W9-AMA-705

Video, Stereo and Optoelectronics

18 Advanced Electronics Projects

To Cindy

with thanks for her patience,
understanding and weekend sacrifices

ACKNOWLEDGMENT

We are most grateful to Mrs. Stella Dillon for her virtuoso performance at the word processor. Her invaluable help is most sincerely appreciated.

Video, Stereo and Optoelectronics

18 Advanced Electronics Projects

Rudolf F. Graf
William Sheets

TAB BOOKS
Blue Ridge Summit, PA

FIRST EDITION
FIRST PRINTING

Library of Congress Cataloging-in-Publication Data

Graf, Rudolf F.
Video, stereo & optoelectronics : 18 advanced electronic projects / by Rudolf F. Graf and William Sheets.
p. cm.
ISBN 0-8306-8358-5 ISBN 0-8306-3358-8 (pbk.)
1. Electronics—Amateurs' manuals. 2. Radio—Amateurs' manuals. 3. Television—Amateurs' manuals. 4. Optoelectronics—Amateurs' manuals. I. Sheets, William. II. Title. III. Title: Video, stereo, and optoelectronics.
TK9965.G712 1990
621.38—dc20 89-29128
CIP

TAB BOOKS offers software for sale. For information and a catalog, please contact TAB Software Department, Blue Ridge Summit, PA 17294-0850.

Questions regarding the content of this book should be addressed to:

Reader Inquiry Branch
TAB BOOKS
Blue Ridge Summit, PA 17294-0214

Acquisitions Editor: Roland S. Phelps
Technical Editor: David M. Gauthier
Production: Katherine Brown
Cover Design: Lori E. Schlosser

The photograph on the cover has been graciously provided by Mr. Brion C. Fenton, Editor of *Radio Electronics,* Gernsback Publications. This courtesy is gratefully acknowledged.

Contents

Introduction

This book describes reproducible projects that are relatively easy to build, but are much more sophisticated than the "run of the mill" experimenter projects, *and* far more useful. Blinking lights, beeping horns, and "tick-tock" projects have been avoided. Also avoided are the old cliches such as digital clocks, rain sensors, and "gadgets" that one is much better off buying than making.

Recently, the digital revolution has brought a wealth of projects involving computers and specialized equipment. We have avoided these, too, because they are quite specialized, in most cases, and possibly require the experimenter to purchase special preprogrammed EPROMs or software that are too "black-box."

We are aimed more at the general experimenter and believe that a useful project is much more rewarding than some "gadget" or exercise that quickly becomes tiresome or otherwise soon wears off in its novelty. Instead, we have concentrated on entertainment-oriented projects in the audio, radio, and video fields, as well as photographic items.

There has been an apparent fall-off in "radio" projects and magazine articles. In fact, we are aware that most of the younger electronic enthusiasts have little or no idea of how a radio functions or how to work with VHF and UHF frequencies. The ready availability of very cheap receivers from the far East, and the near nonavailability of parts has unfortunately discouraged experimentation. This is not only sad, but also a threat to our national security. Rf/Microwave engineers are getting scarce. What good are the computers if they cannot communicate? We live in an analog world, in spite of the digital revolution about us, and we cannot afford to neglect this field of electronics!

Our projects use both digital techniques and rf techniques to produce really useful, modern, and challenging—yet easy-to-build—items. Most of the projects are of the, "We would like to have, but no one makes it" or "That would be great if it was available'' type. Included are wireless headphone projects; FM *stereo* transmitter and TV transmitters of limited range but great educational value and utility; receivers for the longwave and FM broadcast bands that offer a high degree of performance and do things not found on any manufactured unit; video effect devices not generally available to the home video enthusiasts at realistic prices; and photo equipment that is expensive to buy but very easy and inexpensive to build.

We cover the spectrum from VLF to UHF to IR and light. We feel that these projects are, in many cases, unique. They have *all* been tested, built, and are well engineered. Sources of supply for certain specialized items, that are sold to experimenters at reasonable prices, are mentioned. PC board fabrication is covered in detail, as are useful experimenter techniques that the average home photo darkroom enthusiast can use to produce excellent-quality PC boards. For those who would rather purchase ready-made boards and kits of parts, a source of supply is given, covering most of the projects in this book. Also, use is made of many available Radio Shack items to simplify parts procurement. Catalog items are given by their catalog number when possible. Also given is a list of other parts houses with which we have dealt.

The purpose of experimentation is to learn and enjoy, but why not get a useful piece of equipment that is well engineered, ready for years of service, as well as educational? To this purpose we are dedicated.

Part 1
Printed Circuit Techniques for the Hobbyist

Today, most electronic equipment and subassemblies are constructed on thin sheets of dielectric material upon which various conductors are printed or otherwise mounted. These assemblies are commonly called printed circuit boards (PC boards). Insulating materials may include fiberglass, epoxy, phenolics, Teflon, ceramics, and other materials with special electrical characteristics. We will limit our discussion to phenolic and fiberglass reinforced epoxy materials. What is commonly available to the home hobbyist generally will be either phenolic laminate with a wood or paper base, or an epoxy-fiberglass type called G-10. The two materials can be distinguished by their color (but not always). Usually, the phenolic board is a brown color and the epoxy fiberglass is a light blue-green to yellow-green color. The phenolic board tends to crack and be brittle when cut with a fine-tooth saw, while the fiberglass board (commonly called G-10) is more flexible and very difficult or impossible to crack.

The phenolic type board is low in cost and commonly used in entertainment electronics equipment and other low cost applications. It is easy to work with, and does not present excessive wear on drills or other tools. Once properly mounted, it is durable. However, PC boards made of this material tend to be easily cracked. The authors have had extensive electronic servicing experience and have seen many instances of problems caused by the relative ease that this material can crack, causing nasty intermittent problems. It is almost a crime to use this material in an expensive TV set, VCR, or other application involving dense circuitry and rather heavy board mounted components. Yet, it is done to cut costs. In our opinion, this is foolish. One very prominent manufacturer several years ago used this material even in their top of the line TV sets. These TV sets always had a reputation for excellence, but the PC boards tended to crack after several years of heat cycling. As the reader may gather, we do not recommend the use of phenolic PC board materials except in very low cost, small assemblies where durability is not a factor. And for the home experimenter, an accidentally broken PC board is a disaster, since several hours of labor may be wasted. However, Radio Shack has a line of experimenter PC boards and predrilled breadboard assemblies made of phenolic material which are excellent for use as time-saving, ready to use breadboards or one of a kind items, and are great convenience items. We use them for experimental work and highly recommend them where durability and appearance are not especially important. These boards come in several sizes, have predrilled holes and pads on 0.1″ centers, and cannot be made at home nearly as cheaply as you can buy them (currently $2 – 5) when you consider the value of your time and effort. Also, they are available with preetched pads, conductor traces, and also with one side covered with a ground plane for high frequency analog or high speed digital use. But, for home PC fabrications, G-10 materials are the only really practical way to go. For home use, 1-ounce or 2-ounce copper is adequate.

G-10 epoxy fiberglass PC board is available from many sources. It often can be found at surplus dealers, mail order houses, or can be bought from many industrial electronics suppliers. It is generally cheaper to get surplus material, as G-10 can be rather expensive ($7 – 10 a square foot for presensitized material). Surplus material is often easy to get in convenient experimenter sizes (pieces having dimensions several inches on a side) and there is no minimum order requirement, as a rule. Many times, good G-10 PC material can be found at ham fests, flea markets, and usually, the only problem is that it may be somewhat tarnished (red or brown). Avoid any material in which the copper is corroded (usually greenish crusty material). A little work with some steel wool (very fine) cleans this stuff up very nicely. Also, a local PC board house that fabricates PC boards for industry may have scraps available for a nominal charge or even free for the asking. Sometimes, scrap G-10 is sold by the pound. A pound of G-10 material is enough for several projects. We have bought scrap G-10 for as low as $1 – 2 a pound.

If you are going to use the photographic method of PC board fabrication, Kepro Industries sells presensitized G-10 board. A 7″ × 10″ sheet (sold in packages of 5) may cost several dollars in presensitized form, but this is not really bad, when you realize that a PC board is actually another electrical component, and a vital one at that, and possibly $50 – 100 worth of components may be installed on it. So keep this in mind if you balk at the prices of PC board materials. A good, well-made PC board is important to the success of a project and using cheap, cut rate, or otherwise inferior PC materials is like building a quality house on a poor foundation. It doesn't make sense at all.

PC boards may be home fabricated by several methods. They all share the same idea in common—a copper coated insulating material has a pattern imprinted on it, the board is then placed in a chemical bath to etch away all the copper not needed to form the desired pattern. What differs in all the processes to be discussed is the method to form the pattern of conductors on the blank face of the PC board.

Normally, chemicals such as ferric chloride or ammonium persulfate, in a water solution, are used for etching away unwanted copper. Either one may be used, although ferric chloride is generally easier to obtain. Radio Shack sells it in convenient size bottles (16 oz. P/N 276-1535) and one bottle will do several 4″ × 6″ PC boards. It normally takes 15 – 30 minutes with an etchant temperature of about 100 – 120°F, and somewhat longer, up to 45 minutes, at 70°F room temperature. This may vary due to the condition of the chemicals with regard to how fresh they are, and the thickness of the PC board copper (1 or 2 ounce, which is the weight of copper per square foot of board) and also the degree of agitation of the etching bath.

The material that forms the pattern on the PC board and protects the desired copper conductors from being etched away by the solution is called the resist. This resist may be anything that prevents the etchant from getting at the copper and that chemically does not react with the etchant. Suitable materials are waxes, paints, some inks, and photographically deposited resists (presensitized board, to be discussed later). Suitable materials for the home experimenter are lacquer, india ink, crayons, plastic tape or resist inks sold by various companies such as General Cement, etc. (see index). PC board can be cut with hacksaws, jigsaws, or tin snips (G10). The method of PC fabrication used will depend on several factors:

1. Number of PC boards needed
2. Circuit density and complexity
3. Size of PC board
4. Available materials and experience
5. Speed and ease of fabrication

MATERIALS

Basic materials needed, at first, will be as follows:

PC material. Single sided G-10 (one side copper) or double sided G-10 (not recommended for those inexperienced in PC work—do a few single sided boards first).

Tray. Glass, plastic, or fiberglass, large enough to hold PC board sizes anticipated. An 8 × 10 photo darkroom tray is excellent. *Do not use metal trays* for etchant.

Etchant. Ferric chloride 16 oz. available from Radio Shack (P/N 276-1535) or other distributers (see index).

Resist. Any form of lacquer—we have used Testors model airplane dope, hot fuel proof type, preferably black or other dark color.

Fine (#000) steel wool. For cleaning the PC board material.

Rubber gloves or tongs. Plastic for handling PC materials and protecting skin from etchant (may stain or cause skin irritation).

Safety goggles. To protect your eyes if you do not wear glasses or other protective eyewear. You might also want to wear old clothes, since etchant stains are hard to remove.

PC pattern. Depending on your particular project, from magazine article or your own design.

Small Drill Bits. For drilling holes in the PC board. While there are no general "universal sizes," drill sizes #28 and 33 are good sizes for mounting holes; a #57 or 60 drill is commonly used for small trimmer potentiometers and wires, and large electrolytics. A #63 drill is useful for 1/2 watt resistors. A #66 drill is used for most small components such as resistors, disc ceramics, small electrolytics, transistor leads, and IC chips. However, a #70 will be found useful for crowded PC boards or for IC pins which have traces running between them such as commonly encountered on dense digital boards and memory boards. These drill sizes can be obtained at hobby shops and industrial hardware dealers. (If only one size could be called most useful, it would be a #60.)

If you have never made a PC board before, it may be a good idea to try the following experiment. Take a piece of PC board a few inches on each side, say 2″ by 4″. Use steel wool of #00 or 000 (very fine) and polish the copper side until it is bright and shiny. Do not touch this with your fingers after it is polished. Next, using lacquer, paint a pattern on the copper and then set the board aside to dry. When the PC board has dried (a hot air hair dryer can be used to speed this up), place the PC board in a tray of etchant, deep enough to cover the PC board.

You will immediately notice that the copper that is unprotected with lacquer quickly turns a brown color. After one minute remove the PC board and wash it with hot water. The pattern should be definitely starting to etch.

Now examine the PC board and touch up anything that needs it with more lacquer. Then place the PC board (after the lacquer is dry) back in the etchant.

Remove the PC board, wash, and examine it every few minutes. You will get a good idea of how the etching process is proceeding. After about ten minutes, some evidence of etching will be seen, in which light will be visible through cracks and holes that have etched through.

Now leave the PC board in until etching is complete. Any tiny spots that refuse or are stubborn to etch away can be removed with a small knife (X-ACTO Blade) if needed. Actually, on very delicate or dense PC board with very narrow traces, there is the danger of over etching, resulting in loss of the desired conductors, so it is wise to quit while you are ahead and use a knife to remove unwanted specks of copper.

Next, rinse the PC board and remove lacquer resist, either with steel wool or lacquer thinner. Congratulations—you have made your first PC board. You have also gained a good idea of the etching process. Actually, for some very small PC boards, this is the fastest way to make one, where one of a kind is required.

In an actual electronics project, the PC layout may be given in the construction details. The problem then is to get the layout transferred on to a piece of

Fig. 1-1. Basic materials needed for PC fabrication.

PC board material so that a board may be etched. If no PC layout is given then the builder must first lay out the PC board. PC layout is not the main subject of this chapter, but a few hints will be given later.

In order to determine a suitable PC layout, it is best to know in advance what components will be used, physical dimension and mounting details of these components, and detailed terminal connection information. A suitable schematic diagram should be available. In many cases, especially in analog circuits, a schematic diagram may be a useful clue as how to place the components on the PC layout. Signal should flow in a continuous direction (usually left to right) with few as possible crossovers. Circuits having high gain should be laid out so that inputs and outputs are as widely separated as possible. Circuits that operate at high frequencies (above 1 MHz or so) should have short conductors and generous areas of ground conductor (ground plane). Also, PC traces carrying more than a few hundred milliamperes of current should be made wide enough so as to reduce resistance and possible overheating of conductors. If possible, short fat PC traces are preferable to long thin ones. Also, avoid sharp bends in PC rails as they can be susceptible to cracking and breaks. Figure 1-2 shows examples of good and poor layouts for a hypothetical transistor rf amplifier circuit. However, there is no substitute for experience.

Sometimes, especially in digital circuits you may find that the circuit to be laid out has a ''nonplanar'' topology i.e., it is impossible to lay out a PC board without ''crossovers'' of certain conductors. There are several ways around this (see Fig. 1-3).

1. Use of components as ''bridges'' by their placement on the board
2. Jumper leads or wires
3. Use of double sided PC boards.

The first method should be obvious. The second method has the disadvantage of extra connections and to some builders, an esthetically disagreeable situation, and an admission of inability to come up with a ''clever'' layout. For those purists, a 4.7 or 10 ohm resistor can be used as a jumper. In fact, one company makes ''zero ohm'' resistors, which are simply jumper wires with a composition body molded around them so that the jumper looks like a resistor. This is often unnecessary, since many circuits operate at sufficiently high impedance so that 10 ohms or less resistance has no effect on operation. However, the added complication of a double sided PC board merely to eliminate a few jumpers is not always justifiable, especially in a home constructed electronics project.

Double sided PC boards are generally required in many digital circuits or in ultra high frequency applications where a ground plane is necessary, or in those cases where shielding is required. They are best avoided by the home construc-

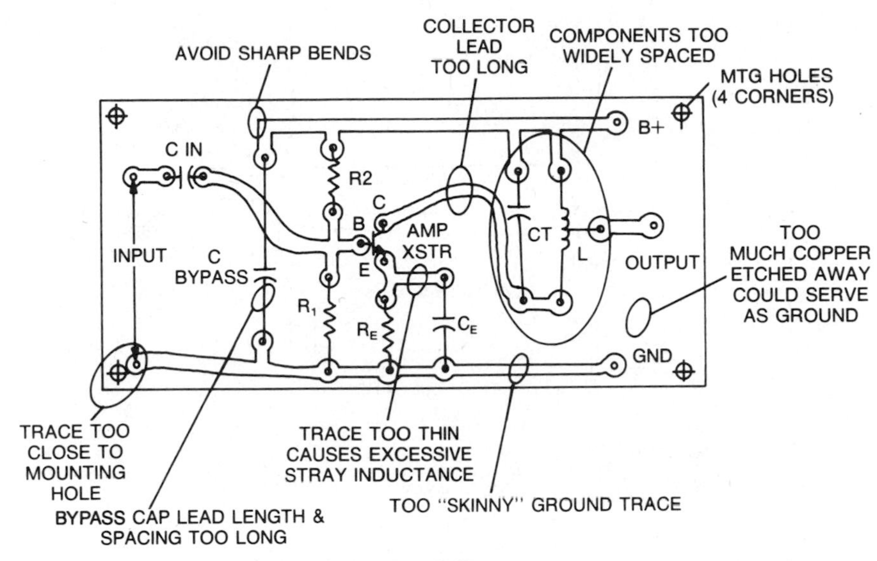

POOR LAYOUT

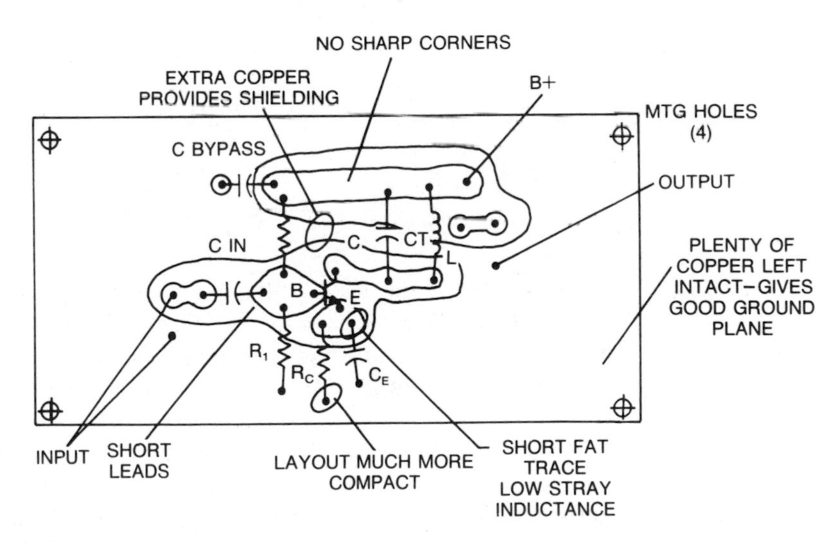

BETTER LAYOUT

Fig. 1-2. Examples of good and poor PC layout.

tor, unless photographic methods are employed, as exact registration of top and bottom patterns is important. Projects in this book mainly employ single sided boards. Also, in double sided boards, connections between top and bottom traces must be made with jumper wires, or else the use of plated through holes, in which the metal is plated inside the component holes drilled in the PC board.

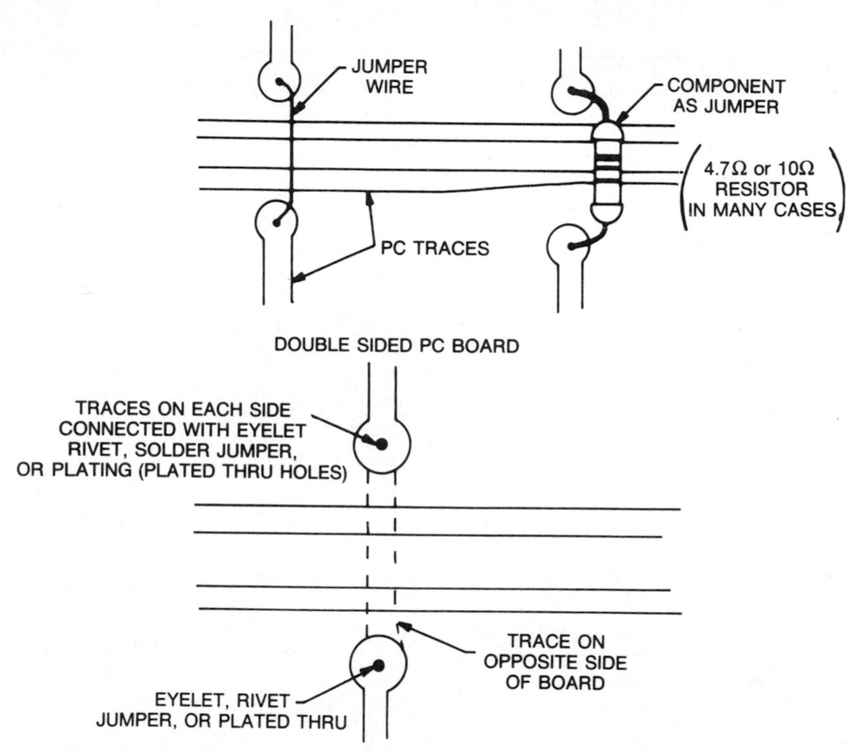

Fig. 1-3. PC crossover and jumper methods.

This technique is generally beyond the ability of the home constructor and will not be discussed here. Most double sided boards used in home construction are generally simple or noncritical for the aforementioned reasons, and if possible should be used only if a few jumpers are either impossible or intolerable for some reason.

When a suitable PC layout is finally obtained, the next step is the transfer of this layout to the PC board. The following discussion will describe a simple, nonphotographic method of doing this. The photographic method will be covered later.

First, cut the PC board to size, and drill any required mounting holes (usually #28 or #33 drill, for use with #6 or #4 screws, one in each corner of the PC board. Larger boards (6″ or more) should have additional mounting holes for best support and to avoid possible damage.

Next, thoroughly clean the PC board material with fine steel wool, until every part is bright and shiny. This is important so as to ensure adherence of resist material and to permit even etching of the board. Be very careful not to get oil or grease on any part of the PC board from this step on. Use of cotton gloves or finger cots is recommended for best results.

Now, using whatever resist materials you choose—lacquer, resist pen, india ink, nail polish or even press-on patterns (Radio Shack P/N 276-1577, or equal), draw, paint or press on the desired pattern. Make sure all copper that is to be retained as part of the PC board is covered with resist. Use alcohol or acetone to remove any accidental fingermarks incurred during handling. A good idea is to use a piece of nylon cord, run through one of the mounting holes, as a means of handling the PC board.

When you are satisfied that the pattern has been correctly and completely placed on the PC board, the next step is to place the PC board in the etching bath. It is best to preheat the etchant up to about 120°F or so. This can be done by immersing the etchant bottle in a hot water bath for 15 minutes or so. A photographic tray (8×10 size) of *plastic or glass, not metal or enamelled metal* is a suitable vessel for holding the etchant. If desired, this tray can be placed in a larger tray (11×14 photo tray) or a sink with hot water in it, to maintain the etchant at a higher temperature. Make sure adequate ventilation is available as strong fumes can be emitted from the etchant at higher temperatures. Alternately, room temperature etchant can be used, but it will take longer to etch the board.

Place the PC board in the etchant, pattern side down, and agitate the etchant tray about once per second. Monitor the etching process every five minutes, after ten minutes have elapsed since placing the PC board in the etchant.

Typically twenty to forty minutes will be necessary depending on temperature, agitation, and PC board size, as well as condition of the etchant. A 16 ounce bottle should etch about 100 square inches or more of copper from a PC board. If etching appears to stop and go no further at some point, the etchant is probably exhausted. In this case replace it and continue.

The etching process can be monitored by carefully lifting out the PC board, washing it off in warm water, and visually inspecting the board. Make sure all resist is intact, and repair any pits, holes, or scratches as required. When the pattern is etched, do not leave the board in any longer than necessary. Very small residues, stubborn specks, etc., can be removed with a sharp knife or razor blade. It is more prudent to do this rather than risk over etching the PC board, which cannot be repaired. An over etched board must generally be done over. This is not too serious when the photographic process is used, but if the resist has been applied by hand, a lot of time and effort has been wasted.

After the PC board is etched, remove the board from the bath and rinse it off in warm or hot water for about a minute or so. Dry the board with a paper towel. Check for any last minute defects. If the board is OK then the next step is to drill all holes for component mounting. When all holes are drilled, the resist can be removed. Use a suitable solvent. Steel wool can be used, but it may clog the holes and cause tiny, hard to see shorts that can be difficult to locate. The PC board is now ready to have the parts installed. When soldering components, be careful not to overheat the PC board. Generally, a small (25 watt) hot (700°F) iron is best. The idea is to quickly heat the joints, solder them, and finish quickly. A too small or low powered iron can cause more damage than you might think, since it may take too long to heat the joints suitably, resulting in excessively long time intervals in which components are exposed to high temperatures. If available, a liquid flux consisting of rosin dissolved in isopropyl alcohol, or other such flux (Kester Solder makes this in prepared form) aids in soldering to the PC board. Use only enough solder to make a good joint.

PHOTOGRAPHIC METHODS

The photographic technique to be discussed is a much better way of fabricating PC boards that are complex (more than say 50 components) or that are small, or when several boards may be required. It is more complex and time consuming. However, whether one or hundreds of PC boards are needed, the amount of effort is the same. Results can be ''professional'' in quality and appearance with a little care. Basically, an ''artwork'' which is a scale (usually 2:1) drawing of the desired pattern is made first. This can be drawn by hand, or ready made PC patterns used. Black or red lithographers' tape can be used as well. The tapes are available from companies such as Bishop Graphics, and come in many sizes and shapes. Among these are IC pads up to 40 pins, tapes in all widths from .020″ to 1″, circles of many different inside and outside diameters, and almost anything you can imagine. Also, press on or dry transfer type PC materials are available. In addition, a material consisting of a plastic sheet covered with a special ruby colored material is available, called ruby lith. Graphics arts supplies and drafting equipment stores generally have this. You simply cut away the red material with a knife, such as a pointed X-ACTO hobby knife, and the result is a ruby and clear PC pattern. Also, ordinary India ink and white, dull surfaced paper can be used. In many instances, such as in this book, a suitable artwork is supplied as a part of the construction article. It is only necessary to photograph the artwork in this case. Further steps use these photographs, the original artwork being put away until changes must be made in the PC board. In this case, changes are made in the artwork and it is rephotographed. In addition, using ''cut and paste'' methods, various parts of the artwork can be used for other applications or as part of another project. And you still have the original artwork, since all that ever happens to it is that it has its picture taken. However, any PC board done this way does require the photo work to be done whether the circuit has one component or thousands. Therefore, for very simple one-of-a-kind boards, the photographic method should be used with discretion if time is a consideration.

In order to use the photographic method, it is best, but not absolutely necessary, to have your own developing facilities. If you are a photo hobbyist and can do your own black and white developing and printing, you already have most of the knowledge, experience, and equipment you need. If not, possibly you know someone who does, or your local camera club may help you locate someone. If these resources are not available to you, it is really not all that difficult to

Fig. 1-4. Commercial tape and tape patterns used in PC layouts.

do it yourself. The most difficult part of the job is obtaining the working negative. A commercial photographer can do this for you, so can a graphics or lithography firm. Consult your local yellow pages for those available in your area.

If you can do it yourself or you wish to try, the following procedure will describe the methods. Basically, you need a camera. An ideal camera would be an 8×10 view camera of the kind used in studios. Few of us have this camera available. For small PC boards, up to 4×5″, a 4×5 "press" type camera, such as a Speed Graphic of the kind that used to be used by press photographers, is also ideal. But again, only a few serious photography buffs have these cameras available. Actually, a very good job can be done with an ordinary 35mm single lens reflex, the common garden variety Pentax, Nikon, Canon, Minolta, or many other fine makes available. We recommend the commonly available Pentax K1000 manual camera with a 50- or 100mm fixed lens.) A manual camera is really best for this job. You will need a tripod, since time exposures are used, and a cable release for actuating the shutter. Zoom lenses are not generally recommended, since many of these lenses have some optical distortion that causes straight lines at the edges of the field to curve slightly at some lens settings. In pictorial work this is not really noticeable, but in PC boards it could be a problem. To test your camera, photograph a drawing of a rectangle (see Fig. 1-5). The rectangle should occupy most of the picture area. Use slide (transparency) film or negatives. When the slides or negatives (do not use prints to judge this) come back from the laboratory, you can check the image of the rectangle. If the sides are reasonably straight, with less than about 1% deviation from a straight line near the edges (1% of the length of the lines) you will be OK, since you will mainly use the central area of the negative. Most 35mm lenses we have seen, of

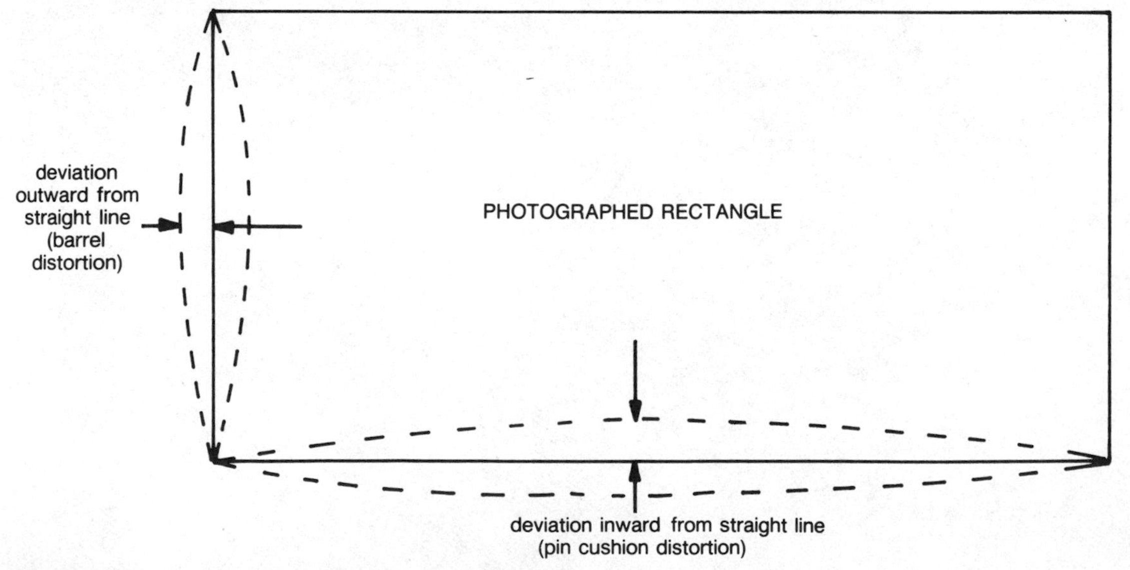

Fig. 1-5. Method of testing your 35mm camera for PC work.

quality makes, will do OK, if their focal length is in the 50–100mm range. Some telephotos up to 200mm are OK but their long focal length can be a problem in actual use, as you will find out, unless you have a very large room area to work in. By the way, a 35mm SLR is a nice item to own, if you do not already have a camera. For PC work, small 110 size or 126 size cartridge cameras are not generally satisfactory, nor are 120 size cameras, since the required films are not generally available in sizes to fit these cameras.

Sensitized material needed for PC work are the kind of films that have high resolution, fine grain, and high contrast. Speed is unimportant, and is not a requirement. Only black and white materials are used, and need be considered. Also, a film that can be developed under a safelight is preferable, which implies a film that is either blue sensitive only, or orthochromatic (sensitive to blue and green, but not red). However, films like Kodak Panatomic X or Ilford FP4, which are slow, fine grain panchromatic films, (sensitive to all visible light wavelengths) are satisfactory if developed for maximum contrast but development must be in total darkness. High speed films such as Kodak Tri X, T max 400, and Ilford HP5 are not satisfactory. Speed is not needed, and they are not contrasty enough. Resolution of fine detail, while satisfactory for pictorial work, is not as good as the slower classes of film. See Table 1-1.

There are three steps needed. The first step consists of copying on 35mm film the artwork pattern to be used on the PC board. The film is developed to yield a negative of the artwork. By the way, the artwork is dark where copper is wanted, and clear or white where copper is to be etched away. The negative formed in the 35mm camera on the original film has clear areas where copper is wanted, and dark or opaque where copper is to be etched away. Actually, this is the way we want our working PC negative, except that the first step yields a negative much too small (typically 1 × 1½″) or less.

The second step consists of using an enlarger to "blow up" the negative to final size. A piece of lithographic film or other very high contrast film is placed on the enlarger baseboard. With the negative obtained in the first step in the negative carrier in the enlarger head, the projected image on the "lith" film is made to conform to the final size of the PC board. This is done by proper enlarger focus, bellows, and the height adjustment of the enlarger. If you are using this method you are familiar with this procedure, as the same method is used to "crop" photographs on the baseboard or easel. Use a lens aperture (f/5.6 to f/8 usually) that gives best lens performance. Exposure times are typically 2 to 5 seconds (Kodak Kodalith 2556). The idea is to go for maximum sharpness. By the way, we use a dichroic color head and a Rodenstock Rodagon lens. The color head is set for white light. A condenser type B/W enlarger is also perfectly satisfactory. The main thing is the lens sharpness and optical quality. However, most "standard equipment" enlarger lenses that are factory supplied for your enlarger will do well.

The lith film is then developed. The result will be a positive image (clear where no copper, black where copper is to be). It will look like a copy of the original artwork, either full size, or reduced size if you are working with a 2:1 (master art twice actual size) or 4:1 (four times actual size). However, you can use any scale you like, as long as the result of this step is *exactly* the size of the PC board you will be using. It is easier in many cases to use 2:1 artwork, as it is easier to work with, especially with dense or small boards. 4:1 is even easier, but for our purposes, unnecessary unless you are working with extremely dense or very small boards with fine detail. We will discuss materials later.

At this time our result is an actual size PC artwork on film. If we were using a positive photoresist material (that type in which after development, *exposed* areas of resist are removed) we would be done. However, our process is geared for *negative* photoresist (exposed resist is hardened, unexposed washed away after development). So, we have to reverse the image so as to obtain a negative (we have a positive in step 2). This is done by simply "contact" printing our actual size positive on another piece of lith film.

While this extra step may seem a chore, actually it is an advantage. Small errors, or even minor circuit changes, can be done on this positive, without going through the whole camera process again. Furthermore, it is easy to make duplicate negatives (step 3) if you wish to make several PC boards at once, or to have extra "insurance" copies.

To make the final negative, place the result in step 2 over another similar piece of lith film. Use a sheet of glass over the films or a contact printing frame to ensure good film to film contact. Expose the film using either light from the enlarger (no negative in carrier, lens at same settings as step 2) or turn the darkroom lights on for about a half second or so. Now develop the film and the result will be one that is black where no copper is wanted, clear where copper is to remain.

Be sure that the image is sharp at each step. Also, minor defects (pinholes, scratches) should be repaired. You can use India ink, black lacquer, or a special paint called film opaque sold in photo stores. for retouching. By the way, black tape strips, or red lithographers tape can be used, since lith films are not sensitive to red light.

Now, using your negative, place it in contact with a piece of PC board that has been previously sensitized (you can buy presensitized board or sensitize it yourself with available spray on photoresist materials (see later). Use a heavy piece of glass (clear) to ensure intimate negative to PC board contact. Expose the PC board per the required method (usually photoflood lamp, UV lamp, or sunlight) as suggested by the manufacturer of either the presensitized board or the photo resist. We use Kepro Corp. type PC boards and these are exposed for two minutes to a light source consisting of five 15-watt T-8 "black light blue" fluorescent tubes (15T8/BLB Philips type) at a distance of six inches. This is given as a typical example.

After development of the exposed PC board (two minutes in Kepro KD developer, which appears to be a Xylene based liquid). The PC board is washed carefully for one minute in tepid (80°F or so) water. The board can now be

STEP 1 -35mm NEG

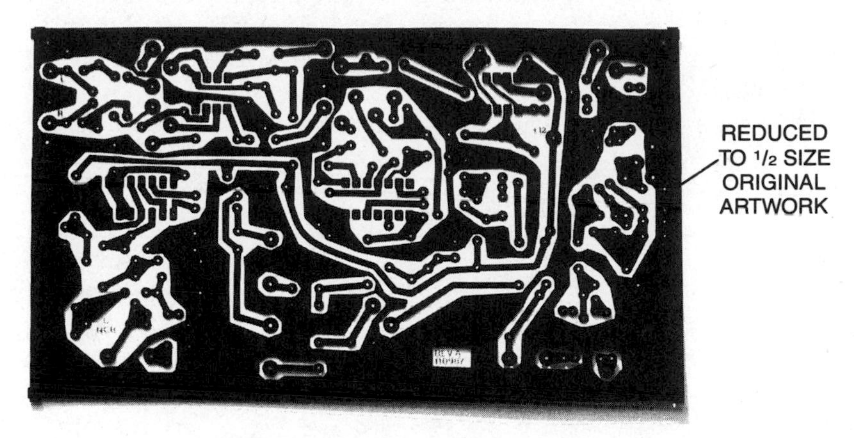

STEP 2 - MASTER POSITIVE

STEP 3 - WORKING NEGATIVE

Fig. 1-6. Results of steps 1,2,3 in photographic method.

etched as mentioned previously in earlier paragraphs. The result is a high quality professional looking board that is equal to those found in factory made items, or better, since we did not try to cut costs.

After etching, the board can be tin plated using electroless tin dip sold by Kepro Corp., or equal. About two to three minutes immersion seems adequate. However, this is not needed, just a luxury. Also, the board can be solder coated with a hot soldering iron if you wish. Now, drill the holes and assemble the PC board.

Next, we will discuss materials used in the photographic process. A list of suppliers is included in the Sources appendix that deal in supplies for PC work. (This list is current (winter 1988 – 1989) but prices and availability are of course subject to change.)

Suitable films are shown in Table 1-1. Note that there are a few different suitable B/W films.

Table 1-1. Suitable Films.

Film	Used For	Availability	Developers
Kodak 2556 Kodalith Ortho (orthochromatic)	Camera work working negatives	35mm 100′ lengths sheets 4x5	Use only Kodalith A&B (Kodak) Developer
Fine grain Positive 5302 Kodak (Blue Sensitive)	Camera work intermediate (1st) reduction Possible working negatives	35mm 100′ rolls sheets 8x10	D11, D19, D76 (Kodak)
Kodak Panatomic X (panchromatic)	Camera work double-sided boards	35mm 36 ex rolls 120, sheets 4x5 8x10 100′ bulk rolls	D76, HC-110 (Kodak) and others
Kodak Tech Pan (Panchromatic extended red sensitivity)	Camera work double-sided boards	35mm 36 ex rolls	D76, HC-110, Technidol (Kodak)

Other firms such as Agfa-Gaveart manufacture equivalents. Do not overlook various films available from mail order houses such as Freestyle Sales in Los Angeles, California, which has a large selection of graphic arts type films suitable for PC work. We are mentioning those that we are familiar with and have tried for actual working negatives. (Those used to expose the PC board with) is a lithographic film with extremely high density, called D max, which is the maximum density expressed as a logarithmic function of the percentage light transmission. For example, a D max of 2.0 has 1% light transmission, a D max of 3 has 0.1% transmission, etc. It is possible to use films such as the 5302 fine grain positive, a material normally used for printing B/W transparencies and slides from B/W negatives by employing a high contrast developer such as Kodak D-19 or D-11, but lith films give the best results.

Many of these films are only sold in lengths of 100′ in bulk, since they are used only professionally. However, it is easy to obtain blank 35mm cassettes (any photo store) and load up your own cartridges. A rule of thumb is to load up 10 inches of film plus 1$^1/_2$″ per exposure. Thus, a 12 exposure roll would need 28 inches (2$^1/_2$ feet) and a 36 exposure roll would need 66 inches (5$^1/_2$ feet). Do this under appropriate safelighting.

Safelights for working with these films are specified by the manufacturer.

Kodak 5302 can be handled under a safelight that is suitable for B/W photo printing, a dim amber light. Kodalith ortho uses a dim ruby red safelight. However, you can probably get by (no guarantee) with a small 4-watt red Christmas tree lamp (GE Glow-Bright) sold in many stores around the Christmas season. (Get the translucent kind.) Keep film at least 4 feet away from the light. A Kodak 1A safelight filter with a 15-watt bulb is best. However, test your safelight by exposing a piece of film 4 feet away for five minutes. Develop the film. If it is clear, you are OK. If not, your safelight is unsafe.

An ordinary 40-watt red or yellow incandescent (bug-light) lamp will do as a safelight for the PC board photo resist, or a smaller 7$^1/_2$-watt clear or white night light can be used. Keep illumination to a minimum.

You can use an ordinary 35mm stainless steel (preferred) or plastic developing tank with a suitable reel to handle the camera film. Working negatives are best developed in 5 × 7 or 8 × 10 plastic or glass trays. Use what is best for your situation.

If you have no darkroom, a closet or bathroom can be used if it is dark enough. It should be dark enough so that after five minutes in the room you cannot readily see anything in the room. A few minor cracks under doors, etc., are OK, especially if you can throw a mat or rug over them. Usually the worst offender is the crack or space at the bottom of the door. Or else, simply do your photo work after dark, with blinds or shades pulled down and room lights off. If in doubt, use the method for safelight testing, leave a piece of unexposed film out for 5 minutes, then develop it. If it comes out clear, you are in good shape. The idea is simply not to unintentionally expose the film being used. If nothing shows on the film, the room is probably dark enough.

Developers for various films are sold in camera stores or photo supply houses. They are inexpensive. Use manufacturers recommended developer for the film you are using. Do not save used developer solutions—they don't generally keep well and best results are had with fresh solutions. Mix up only what you need.

A suitable working temperature for most B/W films is between 68°F and 75°F. For our PC work, temperatures are not very critical but we use 68°F. An inexpensive photo thermometer, accurate to ±1°F will be good enough. These thermometers are only a few dollars and if you are a photo hobbyist you probably have one already.

Ordinary tap water is OK for use in processing these films. If you can drink the water it will do fine for photo work of this nature.

Films should be stored in a cool dry place, preferably between 35°F and 55°F, but your refrigerator will do if you wrap all film in a hermetically tight container. Ziplock bags and/or freezer containers are good for storage packaging. They can be obtained at your local supermarket.

By the way, do not pour these chemicals, or PC etching fluid, in any *metallic* plumbing system. Ferric chloride will eat away brass fittings and traps. It is best to dispose of chemicals by diluting them with large quantities of water, then putting the solution down a toilet bowl a *little* at a time. You might consult local officials as to the best way of getting rid of these chemicals after they are used, if you are in doubt. Keep chemicals away and out of reach of people and animals at all times. Do not throw any bottles of chemicals in the garbage, as someone handling the garbage might break the bottle and accidentally come in contact with the contents.

Make sure you have adequate ventilation especially when using Xylene-based PC board developer, and if possible, avoid skin contact.

As far as PC artworks go, some magazines and publications provide a copy of actual patterns. You can photograph this pattern. Use a polarizer on the camera lens to reduce glare and hot spots due to reflections off the glossy paper. A typical setup is to use two 250W photofloods about 24″ from the artwork, and 45° either side of it (Fig. 1-7). Use a red filter if panchromatic film is used. You will have to try various exposure settings. Panatomic X (Kodak) exposed at E1 50 or 80 and overdeveloped seems to do well. Also, Kodak 2556 Ortho (Kodalith) exposed 1 second at f/11 (use f/8 to f/11 for copy work) works nicely, with no filter needed.

You can also remove the artwork from the magazine and mount it on a copy stand or flat surface, if you do not mind cutting up the magazine. Some magazines (Radio Electronics Magazine, a Greensback publication, for example) print artworks on pages such that the reverse side is blank. In this case, you can coat the paper with salad or mineral oil to improve transparency and then use a light table rather than photofloods. You might even use this kind of artwork as a positive, if it is actual size, and directly contact a working negative step 3 of photo process without doing steps 1 and 2. While we have not tried this, it certainly might have a chance of working.

A word about double sided boards. Two artwork patterns, one for each side, are required. They must register correctly i.e., alignment must be such that the top and bottom patterns of the PC board line up exactly. Separate negatives are made for each side. The photo method is practically a must for this

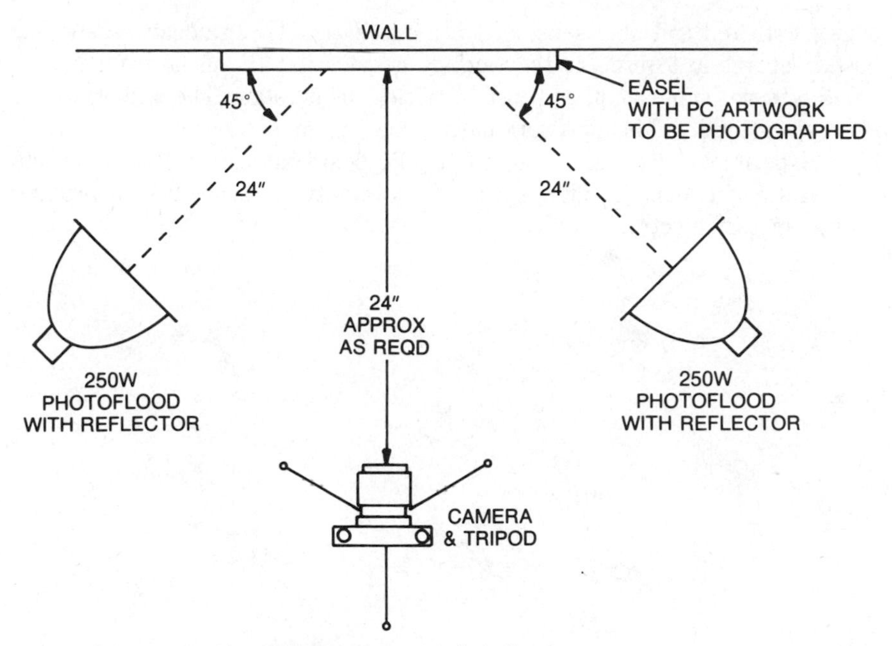

Fig. 1-7. Copy setup using 35mm camera and photo floods.

process. Then, the negatives should be placed in correct registration using a light table, and fastened together such a way as to allow inserting the sensitized PC material between them. Both sides are then exposed, the PC pattern developed, and then etched the usual way. You must be very careful to maintain good alignment and registration.

There is a way to generate double sided artwork using one pattern, that gets around several registration headaches. Basically, one side of the pattern is taped up using a special transparent red tape material (Bishop Graphics has this material available). The other side is taped up, using a blue material in a similar manner. Registration is automatic, since the same base or sheet of paper is used.

The pattern is photographed using a light table to illuminate it from behind. First, an exposure is made using a blue filter (Wratten type 47). The camera sees only the dark lines where the red material is placed, the clear and blue appearing the same. Next, the pattern is photographed using a red (Wratten type 25 or 29) filter. This time, the blue lines are dark, the clear and red "dropping out." The result is two separate images, of each side, that register perfectly, and only contain details of either the red or blue side. Naturally, panchromatic film such as Panatonic X, FP4, or Kodak Tech Pan must be used, since both red and blue sensitivity are necessary. These negatives are then used separately to prepare the working negatives in the usual manner. This method saves some labor, and ensures perfect registration. The disadvantage is that special red and blue artwork taping materials must be used.

Experimenters are encouraged to try PC board fabrication. It is not as difficult as it might sound, and it opens up a wide variety of possibilities important to contemporary electronic work.

Part 2
Audio Amplifiers and Projects

There are many situations when a 200-watt amplifier system is not called for, and where a 25- to 50-watt amplifier will do well. In a typical apartment situation, or in a quiet country home, the use of 25 watts of audio with any kind of decent speakers, will probably drive the user out of the room.

PROJECT 1: AUDIO POWER AMPLIFIER

This section describes a simple power amplifier that is capable of over 40 watts rms output into an 8 ohm load, with low distortion, and is capable of delivering this power continuously for hours, a very severe test. It may be run at lower power outputs by using a smaller power supply. Depending on your requirements up to ±40 volts may be used with the circuit. A simple power supply using a 25V CT 2A transformer and bridge rectifier will give +17 volts and up to 15 watts rms power output into a 4 ohm load. At this power level, the heat sinking requirement for the output transistors is very modest, and the amplifier is practically indestructable.

Most modern solid-state power amplifiers are basically high power, dc coupled, operational amplifiers using dc and ac feedback. The open loop gain is usually extremely high (usually 80dB or more) and the closed loop gain is typically 20 to 40dB. Generally, one volt peak-to-peak will drive the amplifier to full output, which may be 15 to 120 volts peak-to-peak. With feedback, distortion can be extremely low, and frequency response is usually a few hertz to over 20 kHz. What is practically achieved is an amplifier that is far better than the sources feeding it, or the speakers it drives. Therefore, an amplifier of this type will do an excellent job in high fidelity applications.

The hobbyist can learn a lot about amplifier theory and operation by building a simple amplifier such as the one to be described. By using a small power supply, inexpensive power transistors, and keeping the power level around 15 watts per channel, without fancy frills, the danger of "smoking" $50 worth of output transistors or other components during the testing phase is eliminated. Running a 40–50 watt capability circuit at 15–20 volts instead of 40 volts results in a circuit very forgiving of initial wiring errors or accidental short circuits, solder bridges, or other unexpected run-of-the-mill goofups that every experimenter and hobbyist experiences now and then. No "black box" IC devices are used. ICs are fine for production but do little to teach about the "workings" of circuits like this one. If more power is desired later, it is simple enough to rebuild the power supply and use larger heat sinks for the power transistors, this time having experience with the circuit. Use of discrete components keeps costs per part low and gives the builder a closer look at what is happening.

Referring to Fig. 2-1, a simplified diagram of the amplifier, Q1 and Q2 form a differential amplifier pair. The sum of the emitter currents flows through R3. Since the transistors Q1 and Q2 are a high gain type (β about 100), and since the emitter current of each is the sum of the base current and collector current, we can neglect the base current and say that the collector current is equal to the emitter current with a very high degree of accuracy. Since the base voltage of each of Q1 and Q2 is very close to ground (zero) a total current flows through R3 that equals the supply voltage on the negative side divided by the value of R3. This current ideally is constant.

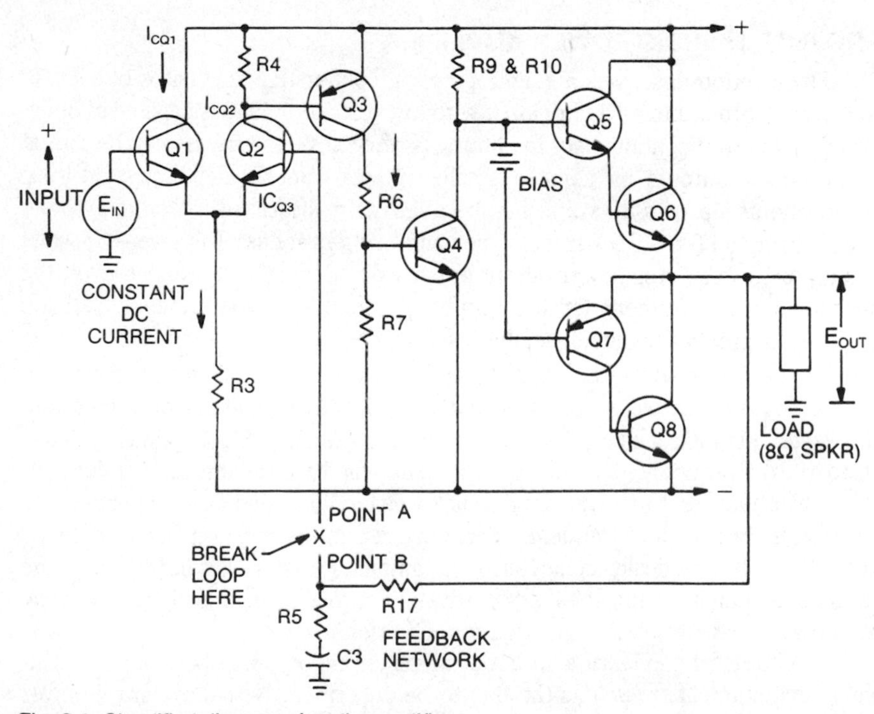

Fig. 2-1. Simplified diagram of audio amplifier.

Assume we get an input signal E into the base of Q1. Assume this signal is positive going. It will forward bias Q1, causing it to draw more collector current (or emitter current). Since Q1 and Q2 have their emitters returned to a source of constant current, if the emitter current of Q1 increases, Q2 must decrease by the same amount, since the total current must be constant. This causes the collector current of Q2 to decrease. The lower resultant current through R4 causes the base to emitter voltage of Q3 to decrease. Note that Q3 is a PNP type. Q3 now conducts less, causing the collector current of Q3 to decrease. Therefore, there is less voltage drop across 7, and Q4, an NPN transistor, also now has less bias. Therefore, Q4 collector voltage will rise, since the collector current of Q4 also decreases, and the voltage drop across R9 and R10 decreases. This collector voltage change passes to the base of Q5, an emitter follower, which is coupled to Q6, also an emitter follower, driving the load. Therefore, a positive going voltage appears across the load.

Q7 is a PNP transistor, and a positive going base voltage will tend to turn it off. Since Q8 gets its bias from Q7, Q8 will tend to turn off, and its collector voltage will go positive.

What we have is a high gain (over 1,000 times) dc amplifier. In order to establish an operating point, we must insert feedback into this system. Note that a positive dc voltage introduced into the base of Q2 will have the opposite effect. It will make the output go negative. Our reference is ground since the base of Q1 (input) is grounded for dc through R1.

If we connect the amplifier output to the base of Q2, as seen in Fig. 2-1, and introduce a positive signal at point A (if we "break the loop") a much larger negative going signal would appear at point B. This would "fight" the input signal, and tend to keep the load dc voltage at zero. This is exactly what we want for *dc*. However, we want the amplifier to amplify ac signals. By making the feedback loop different for ac, we can achieve this. R17 is 10k ohm, R5 is 470 ohm and C3 is 47μF. At audio frequencies, C3 is a "short circuit," and we have about $^1/_{20}$ of the output signal fed back. This gives a closed loop gain of 20 times for ac. For dc the gain is unity since C3 is "open" and the drop across R17 is negligible. Figure 2-2 shows an analysis of the circuit for those interested in the mathematical details.

Note that the gain seen between the bases of Q1 and Q2 is extremely high (approximately 1,000 times). Note also, that if the product of the open loop gain and the feedback ratio is much larger than ten, (it will be, in any practical situation), the open loop gain does not really matter and the closed loop gain is determined by the ratio of the resistors in the feedback network. This property of a feedback amplifier turns out to be very useful in modern electronic circuitry, since one can accurately control the response of an amplifier with a few passive components that can be easily made to close tolerances. The open loop gain can vary all over the place, but as long as it stays above a minimum value of about ten times the desired *closed* loop gain it has very little effect on the closed loop gain.

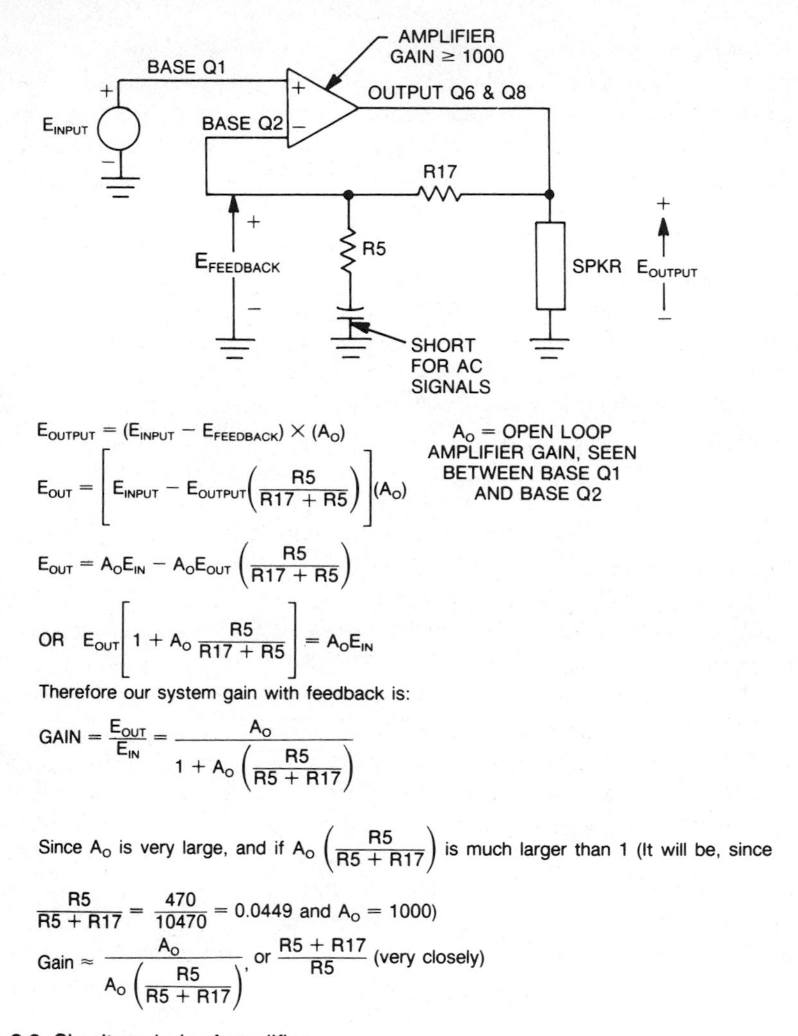

$$E_{OUTPUT} = (E_{INPUT} - E_{FEEDBACK}) \times (A_O)$$

$$E_{OUT} = \left[E_{INPUT} - E_{OUTPUT}\left(\frac{R5}{R17 + R5}\right) \right](A_O)$$

$$E_{OUT} = A_O E_{IN} - A_O E_{OUT}\left(\frac{R5}{R17 + R5}\right)$$

$$\text{OR} \quad E_{OUT}\left[1 + A_O \frac{R5}{R17 + R5}\right] = A_O E_{IN}$$

Therefore our system gain with feedback is:

$$\text{GAIN} = \frac{E_{OUT}}{E_{IN}} = \frac{A_O}{1 + A_O\left(\frac{R5}{R5 + R17}\right)}$$

Since A_O is very large, and if $A_O\left(\frac{R5}{R5 + R17}\right)$ is much larger than 1 (It will be, since

$\frac{R5}{R5 + R17} = \frac{470}{10470} = 0.0449$ and $A_O = 1000$)

$$\text{Gain} \approx \frac{A_O}{A_O\left(\frac{R5}{R5 + R17}\right)}, \text{ or } \frac{R5 + R17}{R5} \text{ (very closely)}$$

Fig. 2-2. Circuit analysis of amplifier.

By the way, hum, noise, distortion, or any other "garbage" generated *inside* the amplifier is reduced by the ratio of the open loop gain to the closed loop gain. So therefore, feedback can "clean up" things inside the amplifier. No wonder today's modern amplifiers are excellent in these respects. It was difficult to use very much feedback with yesterday's amplifiers using tubes and transformers, due to phase shift problems in the transformers. What would happen was that, at certain frequencies, phase shifts would turn negative feedback into positive feedback, causing oscillations. With modern transistor circuitry, phase behavior of the open loop gain can be much more carefully controlled, allowing much more feedback to be used. Clearly, solid state amplifiers have an advantage in this respect, since no transformers are necessary in most audio amplifier designs. Although theoretically, tubes could be used in the same way, it would not be very practical due to the low impedance high current loads used in practice, such as 8 ohm speakers.

A detailed circuit description will now be presented. Referring to Fig. 2-3 (and Fig. 2-1), audio input is applied between the positive lead of C1 and ground. C1 is a coupling capacitor and has negligible reactance at audio frequencies. The audio frequency impedance seen looking into the base of Q1 would be about 3,000 ohms under open loop conditions. However, due to the fact that feedback is used, it is over 100k ohms. R1 is therefore the predominant factor. C1 and R1 have a lower cutoff frequency of around 1.6 Hz. Q1 and Q2 are connected as a differential amplifier, with each transistor biased at about 0.7 milliampere. Emitter bias is fed through R18 from the negative supply (here −17V) and R3. C10 serves to decouple power supply ripple which would appear across R4. R2 and R4 are load resistors for Q1 and Q2. Audio output from the collector of Q2 is directly connected to the base of Q3, a PNP transistor. R6 and C2 form a low pass filter network. C2 is used to establish the 6dB/octave roll-off of open loop gain necessary for loop stability when feedback is used. Without this capacitor, excess phase shifts up around 100 kHz can cause ultrasonic oscillations. R6 and R7 bias Q4, the predriver, and couple audio from the collector of Q3. Note that Q4 is an NPN. The use of NPN and PNP transistors makes it easy to dc couple amplifier stages, which results in excellent low frequency response and stabilizes the dc operating point, since dc feedback can be used. Q4 is run at about 40 milliamperes collector current. R8 is the emitter resistor. About 2 volts drop appears across R8. Q4 drives the output transistors Q5, Q7, Q6 and Q8. R9 and R10, CR1, CR2, R11 and VR1 provide the collector load for Q4. Q4 runs at about 500 milliwatts dissipation and if you intend to operate this amplifier from a supply voltage above 22 volts, a small heat sink would be advisable. Q4 runs somewhat warm (not hot) in this application, but heat sinking is not necessary with the 17 volt supplies used. CR1, CR2, VR1 and R33 form a temperature sensitive bias network for the output transistors. As temperature increases, the drop across CR1 and CR2 decreases, preventing thermal runaway in the output stages Q5 through Q8. Ideally, the diodes CR1 and CR2 should be mounted to the heat sink used for the output transistors. (A little epoxy can be used for this.) However, in our applications it was found unnecessary. If you intend to use higher voltage supplies, it would be a good idea to mount the diodes CR1 and CR2 to the heat sink, and for over 30 volts, it is mandatory.

VR1 serves as a bias adjustment and is adjusted for a 10 millivolt (.01 volt) drop across R14, a 0.22 ohm resistor. This results in a 50 mA idling current in Q6 and Q8. This avoids crossover distortion. R11 sets an upper limit on bias voltage, shunting VR1. Adjustment is smooth and noncritical. C4 provides an audio bypass around the bias network and ensures that Q5 and Q7 have equal audio drive. R9 and R10 form the load resistor for the driver stage.

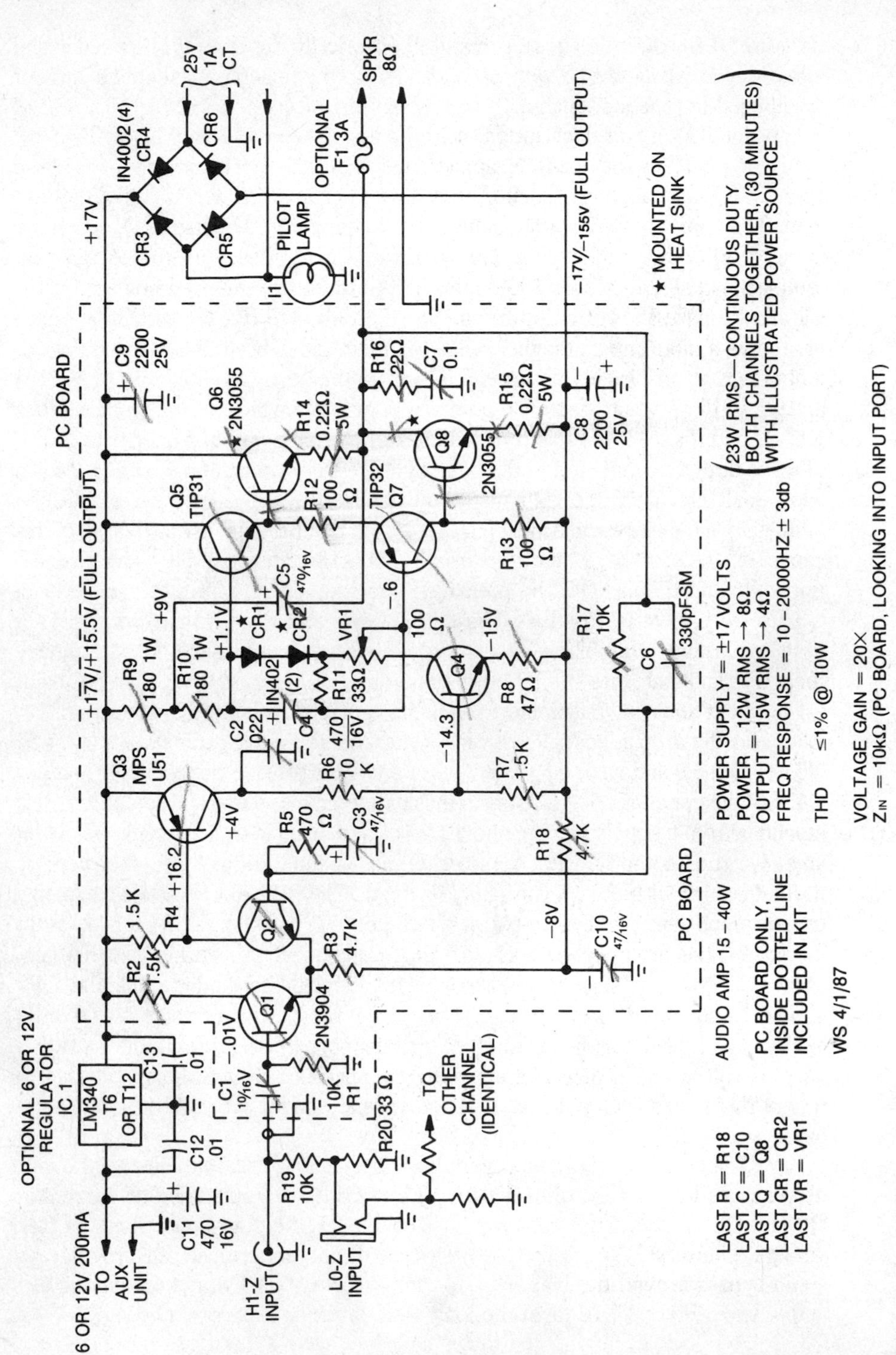

Fig. 2-3. Schematic of amplifier.

They are 180 ohm 1 watt resistors. If more than 22 volt supplies are used, these resistors should be increased to 270 ohm (up to 30 volts) or 330 ohm 2 watts (up to 40 volts). At these voltages, Q4 should be heat sinked and CR1 and CR2 should be in thermal contact (attached to heat sink) with Q5, Q6, Q7 and Q8. C5 provides feedback around the output stages and tends to provide a steady value of collector current in Q4 for positive signal peaks. It provides higher peak power by maintaining available drive to Q5 on positive signal peaks.

Q5 is an emitter follower and drives Q6. R12 provides bias for Q5. Q6 is one of the output transistors. R14 provides emitter feedback and is 0.22 ohm. Q7 is a PNP transistor which drives Q8, the other output transistor. R13 provides a load for Q7. Since Q6 and Q8 are connected in series, opposite drive polarities are necessary. Q7 provides this. On positive going signals, Q6 tends to conduct and Q8 turns off. On negative going signals, Q8 conducts and Q6 turns off. This output circuit is called "quasi complementary." It uses two NPN transistors to do the job of an NPN-PNP complementary pair. It does not require two matched NPN-PNP transistors, and features easy transistor interchangeability. R15 provides emitter feedback for Q8. R16 and C7 form a suppression network for ultrasonic oscillations or high frequency instability. R7 provides dc feedback to Q2 to establish the operating point (zero dc volts across the speaker load with no drive signal). R5 with R17 forms a 20:1 ac feedback network. C3 provides a low impedance audio ground return while permitting full dc feedback, keeping R5 out of the picture for dc. C6 sets the high frequency roll-off. It is chosen for 3dB at 20,000 Hz, where it has a reactance equal to 10k ohm, the value of R17. C3 has a reactance of about 470 ohm at 7 Hz, and determines the low frequency cutoff of the amplifier since C1 and R1 do not cause any significant roll-off above 4 Hz.

C8 and C9 are power supply filter capacitors. They are 2,200μF at 25V. For critical applications they should be increased to 5,000 – 10,000μF at a working voltage 10% higher than the maximum expected supply voltages. However, for utility applications of this amplifier they are adequate.

Figure 2-4 shows a suitable power supply arrangement. A 25.2V CT 2A transformer from Radio Shack does a good job. Suitable rectifier diodes are 1N4000 series types, or Sylvania ECG 125, etc. If desired, a bridge assembly can also be used. Larger filter capacitors can also be used if desired. Use of a speaker fuse is recommended especially in higher power applications. At 15 watts per channel, a suitable heat sink would be a 1/8″ thick (3mm) aluminum plate mounted to the amplifier housing. At 25 watts or higher, a larger, finned heat sink (about 6″ × 4″ by about 1 1/2″ with 4 to 6 fins) would be better. Use of larger filter capacitors than shown in the chart is okay and in fact desirable. They are the minimum recommended. At 30 watts or higher, change R14 and R15 to 0.47 ohm, 5 watts for greater stability. Use of 4 ohm speakers is not recommended above 40 watts, although you probably could get away with it. Also, watch the working voltages of the electrolytics. Use of 50 working volt units for C4, C8, and C9 is desirable for supply voltages above ±25 volts, for 30

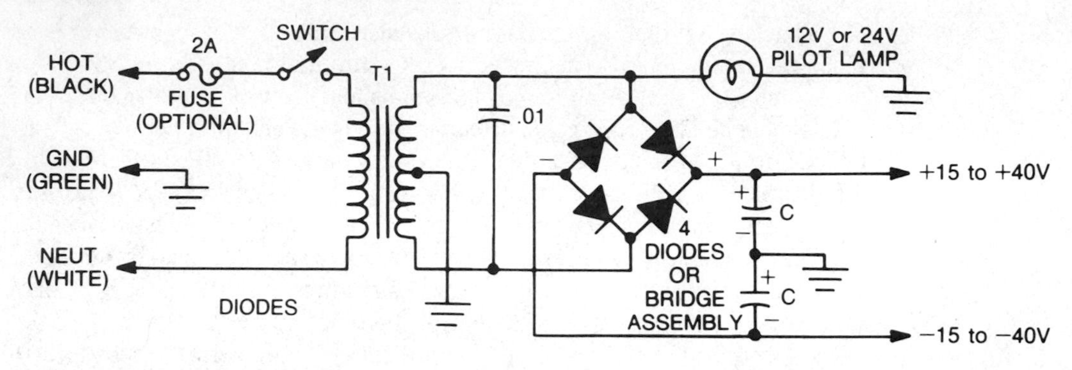

POWER OUTPUT	DIODES	SECONDARY VOLTS T1	MINIMUM* RECOMMENDED C μF (MINIMUM)	R9 & R10	NOTES
15W	IN 4002	25VCT 2A	NONE REQUIRED	180Ω 1W	NONE
25W	ECG 125	36VCT 3A	4700 μF	270Ω 1W	HEAT SINK Q4
30W	5A BRIDGE	40VCT 3A	4700 μF	270Ω 1 OR 2W	— HEAT SINK Q4 — CHANGE C4, C8, C9 TO 50WV UNITS — CHANGE R3 to to10kΩ — CHANGE R14 & R15 TO .47Ω — SPEAKER FUSE RECOMMENDED (3A)
40W	5A BRIDGE	48VCT 3A	6800 μF	330Ω 2W	
50W	5A BRIDGE	60VCT 3.5A	6800 μF	330Ω 2W	

*NOTE
IF CAPACITORS AT LEAST 2200 μF MORE THAN RECOMMENDED ARE USED, C8 & C9 MAY BE OMITTED

RECOMMENDED POWER SUPPLY CONFIGURATIONS

Fig. 2-4. Power supply for amplifier.

watts and higher. Also, note that C8 and C9 are in parallel with the main filter capacitors in the power supply. If capacitors of 8,000 to 10,000μF are used (as in higher power applications) in the power supply, C8 and C9 can be omitted. They are minimum size capacitors used in the 15 watt versions, since they conveniently mount on the PC board and simplify construction.

Construction of this amplifier is straightforward and noncritical. Just keep input audio leads away from ac power leads. Also, use adequate heat sinks—the larger the better. Figure 2-5 shows a *minimum* heat sink for a dual 15 watt amplifier, as a guide. Install resistors first, then capacitors, then lastly, install transistors. Make sure all capacitors are polarized correctly, and all transistors are correctly installed. Q6 and Q8 are mounted on the heat sink and wired to the PC board with No. 20 or 22 gauge hook-up wire. Q6 and Q8 can be installed on the heat sink with sockets, if desired. Make sure that the cases of Q6 and

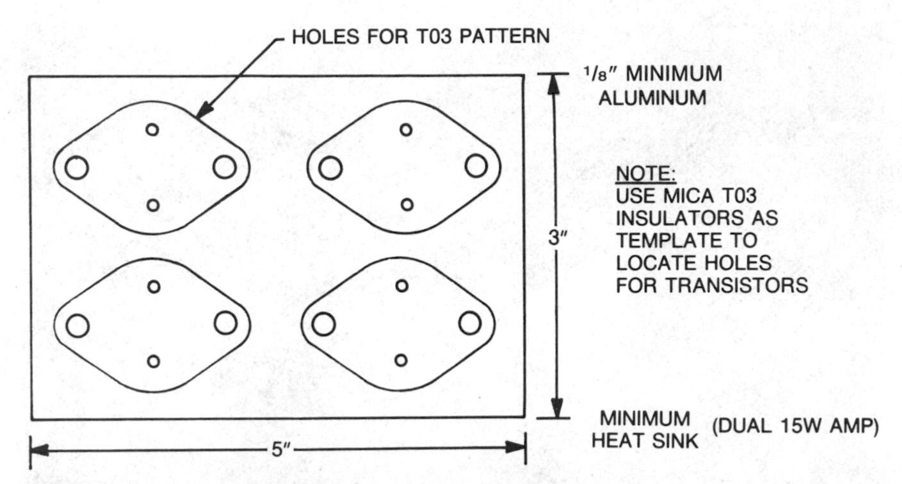

Fig. 2-5. Heat sink placement.

Fig. 2-6. 25W amplifier for pocket stereos, front view.

Q8, which are the collectors, are insulated from the heat sink. Mica washers with a *small* amount of silicone grease (to aid in heat transfer) should be used. Do not overtighten mounting screws, and check with an ohmmeter to make sure that insulators have not been punctured.

Fig. 2-7. 25W amplifier for pocket stereos, rear view.

When construction is complete, check all components for correct position and installation. Check for shorted PC traces, solder bridges, poor solder joints, and incorrect wiring. When you are sure everything is okay, it is time to apply power to the amplifier. In order to protect the components from damage, the following checkout procedure is recommended.

1. Connect a 60 watt 120V ordinary household light bulb in *series* with the primary of the power transformer. See Fig. 2-9.
2. Connect either a double (red-green) polarity LED, or two back-to-back LEDs, through a 2.2k 1/4 watt resistor, to the speaker terminals. See Fig. 2-9.
3. Apply power. The 60 watt lamp may flash momentarily as the power supply filter capacitors charge. It should then *go out*. NO glow should be seen (a very faint *dull* red is okay for higher power versions, if a large transformer is used). The LEDs may also flash momentarily but *must* go out. If one lights continuously, something is wrong. Find the problem before going to the next step.
4. Now, remove the LEDs and the 60 watt bulb. Connect an 8 ohm 25 or 50 watt resistor, or a speaker, to the amplifier. Apply 120Vac power.

Fig. 2-8. Internal view of amplifier.

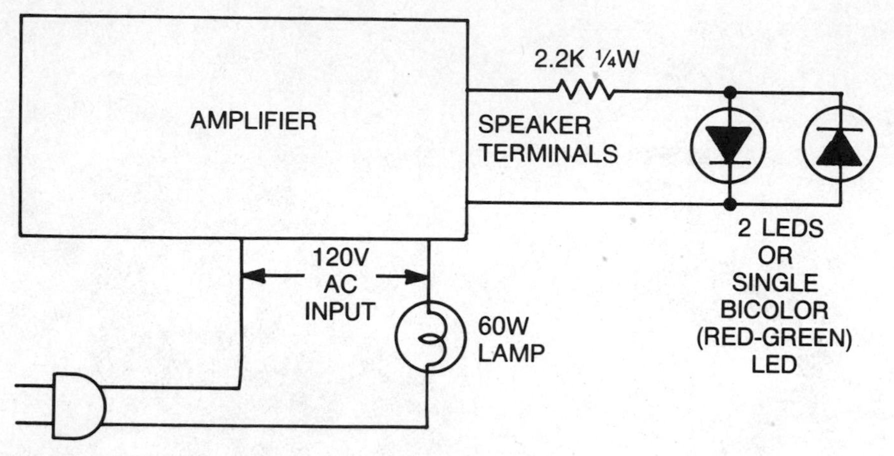

Fig. 2-9. Initial amplifier test.

RESISTORS

R1,R6,R17,R19	10k ohm
R2,R7,R4	1.5k ohm
R3,R18	4.7k ohm
R5	470 ohm
R8	47 ohm
R9,R10	180 ohm* 1W
R11,R20	33 ohm
R12,R13	100 ohm
R14,R15	0.22 ohm* 5W
R16	22 ohm
VR1	100 ohm pot PC board mount

CAPACITORS

C1	10μF/16V
C2	.22μF Mylar ±20%
C3,C10	47μF/16V*
C4,C5,C11	470μF/16V*
C6	330pF Mica ±10%
C7	0.1μF Mylar ±10%
C8,C9	2200μF/25WV*
C12,C13	.01 disc GMV 50V

NOTE

C11,C12,C13 not included in kit, part of optional regulator CKT

DIODES

CR1,CR2 CR3,CR4, CR5,CR6	1N4002

TRANSISTOR & SEMICONDUCTORS

Q1	2N3904
Q3	2N3906 or MPS US1
Q4,Q5	TIP31
Q6,Q8	2N3055
Q7	TIP32
IC1**	LM340T-XX or 7805-XX as required

MISCELLANEOUS**

PC board
Mounting hardware for Q6 & Q8:
Insulators (T03)
Shoulder washers
Jacks, connectors, heat sink switches, pilot lamp, wire cabinet, ac line cord, etc.
**Items not included in kit

NOTE

Suggested transformer and cabinet for dual (stereo):
Transformer – Radio Shack P/N 273-1512
Cabinet – Radio Shack P/N 270-274

*See Text

Fig. 2-10. Amplifier parts placements list, Project 1.

5. Quickly measure the voltage drop across R14 (emitter resistor of Q6). Adjust VR1 for a 10 millivolt (.01 volt) drop. If R14 has been changed to 0.47 ohm, adjust for 20 millivolts (.02 volt). This sets the idling current in Q6 and Q8 to 40 to 50 milliamperes.
6. Check the dc voltage across the speaker or load resistor. No more than about ±250 millivolts (1/4 of a volt) should be present. Usually, it will be less than 1/10 of a volt. If more, something is wrong. Check transistors, R1, R17 or C3 and repair as required.
7. Check the power supply voltages. They should be equal in magnitude, opposite in polarity, and about 70% of the transformer secondary voltage (17 volts for a 25 volt transformer or 26 volts for a 36 volt transformer, etc.).
8. If a scope is available, connect it across the 8 ohm load resistor or speaker.
9. Apply about 100 millivolts p-p audio to the input. It should produce output. If a speaker is used, sound should be heard.
10. Increase audio drive (an audio generator with a 400 Hz or 1,000 Hz sine wave is handy here if available) and check for any visible (on scope) or audible distortion (speaker). If possible, run the amplifier at

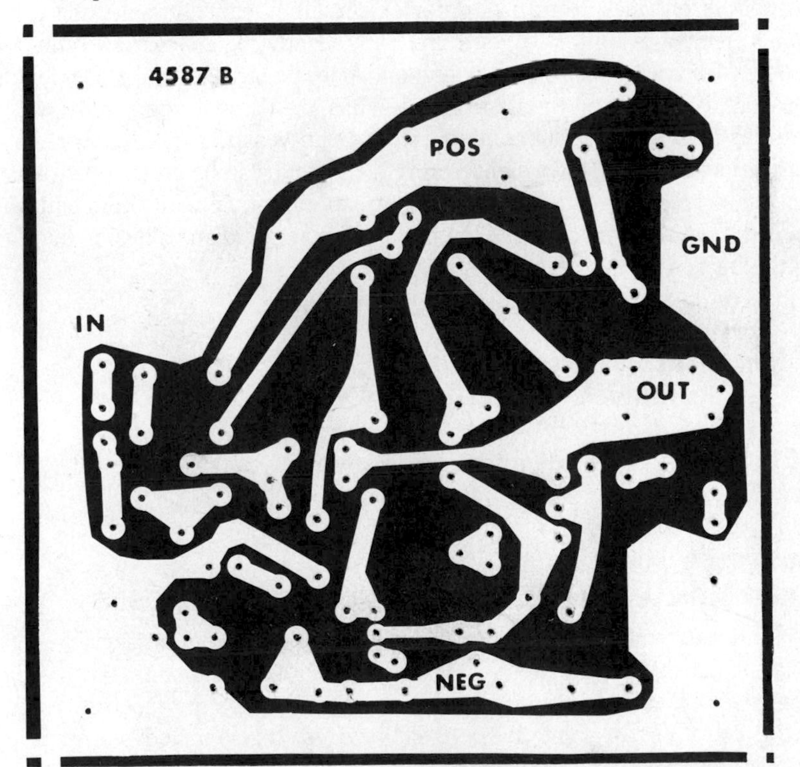

Fig. 2-11. Amplifier PC artwork.

full output. Check Q6 and Q8 for overheating. (You should be able to hold your hand on them) and check the power supply voltages. No more than a 20% drop from that measured in Step 7 should be evident. If there is more drop than this, a larger (higher current) power transformer is indicated.

11. You should be able to run this amplifier at full rms power output for one hour without anything overheating. The heat sinks should get no warmer than you can comfortably touch (65°C or 150°F). The same goes for Q4, and the power transformer. Make necessary modifications so this severe test can be met. This will ensure long, trouble-free service.
12. Now disconnect all test equipment, and the amplifier is completed.

Figure 2-3 shows a 33 ohm resistor, stereo (3 conductor) jack, and 10k resistor used to provide an auxiliary low impedance input. Our prototype was used as an amplifier for a small "walkman" type pocket stereo. It sounds great! We used a Radio Shack 25V CT 2A transformer for a power supply.

Your local Radio Shack carries a suitable cabinet (P/N 270-274) and power transformer (P/N 273-1512) for a dual 15 watt version. Other, larger transformers can be obtained from mail order or surplus houses. In selecting a power transformer, it is better to get a unit with slightly higher current ratings than needed. This will ensure cool operation. Also, larger heat sinks are desirable. Our prototype worked the first time, and has been trouble-free. A 40 watt version has been operational continuously for one year (24 hours a day) in a PA application. This amplifier should prove useful and also be an excellent learning tool for anyone interested in servicing or experimenting with audio amplifiers.

A kit of parts consisting of the printed circuit board and all parts that mount on the board is available from:

North Country Radio
P.O. Box 53, Wykagyl Station
New Rochelle, NY 10804

Price: $32.75 plus $2.50 postage and handling.
New York residents must include sales tax.

PROJECT 2: AUDIO SIGNAL TRACER/INJECTOR FOR TEST PURPOSES

This device will undoubtedly prove handy in many routine servicing operations. It consists of an audible signal monitor for "listening" to the signals present in an electronic device, such as an audio system, receiver, amplifier, tape deck, at circuit points inside these devices. It also includes an rf detector probe for use with high frequency modulated signals, such as those found on an antenna, rf amplifier, or i-f section of a receiver. In much service work, the task of finding the problem is greatly simplified by putting into the item being serviced some form of input signal and checking to see what happens to the signal as it goes in and out of various stages or circuits. If there is something amiss, eventually the signal will disappear or suffer a high loss, such as in an audio stage with a defective IC, transistor, or other component. At this point, another signal from an external source can be "patched in" and followed through around the "roadblock" to check to see if subsequent stages or circuits have any other problems. If not, it is safe to assume that you have located the problem. At this point, testing of individual components, voltages, or currents will reveal the defective component that is causing the problem at hand. See Fig. 2-14.

The signal tracer is simply a high gain, low noise audio amplifier, with a loudspeaker for audible monitoring. A diode detector probe is furnished for checking rf and i-f signals. In addition, a signal injector circuit, which is a simple audio oscillator, is provided to act as a signal source if no input signal is available or practical to obtain. This feature makes for convenience in field service work, since it eliminates the need for an extra audio oscillator. Both the audio amplifier and oscillator operates from either an ac source or a +9 to +12 volt battery pack if isolation from the ac lines is desired. A suitable PC layout is illustrated in Fig. 2-15, or the circuit can be constructed on a piece of G-10 material breadboard style, or if preferred, vectorboard. Adequate grounding and shielding should be provided since this is a high gain circuit. It is best to enclose the finished instrument in a metal enclosure, or in an aluminum foil lined plastic case.

Referring to Fig. 2-14, the signal tracer portion operates as follows. Input signal at J1 couples through C1, a dc blocking capacitor, to R2, a current limiting resistor to the gate of field effect transistor Q1. R3 provides a dc return to ground for the gate of Q1. C2 and R4 provide source bias and set the dc operating point for Q1. Amplified signal appears at the drain of Q1 across R5. R5 provides dc bias to Q1. C4 provides an ac path to ground (bypass) for any power supply stray signals appearing at the bottom of R5. S2 and R1 provide a means of injecting dc bias to J1 for the purpose of noise testing components connected to J1 (input) or as a means of providing dc bias to the external rf probe (see later) to increase its low level rf detection sensitivity. R9 provides B+ decoupling to C4.

Audio signal across R5 is coupled to R6, the amplifier gain control, through C3. The setting of R6 determines the overall amplifier gain. Signals as low as 10 microvolts at J1 can be plainly heard at full gain setting. The signal from the wiper of R6 is connected to the input of IC1, a small audio power amplifier with about 250 mW (1/4 watt) audio capability, sufficient to drive LS1, a 16 ohm to 32 ohm speaker, transistor radio type, not critical as to impedance. A 3″ size is adequate. R7 and C5 suppress a tendency for IC1 to oscillate at high frequencies. C6 is a dc blocking capacitor for coupling audio from IC1 to LS1. C7 provides dc supply bypassing for IC1. R8 and LED1 serve as a power on indicator.

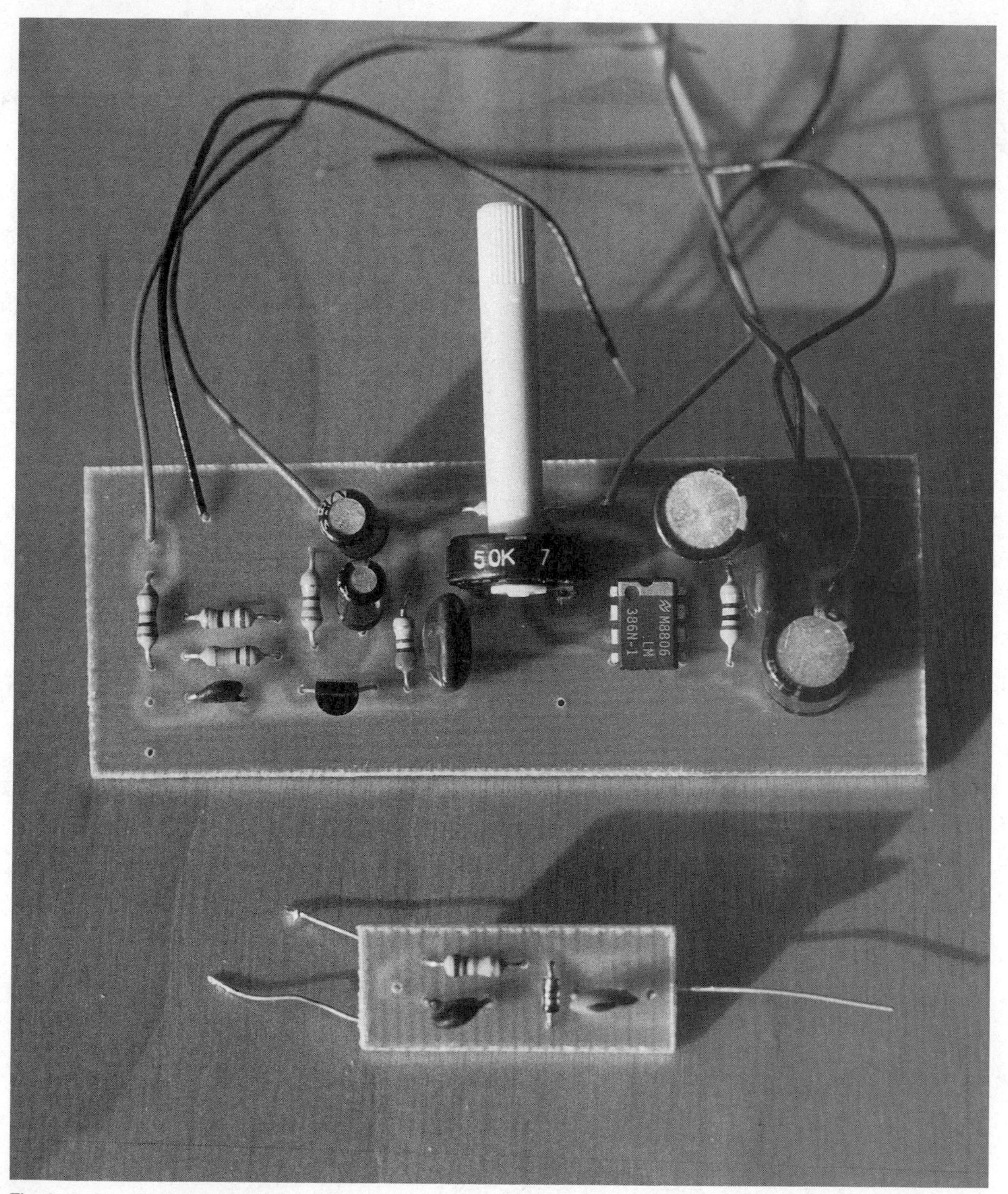

Fig. 2-12. Signal tracer and detector PC board.

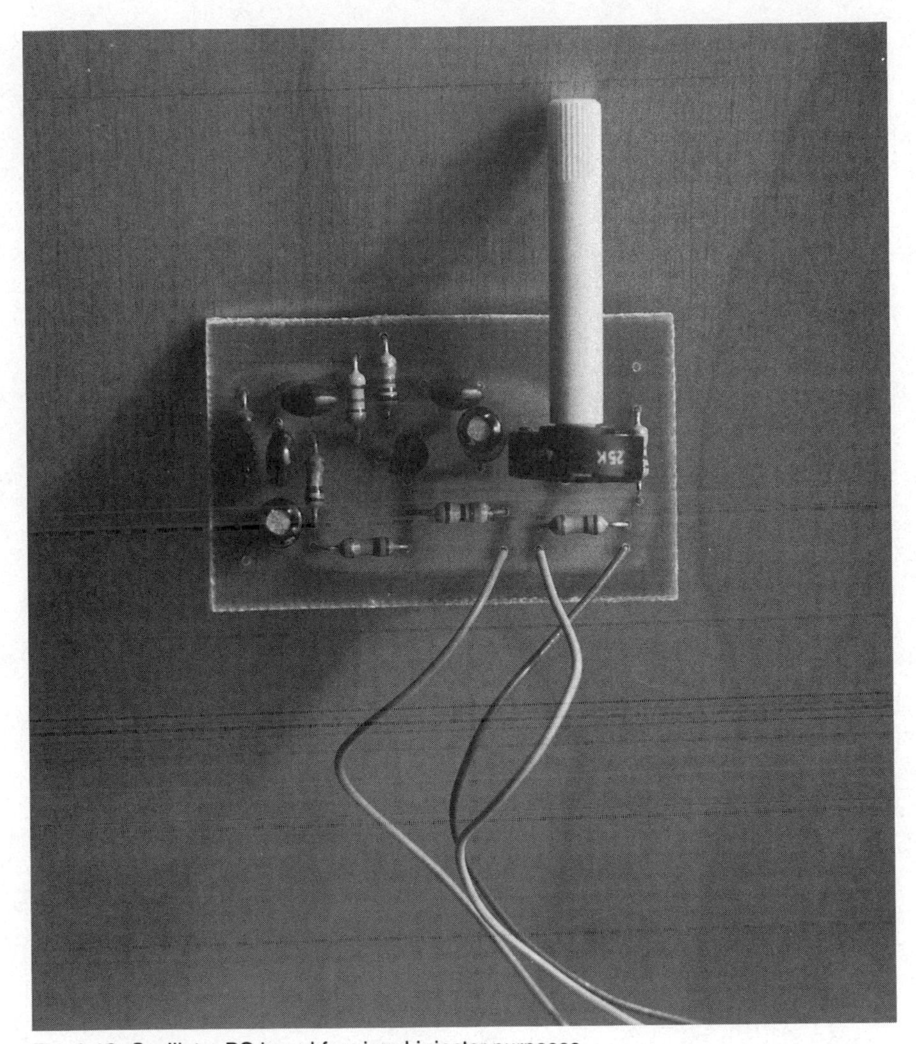

Fig. 2-13. Oscillator PC board for signal injector purposes.

S1 is a dc on-off switch. If convenient, it may be part of R6, the gain control, although a separate mounted switch for S1 is probably preferable to save wear and tear on R6. D1 is a diode to protect IC1 against accidental reversal of power supply. It also serves as a rectifier if an ac power source is used to power the unit. In this case, C7 should be changed to a 4700μF unit to reduce 60 Hz ac hum, and C4 should be increased to 470μF to further reduce ac hum.

The detector probe can be constructed inside a plastic tube, medicine bottle, or a piece of 1/2″ PVC plastic pipe obtained at a plumbing or hardware store (see Fig. 2-17). A ground wire 6 to 8″ long fitted with an alligator clip is suitable but make sure flexible stranded wire is used to reduce breakage. A 5 foot length

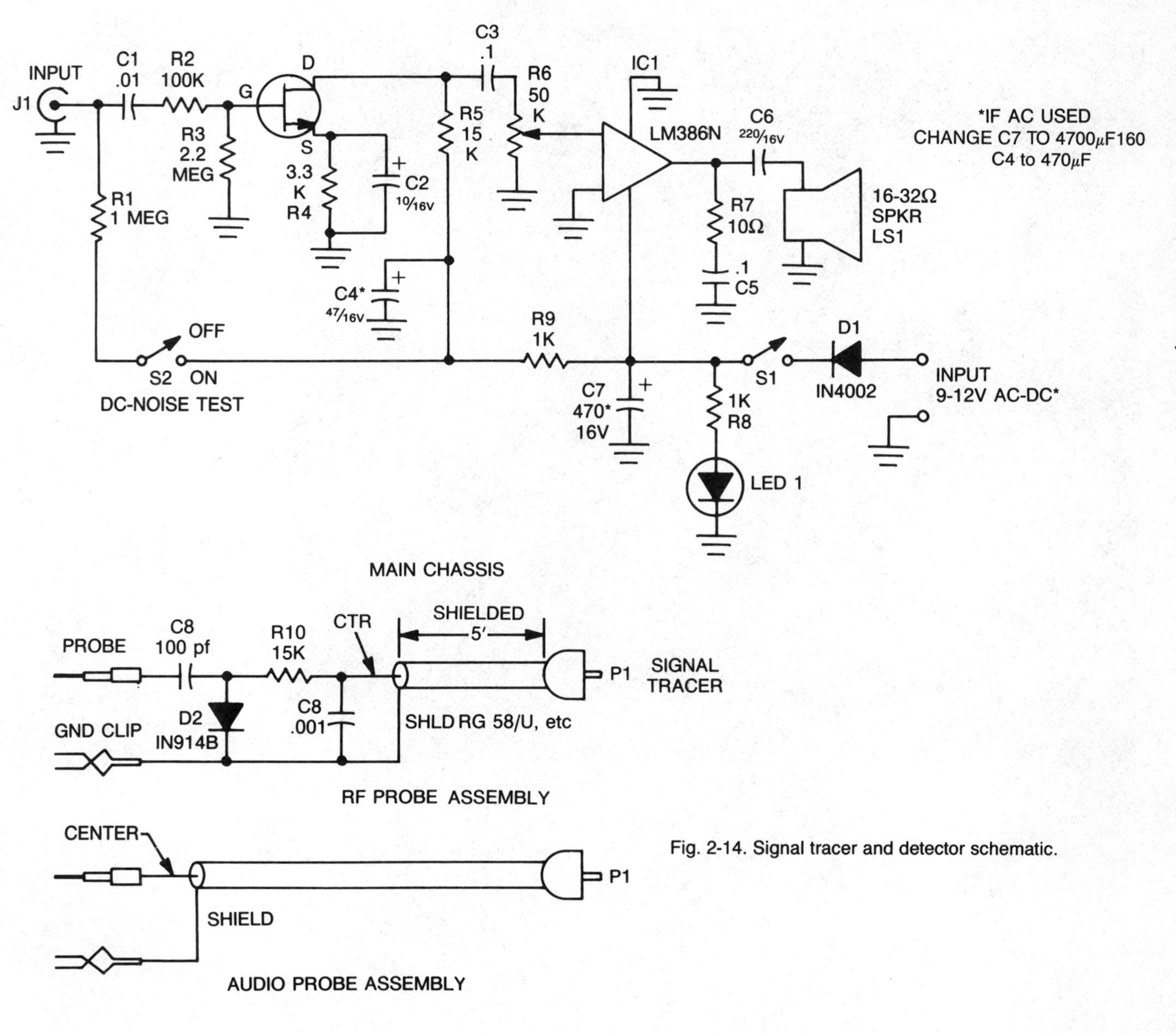

Fig. 2-14. Signal tracer and detector schematic.

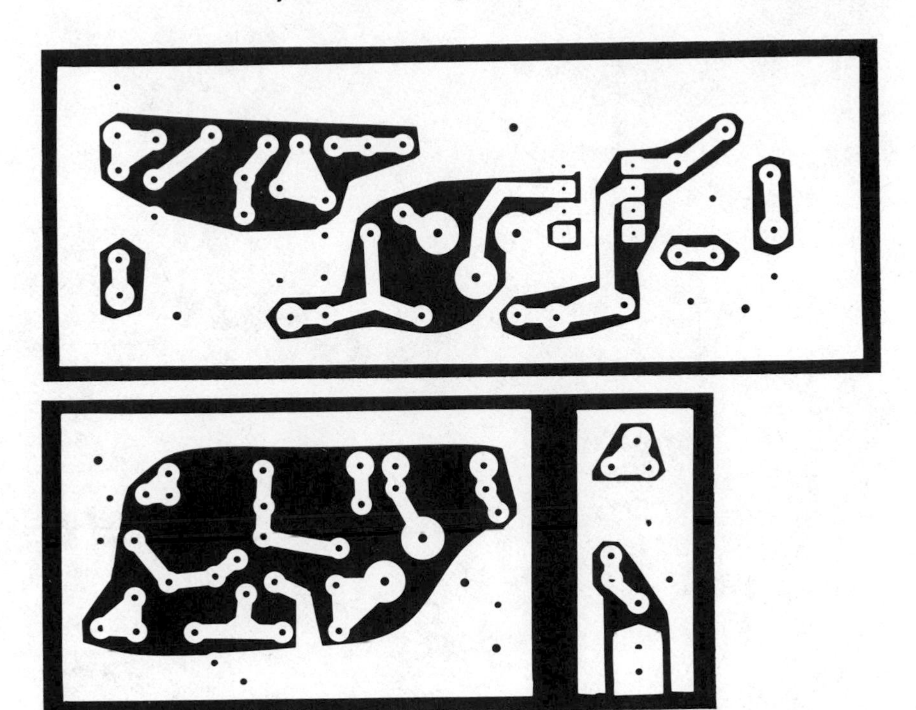

Fig. 2-15. Artworks for signal tracer, detector, and oscillator.

of RG 58/U coax cable can be used as a test lead to the signal tracer. Either a BNC or phone plug (RCA style) can be used at J1 and P1. We prefer a BNC, but this is up to the discretion of the builder. If only audio work will be done, just the probe assembly shown in Fig. 2-15 is needed. However, having both probes is desirable in any case.

The signal injector schematic is shown in Fig. 2-18. Q1 is a 2N3565 low noise, high gain and NPN transistor. R14 provides dc bias for Q2 through switch S3, the oscillator on-off switch. R11, R12, and R13 provide base bias for Q2. C10 prevents loss of ac gain via feedback through R12 and R13. C11 provides dc blocking and ac coupling of oscillator signals from the collector of Q1. C12, R18, C13, R19, and C14 and the input impedance of Q2 (around 10k ohms) form a phase shift network. At the frequency of oscillation, there is a net 180° of phase shift and a loss of 29× in the network. Therefore, a minimum amplifier gain of greater than 29 is necessary for oscillation to occur. The theoretical frequency of oscillation is given by

$$f = \frac{1}{2\pi \sqrt{6}\, RC} \text{ or } \frac{1}{15.4\, RC} \text{ approximately}$$

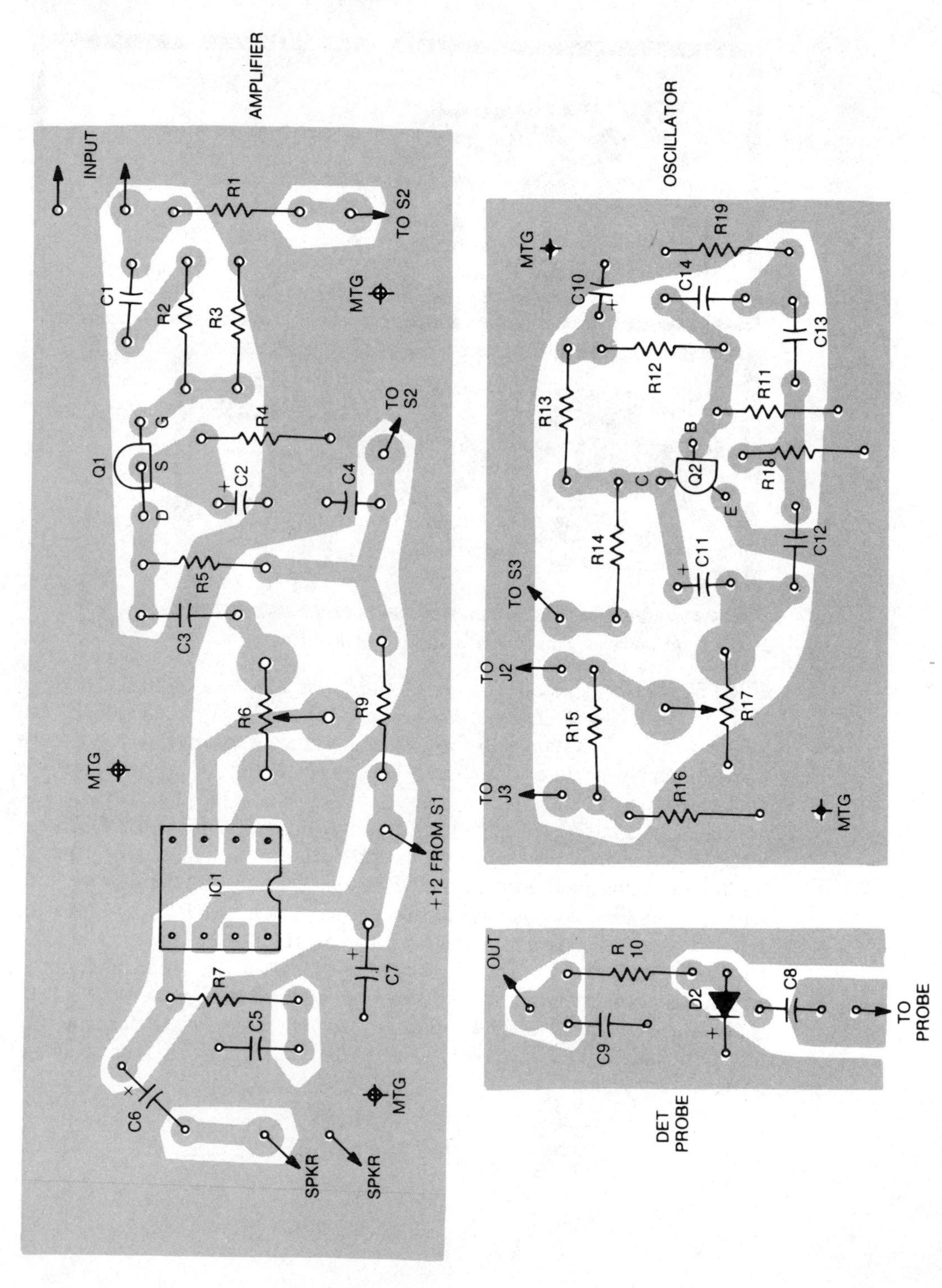

Fig. 2-16. Layouts for signal tracer, detector, and oscillator.

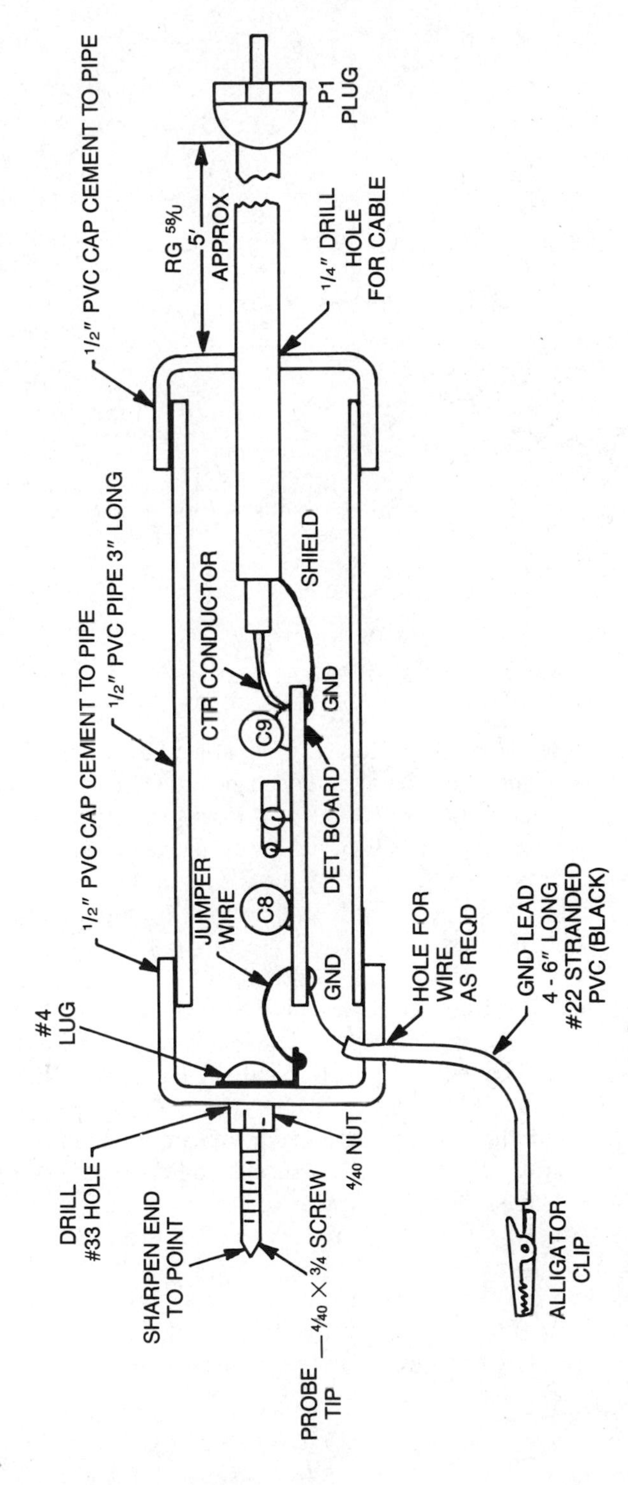

Fig. 2-17. Simple detector probe using PVC pipe.

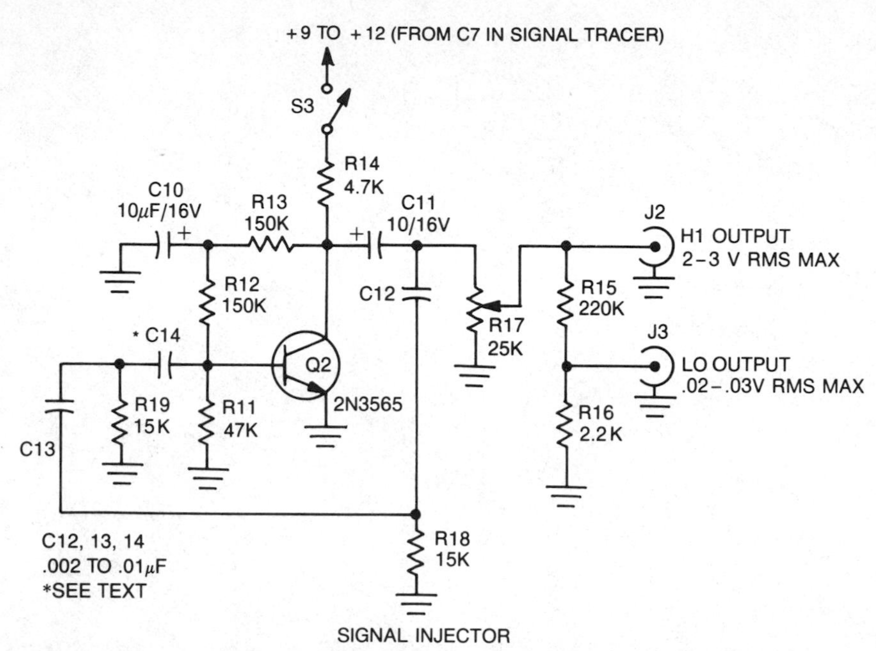

Fig. 2-18. Signal injector (oscillator) schematic.

This formula predicts C12, C13 and C14 should be .0043μF if f = 1,000 cycles, but also assumes that the input impedance of Q2 is infinite and R11 in parallel with R12 is also equal to R13 and R14. However, this is not so, but only approximately true. Actually, 1,000 cycles is not mandatory but is customarily used for tests. You can use .002 to .01 mF for C12, C13 and C14 to get frequencies between about 400 Hz to 2,000 Hz, as desired if you prefer, but 1,000 or 400 Hz are customarily used for most audio testing.

R17 is a level control to vary the oscillator output. R15 and R16 provide a means of obtaining low level output signals below 30 millivolts or so without crowding at the low end of adjustment pot R18. J2 has a full 0 – 2 volts (approx.) output, while J3 has low level output. Either RCA phono or BNC connectors may be used as the builder prefers.

The instrument should be constructed to minimize unwanted coupling of audio from Q2 into the amplifier Q1-IC. A suggested PC layout is shown in Figs. 2-15 and 2-16. If desired, either the tracer or injector circuit can be built separately by cutting the PC board as shown. R8 and LED1 is a pilot light and can be omitted if desired.

Checkout is simple. First, test for B+ shorts. The B+ line should read 300 ohms or more to ground. If not, reverse the ohmmeter probes. If still low, check for a short or a defective IC1. Next, apply power and check voltage at the drain of Q1. It should be +4 to +8 volts (to 10 volts is OK if a 12V supply is used. Also, pin 5 ±C1 should read half the supply voltage (+9.5 to +6 volts or so). IC1 should *NOT* get hot.

With R6 set halfway open, a hum or click should come from LS1 as the center conductor if J1 is touched with a screwdriver blade. Next, connect the rf probe assembly. Set S2 to "ON." Touch the probe to a 10′ length of wire or an outside antenna. A loud buzz should result, and possibly you might hear radio stations.

Next, turn on S3. About +3 to +8 volts should be present at the collector of Q2, and +9.6 or +0.7 volts on the base of Q2. Connect a wire from the center conductor of J3 to the center conductor of J1 (tracer input). With R6 set halfway, a tone should be heard at all or most settings of R18 (except at extreme low end, possibly). If not, check all components, C12 to C14, R18, R19, and Q2.

RESISTORS 1/4W 5% or 10% Tol.	
R1	1 Megohm
R2	100k ohm
R3	2.2 Megohm
R4	3.3k ohm
R5, R10, R18, R19	15k ohm
R6	50k ohm pot
R7	10 ohm
R8, R9	1k ohm
R11	47k
R12, R13	150k
R14	4.7k
R15	220k
R16	2.2k
R17	25k pot

SEMICONDUCTORS	
Q1	MPF102 FET
Q2	2N3565
D1	1N4002
D2	1N914B or 1N4148
IC1	LM386N
LED1	Red LED (any)

CAPACITORS	
C1	.01 disc 50V
C2, C10, C11	10μF/16V elec
C3, C5	.1μF Mylar
C4	47μF/16V elec
C6	220μF/16V
C7	470μF/16V
C8	100pF ceramic 500V
C9	.001 ceramic
C12, C13 C14	See text

MISCELLANEOUS*	
J1, J2, J3	Phono jack RCA type or BNC
P1	Phono plug RCA type or BNC
S1, S2, S3	SPST switch, any type
LS1	Any small speaker 16 to 32 ohm
Other	Chassis, PC boards, hardware to suit 1/2″ plastic pipe fittings (for probes) alligator clips, TIP jacks as required

*Not included in list

Fig. 2-19. Parts list, Project 2.

This completes checkout. A suitable cabinet and mounting arrangement to produce a finished instrument is given in Fig. 2-20.

A kit of parts consisting of the (3) printed circuit boards and all parts that mount on the boards is available from:

North Country Radio
P.O. Box 53, Wykagyl Station
New Rochelle, NY 10804

Price: $29.50 plus $2.50 postage and handling.
New York residents must include sales tax.

PROJECT 3: FUNCTION GENERATOR

A function generator is a signal generator that can generate waveforms of various types rather than simply a sinusoidal or near sinusoidal waveform of a given frequency. Generally, the generator has capability of generating fairly accurately shaped sinusoidal waves (typically 1% distortion), square waves free of spikes, overshoots, etc., and ramp type waves (triangle waveforms) with good linearity. Also, it is desirable to be able to add a given dc level to these waveforms. For example, a square wave that can have levels with respect to ground that are one polarity (unipolar). This allows the use of the generator for digital logic, (TTL, CMOS, etc.), that may require definite limits on high and low levels. Sufficient output should be available without external interfaces, to directly drive many devices, such as speakers, small motors, magnetic devices, etc. Frequency response down to 0.01 hertz is useful for servo work, and capability to 100 kHz for ultrasonic work is also desirable. See Fig. 2-21.

A function generator will be described that meets these needs. It is built around the Intersil 8038. It covers 0.01 Hz to 100 kHz in several* ranges, can drive a 50 ohm load with 8V rms signal at any frequency in this range, and has ±6Vdc offset, adjustable with a control. Means of controlling the oscillator is from an external sweep source for FM applications, or with a pulse for producing bursts. The signal always starts at zero crossover in this mode. Sinewave distortion is typically less than 1% over this frequency range. A dc level shift control is provided which gives ±6V offsets. Note, however, that the maximum peak output voltage of ±12V may not be exceeded. Trying to offset a 20V p-p waveform with +6V, for example, will cause clipping, since a +16 to −6V swing would be required, which is out of range of the circuit. In this case, a clipped +10 to −4V swing would result.

---

*.01 and .1 Hz range can be deleted if desired.

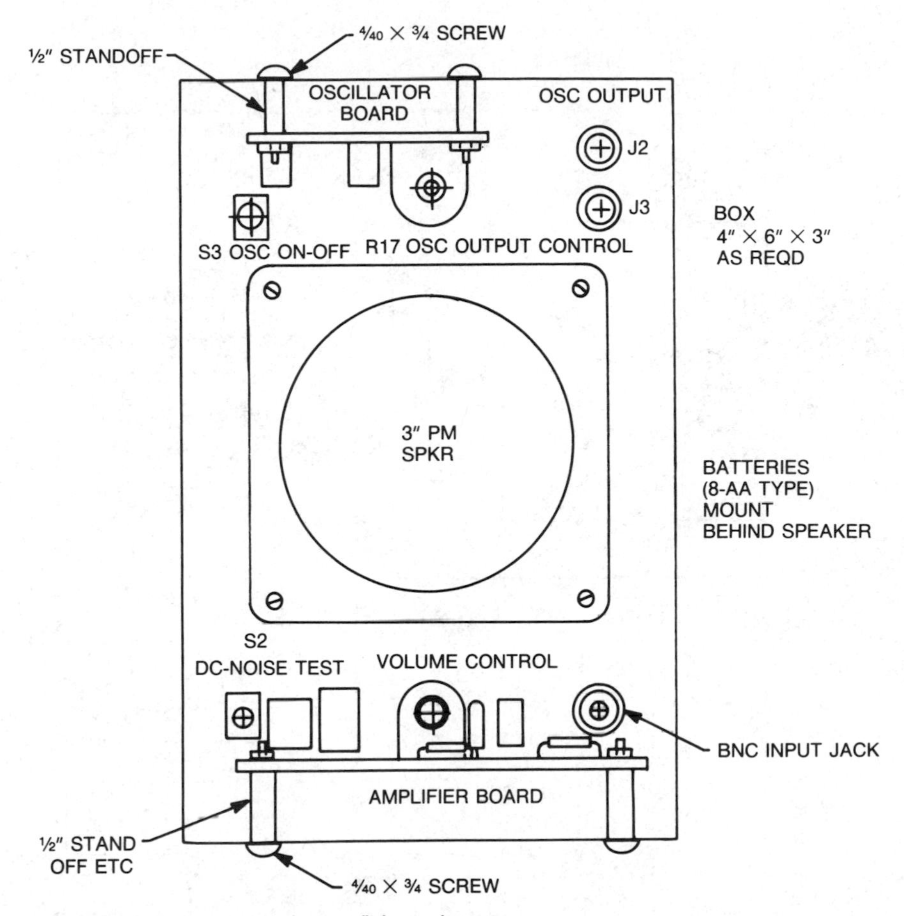

Fig. 2-20. Suggested signal tracer/injector layout.

With loads less than about 30 ohms, somewhat less output swing is to be expected. Remember that 7.07 volts rms into 50 ohms is one watt of power, more than most function generators will deliver. This is enough to light a small bulb or run a small motor and at audio frequencies, will make a lot of noise with a speaker.

The function generator schematic is shown in Fig. 2-21. In order to provide reliable calibrating and easy use, the claimed 100:1 or 1,000:1 frequency swing in one range is not attempted. Such a unit would be difficult to calibrate, set up, and other than as an experiment, would have little value. It is better to break the frequency range into 10:1 increments. One calibrated dial scale can be used for all ranges, since all the ranges are 10:1 multiples of each other. Trim controls are used to provide calibration at both the high end and low end of the frequency range. Only one range need be calibrated, since the other ranges are obtained by switching in capacitors, the same resistors being used on all ranges.

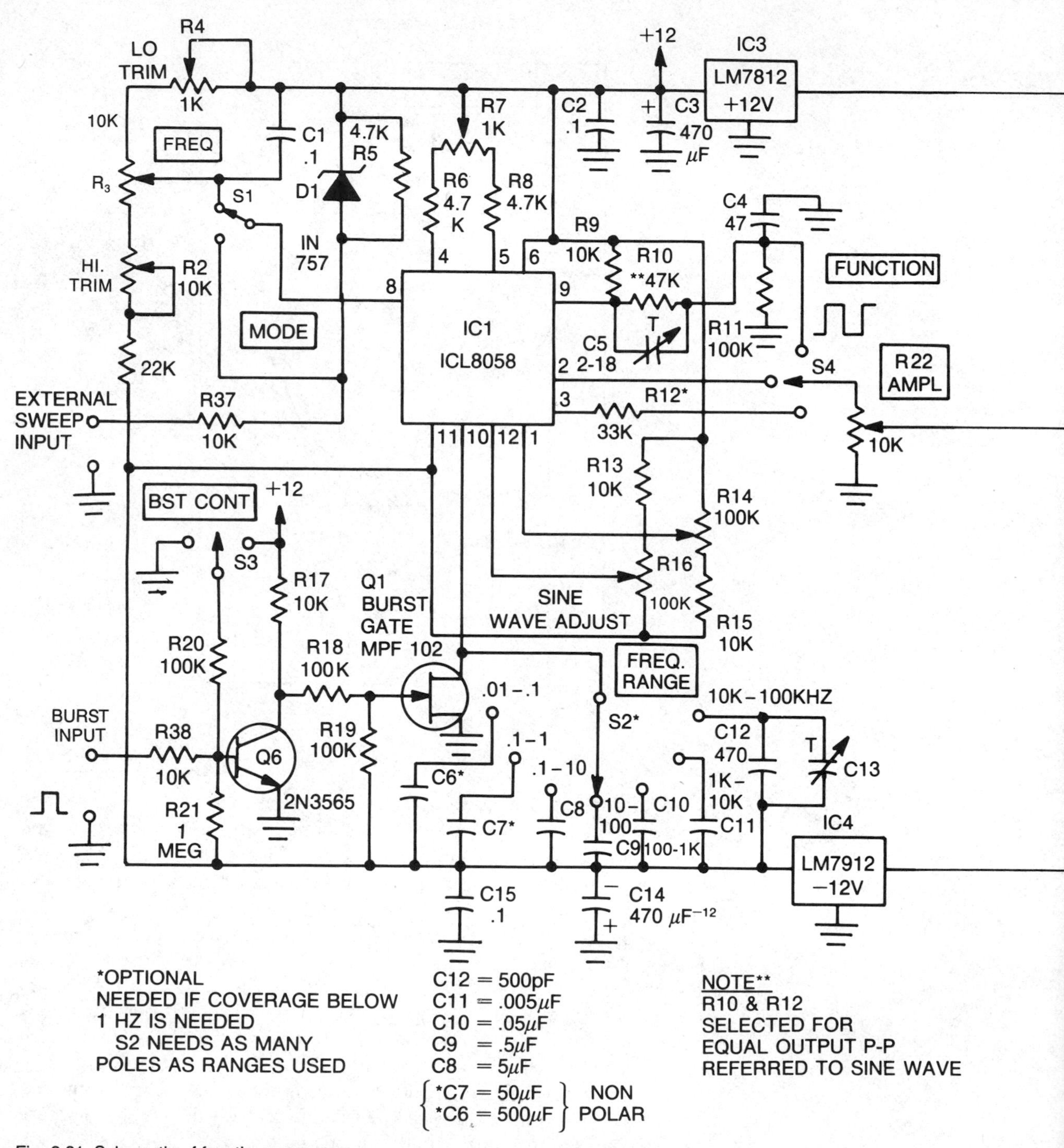

Fig. 2-21. Schematic of function generator.

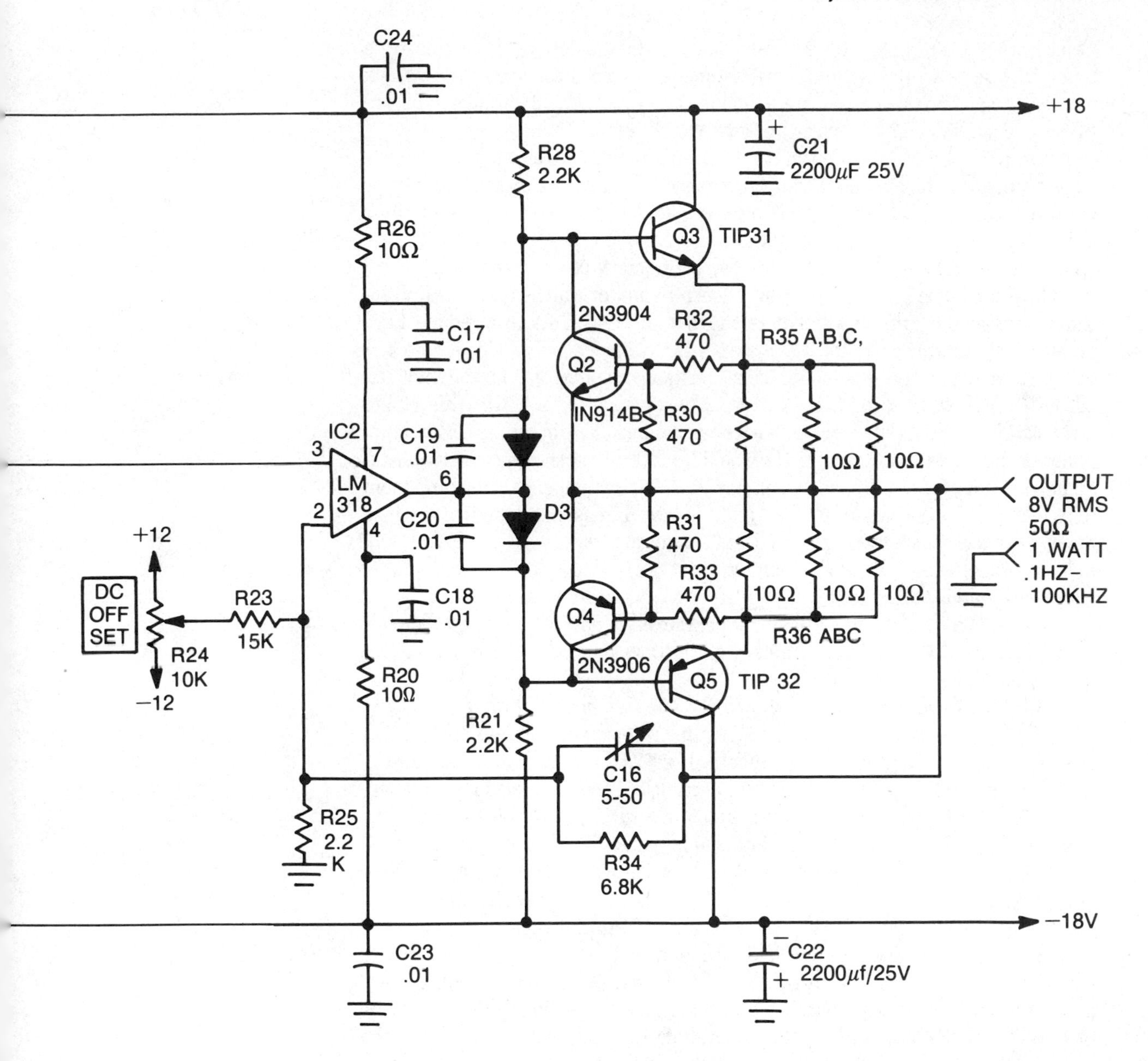

C24
.01
+18
R28
2.2K
C21
2200μF 25V
R26
10Ω
Q3
TIP31
C17
.01
2N3904
R32
470
R35 A,B,C,
Q2
IN914B
R30
470
IC2
3
7
C19
.01
6
LM
318
2
4
C20
.01
D3
10Ω
10Ω
OUTPUT
8V RMS
50Ω
1 WATT
.1HZ-
100KHZ
+12
DC
OFF
SET
R23
15K
R24
10K
−12
C18
.01
R31
470
R33
470
Q4
10Ω
10Ω
10Ω
R36 ABC
2N3906
Q5
TIP 32
R20
10Ω
R21
2.2K
C16
5-50
R25
2.2
K
R34
6.8K
−18V
C23
.01
C22
2200μf/25V

Capacitors C6 through C13 are individually selected 5% or 10% tolerance types. If exact calibration is not a requirement, then 5% capacitors from stock will be adequate. C6 and C7 are tantalum types, while C8 through C11 are Mylar capacitors. C12 is a silver mica or NPO ceramic.

The IC1 8038 drives an op amp with wide bandwidth, which in turn drives a power amplifier stage using a complementary pair of power transistors. All dc coupling is used, since frequency response must be down to 0.01 Hz (one cycle in a hundred seconds) and dc offset is to be provided. A wideband LM318 op amp is used, with a pair of TIP 31/32 complementary power transistors. Short circuit protection is provided by a pair of current limiter transistors. These transistors sense the emitter current of the output transistors and reduce base drive at high emitter currents. Regulated ±12V voltages are provided, by a pair of regulator ICs, to the ICL 8038. This assists in obtaining good frequency stability. The ICL 8038 is rated at −250 p-pm/°C or ±2.5 Hz at 1 kHz over a 15 to 25°C ambiant temperature range. This does not include drift due to circuit components, but conservatively ±5 Hz would be a reasonable figure. This is adequate for most test purposes. Calibration accuracy depends on component tolerances and care in setup, but ±2% is not unreasonable and more than enough for the usual tests that would use a generator of this nature. If you want high accuracy, use a frequency counter.

Many frequency counters have a 10 second gate time available. This will permit measurements to ±0.1 Hz. However, especially on the ranges 0.01 – 0.1 and .1 – 1 Hz, measurement of the waveform period rather than frequency is the best approach.

A PC artwork Fig. 2-22 is shown for the function generator to be described. In addition, a source of ±18V, with 1V p-p ripple, or less, is needed. Current is about 250 mA each side (peak). No reliance on peak current capability can be made here, since at 0.01 Hz, each half of the waveform lasts 50 seconds. A square wave with 12V p-p amplitude and a +6V offset loads the positive supply to 250 mA if a 50 ohm load is used. The output transistors Q3 and Q5 must be adequately heat sinked and the power supply must deliver ±18V at 250 mA continuously as well, without overheating. Also, good mechanical construction with regard to S2 and C6 – C13 is important, since the capacitances determine the oscillator frequency. Use of an adequate power transformer (25V CT at 450 mA), and a metal enclosure for good shielding will go a long way toward insuring reliable performance. Referring to the schematic of Fig. 2-21, a description of circuit operation will be given. Also refer to the data sheet (see appendix) for the ICL 8038 for specific details as necessary.

The frequency of oscillation of IC1 is determined by the charge and discharging of a capacitor connected to pin 10 and the currents generated by two internal current sources used to charge and discharge this capacitor. In turn, these currents can be controlled by the dc voltage at pin 8 with respect to the positive supply voltage. R1, trim pot R2, frequency adjust pot R3, and turn pot

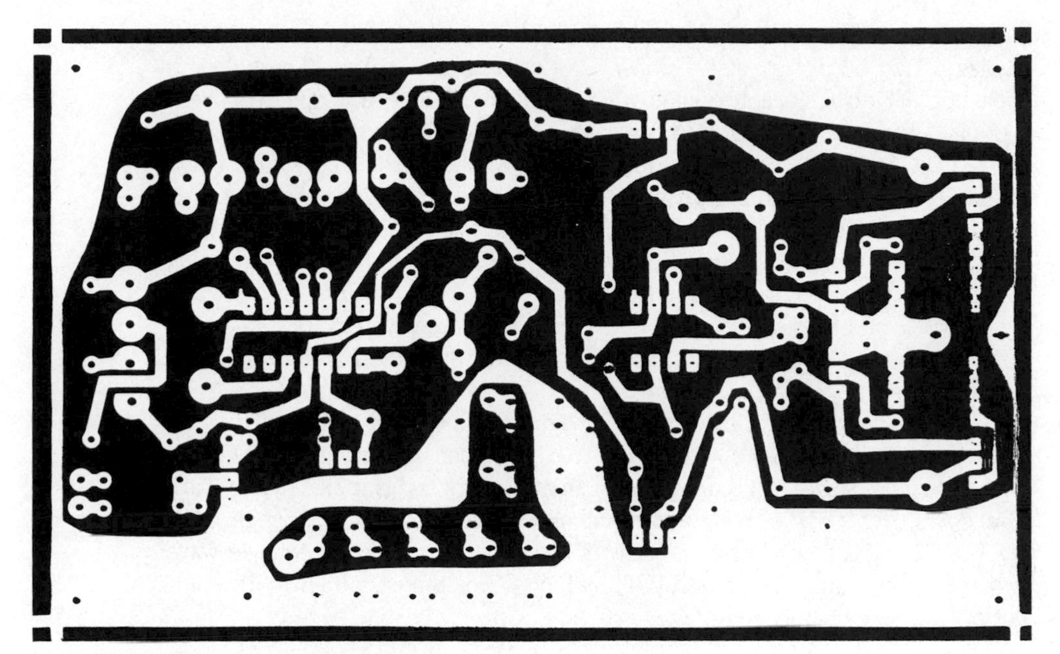

Fig. 2-22. PC artwork for function generator.

R4 form a voltage divider with the output taken from the wiper of R3, the frequency control pot. R1 and R2 determine the most negative voltage available (corresponding to highest frequency end of the range) and R4 determines the most positive voltage available (lowest frequency of range). R2 and R4 are alternately adjusted to provide a 12:1 frequency range to permit overlapping $\pm 10\%$, of each range with its lower and higher adjacent ones. The voltage from this divider goes via S1 to pin 8 of the ICL 8038. C1 is a bypass capacitor.

S1 is used to select either the frequency control voltage for IC1, or an external control voltage if external control is desired. D1, R5, and R37 limit this voltage and protects the ICL 8038 against damage by external sources, etc. If no external control is needed, S1, R37, R5 and D1 can be omitted and the wiper of R3 connected to pin 8 directly.

The frequency determining capacitors C6 through C13 are connected through range switch S2 to pin 10 of the ICL 8038. Q1, a normally cut off FET, connects between pin 10 and ground. In the burst mode, an input pulse is applied to Q6 through protection resistor R38. S3, normally connecting bias resistors R20 and R21 to the positive supply, is connected to ground. A pulse of at least +2V amplitude is inverted by Q6 and cuts off turn-on bias to Q1 through resistor R17 and R18. Q1 is now cut off and removes the ground path at pin 10 of IC1. IC1 oscillates for the duration of the burst pulse, which ceases when the pulse disappears.

In the continuous mode, S3 supplies bias to R20, turning on Q6, which causes Q1 to be cut off at all times, allowing IC1 to oscillate. The selected capacitor (C6 to C13) determines which frequency range IC1 operates over. C12 has a trimmer capacitor C13 across it, since the highest range is most subject to calibration errors due to wiring and somewhat unpredictable stray capacitance. A total of seven overlapping ranges are provided, with a calibrated 10:1 spread (actually 12:1 at extremes) to allow complete coverage of all frequencies from .01 Hz to 100 kHz. Ranges may be added or deleted as desired, using more or fewer capacitors and switch positions.

R6, R7, and R8 set the values of the current sources within IC1. The ratio of R6 and R8 should be unity for a symmetrical waveform. R7 is used to adjust symmetry of the generated waveform.

R9 provides a load for the square wave output. R13, R14, R15 and R16 are distortion adjustments for sine waves, R10 is a compensator trimmer C5 (adjust for best square wave). R11 and C4 form an attenuator to bring the square wave level down to that of the triangle wave and sine wave. R12 sets the triangle wave level to that of the sine wave. S4 selects either square, sine, or triangle wave and applies it to amplitude control R22, which varies the output level. The output of R22 feeds amplifier IC2.

Power to IC1 is supplied by regulators IC3 (+12V) and IC4 (−12V). C2, C3, C14, and C15 are bypass capacitors for both noise reduction and stability purposes.

IC2 is a wide bandwidth op amp, IC2 drives output emitter followers Q3 and Q5. R26, R27, C18, and C19 supply operating voltage to IC2. The output (pin 6) dc couple to outputs Q3 and Q5 through coupling diodes D2 and D3. C19 and C20 bypass D2 and D3 for high frequency signals and improve high frequency response. R28 and R29 supply bias to Q3 and Q5. Emitter resistors R35 ABC and R36 ABC (parallel connected 10 ohm 1/4W) limit current in Q3 and Q5 and add dc stability to the amplifier.

Q2 and Q4 protect Q3 and Q5. If excessive current (more than 500 mA) flows on either Q3 or Q5, R32 and R30 (or R31 and R33) will provide sufficient bias to turn on Q2 (or Q4). These transistors shunt some of the base drive away from Q3 (or Q5) and limit the resultant emitter current to about 500 to 700 milliamperes.

Output is taken from the junctions of R35 ABC and R36 ABC (each is 3.33 ohm, 3−10 ohm resistors in parallel). About 12 volts peak is available at 250 mA (for good linearity) in both the positive and negative directions. C16 and R34 form a feedback network. Overall amplifier gain is about 3−4×. C16 is adjusted for best waveshape of the square wave, and will be about optimum for the sine and triangle as well. R25 provides a dc return and also is part of the feedback network.

Dc offset of the output signal is obtained by a variable dc bias into R23. R24 connects between +12 and −12 volts, and any voltage in between can be obtained at the wiper of R24 by adjustment. This voltage causes IC2 to shift its bias point either positive or negative, and thus the output bias point shifts with it. About ±5 to ±6 volts shift is available with the stated value of R23.

C21 and C22 are bypass capacitors. They are 2200μF 25V units. Q3 and C5 should be heat sinked since there is a possible 10 watts dissipation under short circuit conditions. The cabinet that the PC board is mounted to can serve as a heat sink in some instances. See Fig. 2-23.

Fig. 2-23. PC board for function generator.

A suggested PC layout for the function generator is shown in Figs. 2-22 and 2-23 and a layout using a commercially available cabinet (Radio Shack P/N 270-274A) is shown in Fig. 2-25. While these are only given as examples, they should be used if possible, since our prototype worked well with them. Standard components are used for all circuitry, except for C6 to C11 which are hand selected. It is more important that the capacitors C6, C7, C8, C9, C10 and C11 have a 10:1 ratio between them rather than being the exact value shown in the schematic, since R2 and R4 can correct for this. Use a capacitance bridge, meter, or other method to select these capacitors from stock ±5% or ±10% tolerance parts. If no means of measuring capacitance is available, then you can either just use ±5% and accept possible calibration errors, or use the finished oscillator circuit with a frequency counter to "hand trim" the values by substitution. It is permissible to make up C6 through C11 by using paralleled combinations of smaller values that add up to the required values, if you wish, although it may appear somewhat messy on the PC board. C12 needs no selection since C13 in parallel with it adjusts the value.

Use a good quality pot (Ohmite or AB) that is sealed against moisture for R3. A mil spec type is a good idea here. A dial can be made from plastic to fit the shaft. Calibrations can be made on the dial using a dry transfer letter sheet. A rough dial can be made from cardboard. This can be calibrated using a frequency counter, and a final dial copied from this, on plastic. See Fig. 2-26.

To check out the finished unit, use the following procedure. First set all controls as follows:

S1 — internal input	R22 — midway
S2 — 100–1000 Hz range	R24 — midway
S3 — continuous	C5 — 1/2 meshed
S4 — sine wave	C16 — 1/2 meshed
R2 — midway	R14 — midway
R4 — midway	R16 — midway
R3 — center of rotation	

Next, check for shorts to ground at inputs and outputs of IC3 and IC4. At least 200 ohms should be measured. If not, first find the problem. Reverse ohmmeter probes and take the highest reading.

Now apply ±18 volts to C21 and C22. Check to see if there are the following readings. Do this quickly. (If anything is getting hot, there is something wrong. Immediately remove power and troubleshoot.)

Next, adjust R7 for a symmetrical square wave at output, with R3 in mid position. Now place S4 in the sine position and adjust R14 and R16 for minimum sine wave distortion or for best appearance on the scope. See Fig. 2-27. Next, check for operation overall frequency ranges, each setting of S2. You may have to watch the output for a while on the .01 to .1 Hz range as the waveform changes slowly.

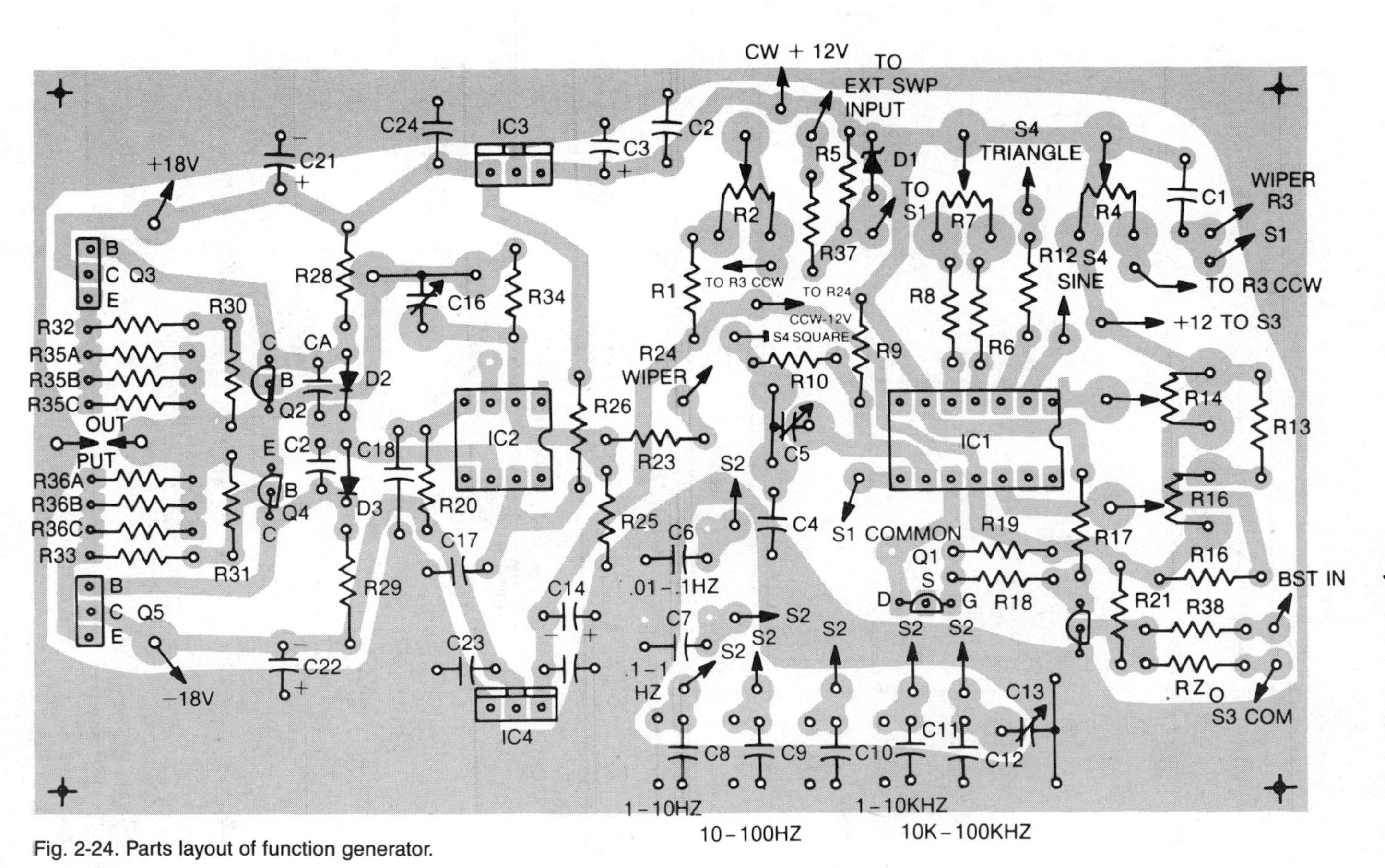

Fig. 2-24. Parts layout of function generator.

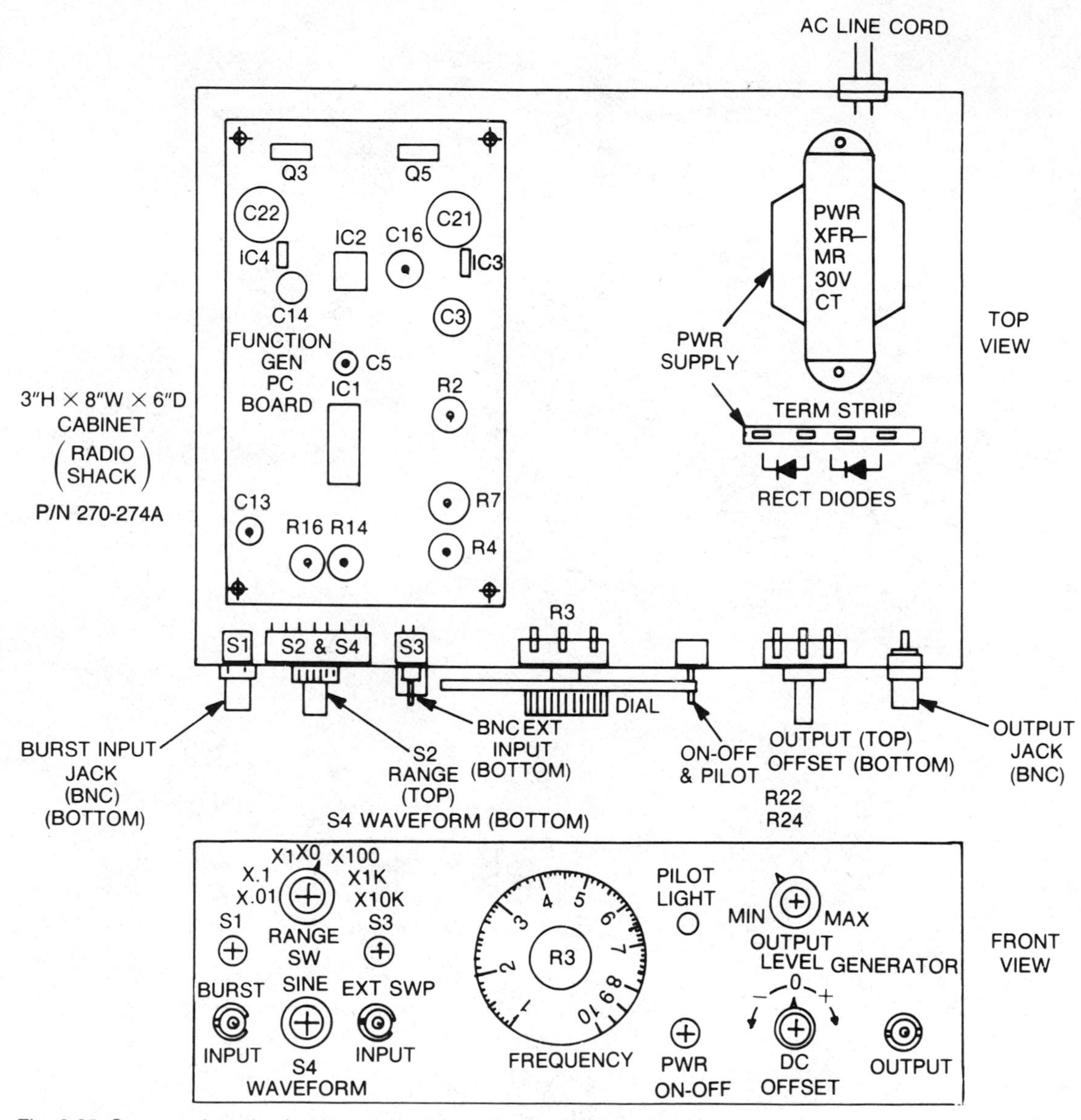

Fig. 2-25. Suggested mechanical layout and packaging for function generator.

Adjust C5 and C16 on the highest range (use about a 10 to 100 kHz square wave) for best looking square wave. C16 can be optimized for triangle, C5 for square if desired. If available, a distortion analyzer can be used to do this. Adjust C5 and C16 for minimum distortion.

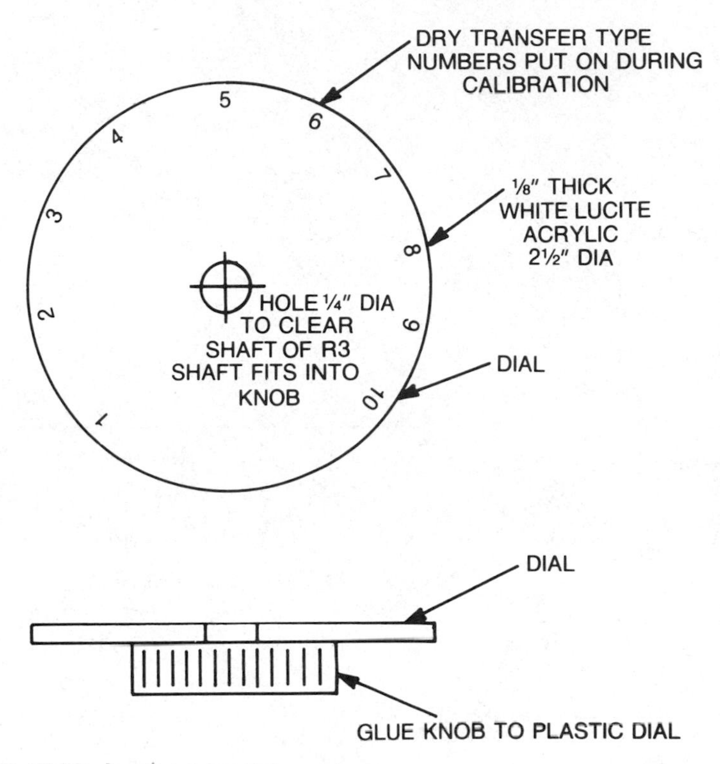

Fig. 2-26. Dial for function generator.

This completes checkout. Now repeat all adjustments for optimum results if desired. The function generator is now complete.

Figure 2-28 shows a suitable power supply. We used a 30V (surplus) 500 mA CT transformer. A Radio Shack P/N 273-1366 (25.2V 500 mA) will do very well as a substitute. Maximum output will be slightly less, but this is not a problem in most cases. See Fig. 2-28.

A kit of parts consisting of the printed circuit board and all parts for coverage of 1 Hz to 100 kHz that mount on the board is available from:

North Country Radio
P.O. Box 53, Wykagyl Station
New Rochelle, NY 10804

Price: $49.50 plus $2.50 postage and handling.
New York residents must include sales tax.

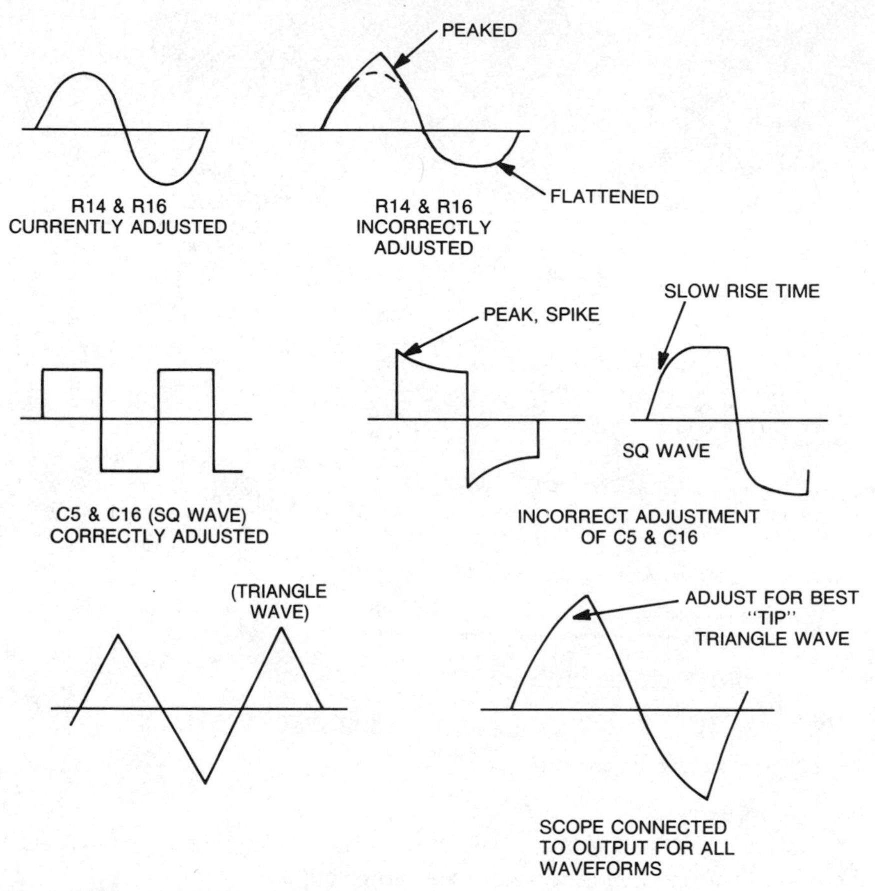

Fig. 2-27. Function generator waveform.

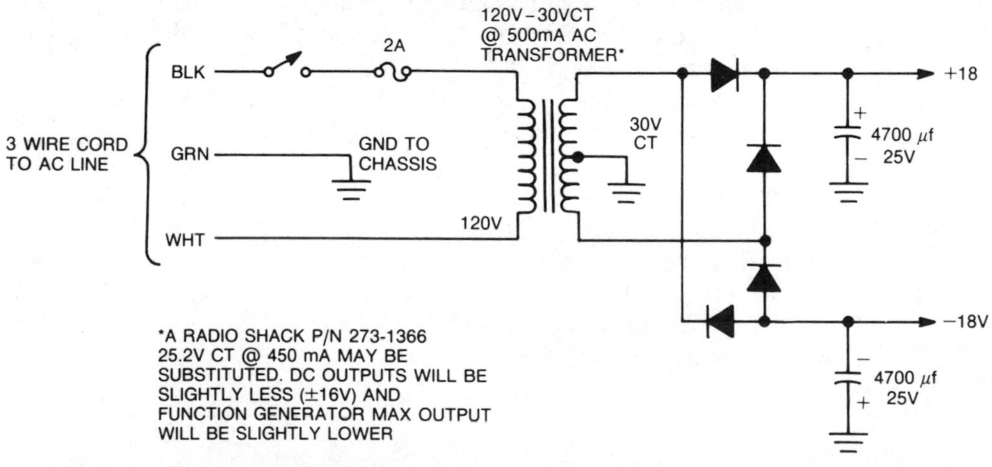

Fig. 2-28. Schematic of a suitable power supply for function generator.

RESISTORS 1/4W 5% or 10%

R1	22k
R2	10k pot, trimmer
R3,R22, R24	10k pot, w/shaft
R4,R7	1k pot, trimmer
R5,R6,R8	4.7k ohm
R9,R13,R15, R17,R37, R38	10k ohm
R10	47k ohm
R11,R18 R19,R20	100k ohm
R12	33k ohm
R14,R16	100k ohm pot, trimmer
R21	1 Megohm
R23	15k ohm
R15,R28, R29	2.2k
R26,R27, R35 ABC, R36 ABC	10 ohm
R30,R31, R32,R33	470 ohm
R34	6.8K1

MISCELLANEOUS*

S1,S3	SPDT switch
S2	Rotary 1P 4 to 8 positions as required
S4	Rotary 1P 3 positions
1	PC board
1	Case as required

Hardware as required
3 sets binding posts or jacks for inputs and outputs (BNC suggested)

*Not included in list

CAPACITORS
(See Fig. 1)

C1,C2,C15	.1 Mylar 50V
C3,C14	470μF/16V elec
C4	47pF NPO/SM
C5	2 – 18pF trimmer
C6-C12	See schematic (Fig. 1) and text—specially selected, depends on application
C13,C16	5 – 50pF trimmer
C17 – C20, C23,C24	.01 disc 50V
C21,C22	2200μF/25V elec

SEMICONDUCTORS

IC1	ICL8038
IC2	LM318
IC3	LM7812
IC9	LM7912
Q1	MPF102
Q2	2N3904
Q3	TIP31
Q4	2N3906
Q5	TIP32
Q6	2N3565
D1	1N751A
D2,D3	1N914B

Fig. 2-29. Function generator parts list, Project 3.

Part 3
Rf Projects: Transmitters

Many times a need arises for a small, low power transmitter to transmit a complete baseband video signal, either color or B/W, with its accompanying audio, either over a coax cable or over the air to a remote receiver. Such applications include:

1. Amateur TV transmissions;
2. Certain video installations where cable hookups are not feasible or possible, such as robotics;
3. Security and industrial work;
4. Simultaneous viewing of program material by several remote TV receivers;
5. Remote sensing application, wildlife viewing, etc.;
6. Cable transmission;
7. Wireless Camera/TV receiver or VCR link.

PROJECT 4: MULTI-PURPOSE VIDEO LINK TRANSMITTER

The unit described here is capable of two levels of rf power output. For low power wireless video applications (covering a house or office, for example) where simultaneous monitoring of program material is desirable without annoying or cumbersome hookups, a level of 1 to 30 milliwatts is available. For longer ranges up to several miles, as for amateur (ham) TV, or security and surveillance purposes, remote sensing, etc., a level of two watts of rf into a 50 ohm load is available. It accepts both color and B/W video and sound and has inputs for video and audio levels compatible with VCRs, camcorders, small TV cameras, and microphones. The unit runs on +12V (nominal) dc, and draws 100 mA nominal (low power version) or 500 mA nominal (2 watt version). The PC board is quite small ($2^1/_4'' \times 4'' \times 1''$ high). It contains everything needed, except power supply and connectors. It employs both subminiature components and surface mount components, for excellent rf performance, high efficiency, ease of reproducibility and construction, and simple tune-up without complex test equipment. In fact, a good tune-up job can be done with only a VOM and a TV receiver. The first four prototypes tuned right up and worked well the first time, indicating good reproducibility.

The transmitter to be described is easy to tune up, uses three-slug tuned coils rather than air wound types, and employs a double sided PC board for shielding and grounding improvements. It employs an integral power amplifier, and integral video gain and audio gain controls for easy interfacing and use. Also, a linearity control optimizes picture qualities.

Be warned that this two-watt transmitter is intended for educational purposes and serious legitimate TV broadcasting such as amateur TV, and industrial and scientific purposes. Its two-watt level is capable of several miles transmission range and those intending to use this design *must do so under appropriate legal circumstances and suitable licensing*. For most of us *this means at least amateur TV and a technicians class amateur radio license*. However, the lower power version should be OK if caution is taken to avoid interfering with other transmissions and that the signal is confined to ones home and/or property. *If a signal is detectable more than several yards away in any direction, you may be in violation of the law*. So be warned. The authors are not responsible and no warranties, implied or otherwise, of any kind, are offered in this matter. You are responsible for your own situations and particular application.

The transmitter is illustrated in block diagram form in Fig. 3-1. Twelve transistors are employed and a twelve volt (11 – 14V is OK) supply such as nicads, auto electrical system, or ac operated supply is assumed. Q1 is a crystal oscillator operating at one eighth ($^1/_8$) of the picture carrier frequency. It can be operated between 52.5 and 62.5 MHz with the current constants in Fig. 3-2. This corresponds to 420 to 500 MHz output frequency which covers 430 MHz Ham TV, and the lower UHF TV channels 14 through 18. With modification of circuit constants, higher or lower frequencies are possible, with reduced performance above 500 MHz, and possibly somewhat enhanced performance below 420 MHz (more rf output), depending on the particular transistors employed. This is left to the reader.

Q2 is driven by the output of Q1 (about 2 – 5 milliwatts) and acts as a doubler. Q2 feeds Q3 with a signal twice the frequency of the crystal. Q3 doubles the frequency to four times crystal frequency. This is typically 210 to 240 MHz. Double tuned interstage networks are used to suppress unwanted harmonics. Q4 doubles the output frequency of Q3 to the final output frequency. About five milliwatts of rf power are present here.

Q5 is an amplifier tuned to the output frequency. Its function is to amplify the signal at the output frequency present at the output of Q4, and also act as a modulated amplifier in the low power version. Q5 is fed either 12 volts (high

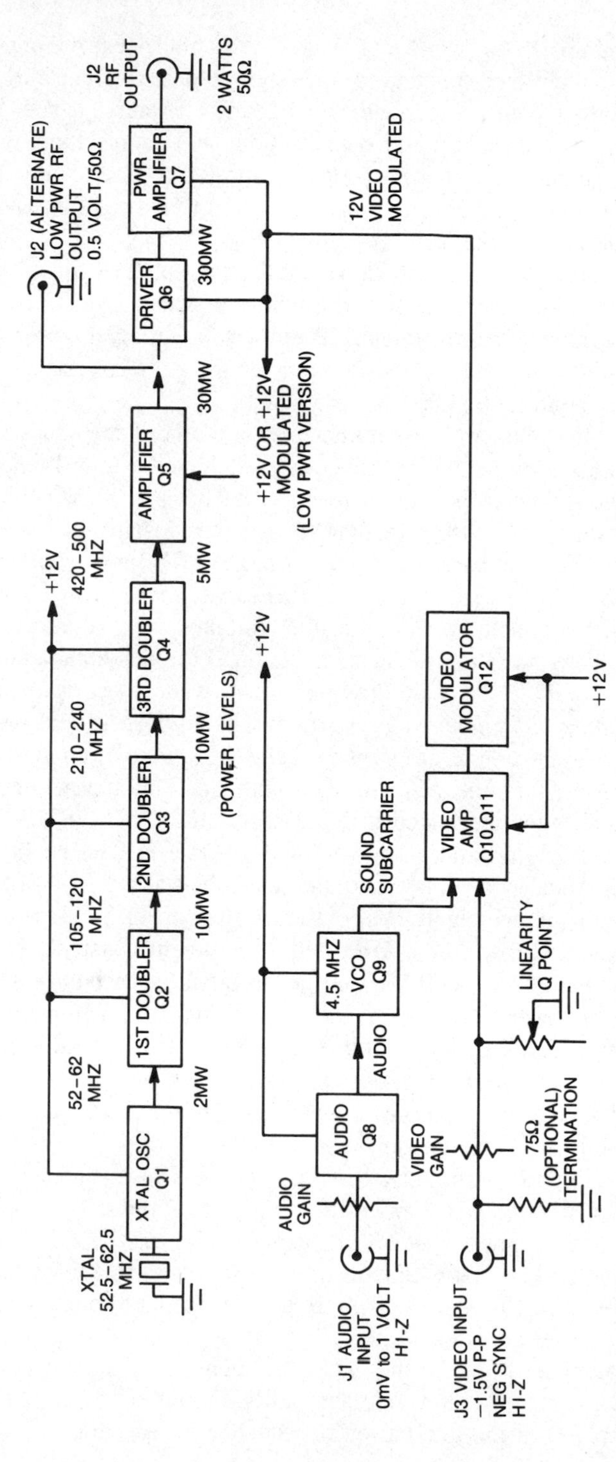

Fig. 3-1. Video transmitter block diagram.

power version) power supply, or is fed with a video modulated (0 to 12 volts) supply voltage in the low power version.

In the low power version rf signal (1 to 30 milliwatts, depending on coupling) is taken from the collector of Q5 and fed to either a cable or a small 6″ whip antenna, for coupling to nearby TV receivers. The TV receivers are tuned to the appropriate frequency (channel) which is an unused low UHF channel preferably two or more channels away from any other busy channels.

For the higher powered versions only, Q6 and Q7 form a power amplifier. The power amplifier employs high gain rf semiconductors, and matching networks are adjustable for optimum tune-up. A fixed tuned strip-line design was contemplated, but at 420–500 MHz would have occupied too much PC board area to comfortably fit on a 2½×4″ PC board. Use of broadband chokes and surfacemount (chip) capacitors, and careful design to avoid possible low frequency spurious oscillation problems, have resulted in a very stable, efficient, reproducible circuit and no UHF "horrors" should be encountered if the design is reproduced exactly as shown, using the specified components. For those who prefer, two complete kits of components are available from the supplier at the end of this article. One kit is for the low power version, the other for the high power version. These kits include the PC board and everything on it. Jacks, connectors, batteries, power supply components, and case are not included, since individual readers may have their own preferences and different interface requirements. These components are best obtained at a supplier such as Radio Shack, etc.

Audio input at J1 from 10 millivolts (microphone) to 1 volt (line inputs, etc.) is fed to audio amplifier Q8. A level control is provided for optimum modulation adjustment. The audio modulates VCO circuit Q9, which produces a 4.5 MHz FM signal. This is the sound subcarrier. It is fed to video amplifier Q10 where it is combined with the input video signal from J3. The video input may be 0.5 to 1.5 volts p-p, negative sync Q10 and Q11 form a video amplifier which feeds modulator Q12. Q12 is capable of producing a video signal which has 0 to +12V level swing, and can drive a load up to 1 ampere. Bandwidth at −3dB is in excess of 10 MHz, assuring crisp picture detail. Q12 acts as a power supply to Q6 and Q7, effectively AM modulating the rf power output. In the low power version, Q5 is modulated in the same manner. A linearity control adjusts the operating point of Q12 for optimum modulation linearity. The Q point must be properly set otherwise clipping of the video signal will occur. This will produce "burned-out" picture highlights (white areas) with loss of detail, and/or sync "buzz" in the audio, as well as loss of picture stability in extreme cases.

When building this project, the reader is cautioned to use only the parts specified in the parts list. UHF circuits can be quite critical as to both component type and value. Also, proper parts placement is very important. Lead lengths should be *very* short. Anything longer than *zero* may cause problems.

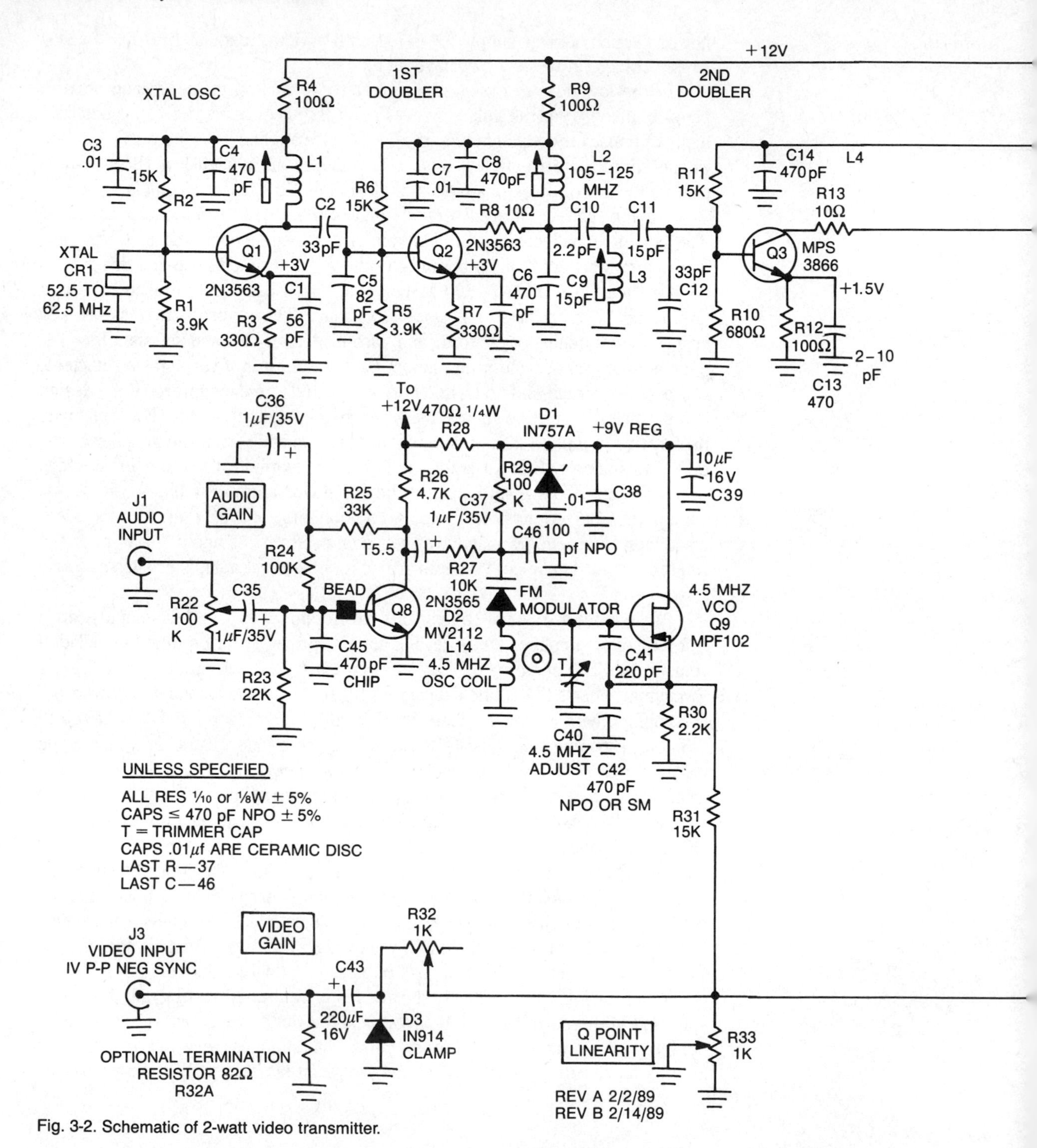

Fig. 3-2. Schematic of 2-watt video transmitter.

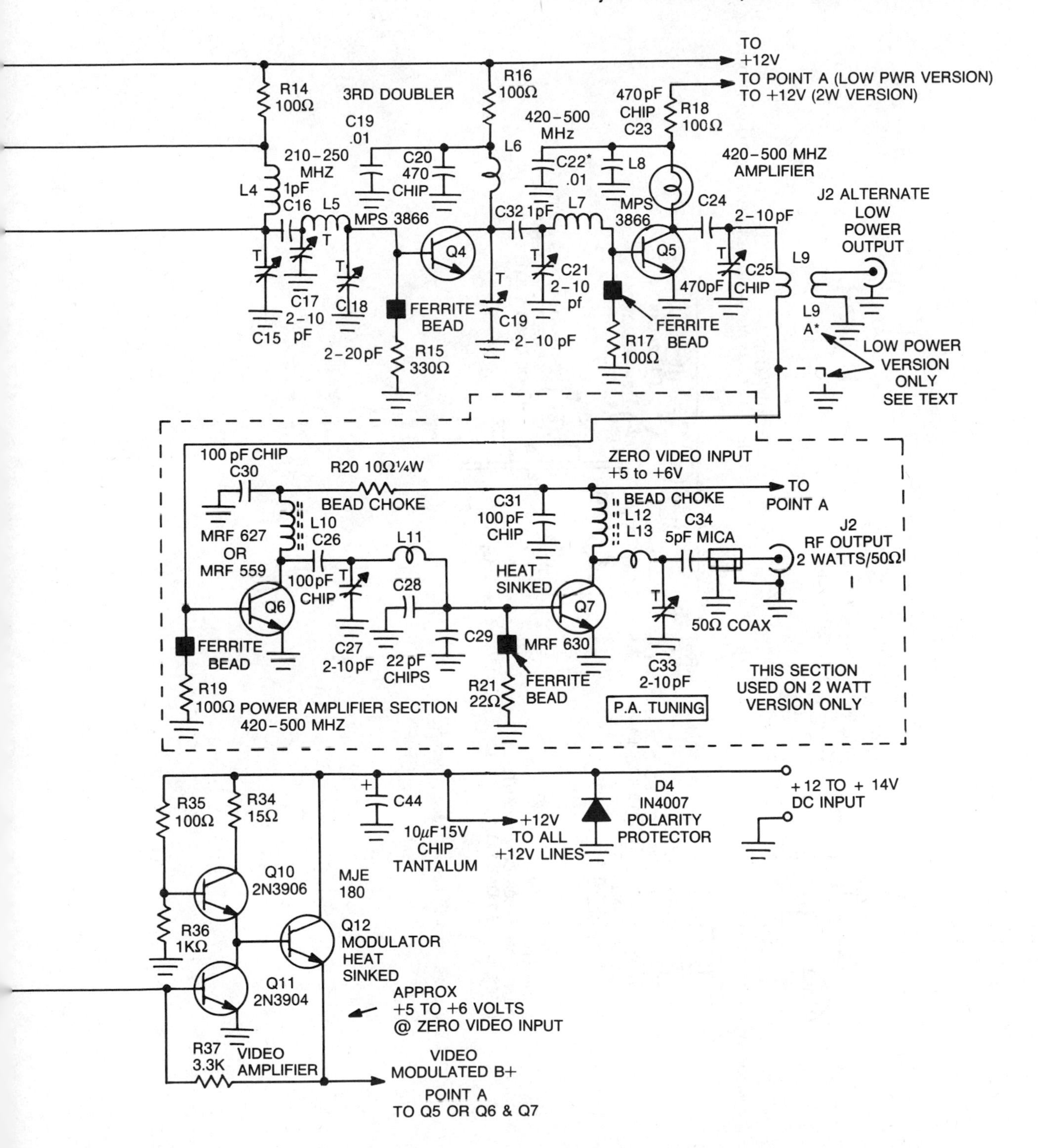
TO
+12V
TO POINT A (LOW PWR VERSION)
TO +12V (2W VERSION)
R14
100Ω
3RD DOUBLER
R16
100Ω
470pF
CHIP
C23
R18
100Ω
420-500
MHz
C19
.01
210-250
MHZ
C20
470
CHIP
L6
C22*
.01
L8
420-500 MHZ
AMPLIFIER
L4
1pF
C16
L5
MPS 3866
C32 1pF
L7
MPS
3866
C24
2-10pF
J2 ALTERNATE
LOW
POWER
OUTPUT
Q4
Q5
C17
2-10
pF
C18
C15
FERRITE
BEAD
2-20pF
R15
330Ω
C19
2-10 pF
C21
2-10
pf
FERRITE
BEAD
R17
100Ω
470pF
C25
CHIP
L9
L9
A*
LOW POWER
VERSION
ONLY
SEE TEXT
100 pF CHIP
C30
R20 10Ω¼W
ZERO VIDEO INPUT
+5 to +6V
TO
POINT A
BEAD CHOKE
L10
C26
MRF 627
OR
MRF 559
C31
100pF
CHIP
BEAD CHOKE
L12
L13
C34
5pF MICA
J2
RF OUTPUT
2 WATTS/50Ω
L11
100pF
CHIP
C28
HEAT
SINKED
Q6
Q7
FERRITE
BEAD
C27
2-10pF
22 pF
CHIPS
C29
MRF 630
50Ω COAX
R19
100Ω POWER AMPLIFIER SECTION
420-500 MHZ
R21
22Ω
FERRITE
BEAD
C33
2-10pF
P.A. TUNING
THIS SECTION
USED ON 2 WATT
VERSION ONLY
R35
100Ω
R34
15Ω
C44
10μF15V
CHIP
TANTALUM
+12V
TO ALL
+12V LINES
D4
IN4007
POLARITY
PROTECTOR
+12 TO + 14V
DC INPUT
Q10
2N3906
MJE
180
R36
1KΩ
Q12
MODULATOR
HEAT
SINKED
Q11
2N3904
APPROX
+5 TO +6 VOLTS
@ ZERO VIDEO INPUT
R37
3.3K
VIDEO
AMPLIFIER
VIDEO
MODULATED B+
POINT A
TO Q5 OR Q6 & Q7

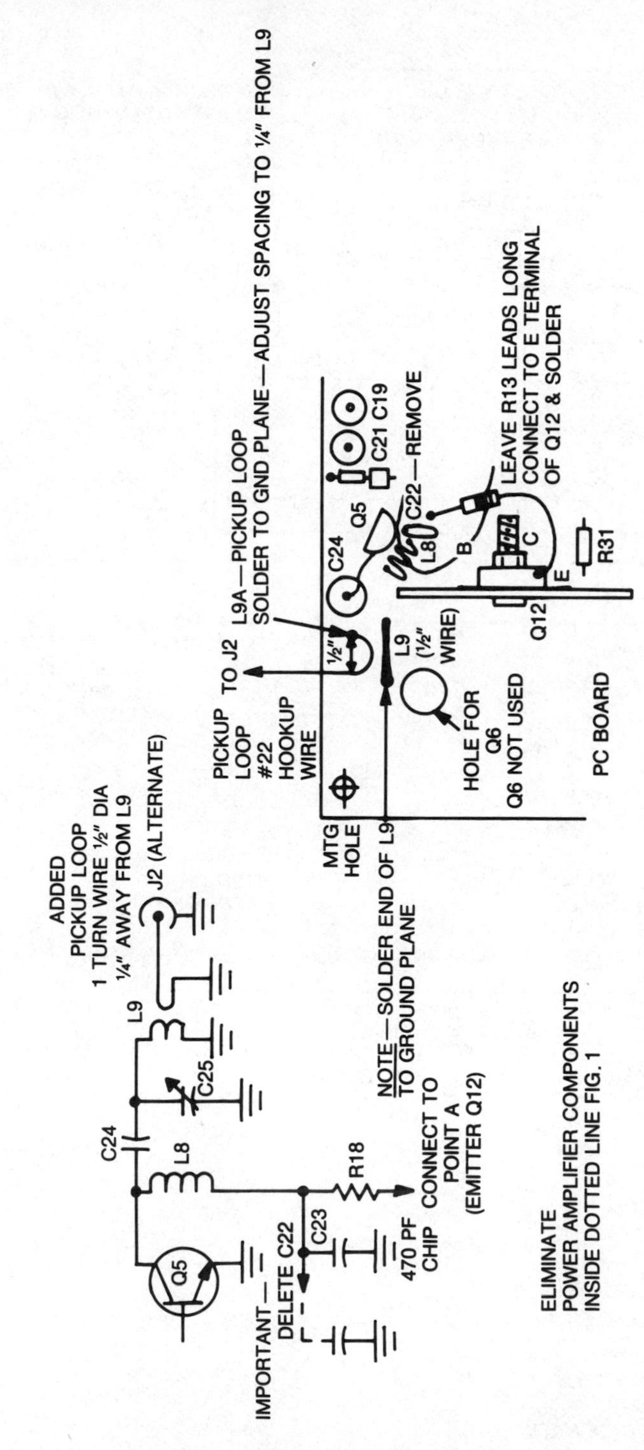

Fig. 3-3. Low-power modification for video transmitter.

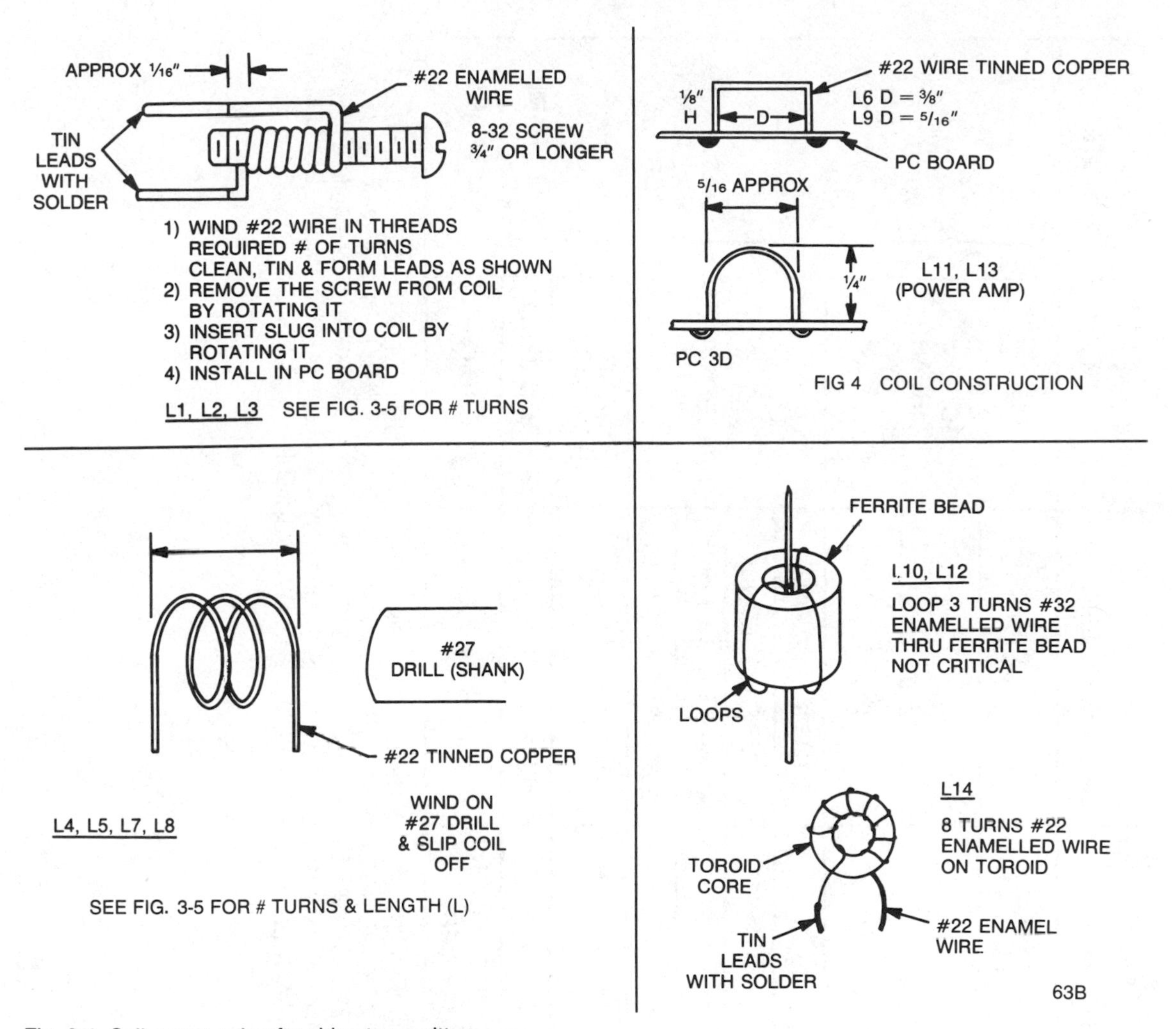

Fig. 3-4. Coil construction for video transmitter.

Note that ten surface mount capacitors (chips) are used, as well as ferrite beads. Also, tenth watt resistors and miniature NPO ceramics are employed for very short leads and close component spacing. Tiny slug-tuned coils, easily made by the constructor (see Figs. 3-4 and 3-5) using readily available materials are used rather than commercial, hard to get, large shielded factory made types. This gets rid of the coil headaches. If the dimensions are followed, no problems should result. In particular, supply bypassing is very important. We have incorporated a chip tantalum capacitor to guarantee this. By keeping everything small, compact, and by using a shielded, double sided PC board, with good bypassing techniques, all the possible "horrors" associated with VHF and

COIL	FREQ RANGE	# TURNS & LENGTH	WINDING FORM	NOTES
L1	420–450 (HAM TV) 450–500 (VIDEO LINK)	9½ 8½	8-32 SCREW THREAD	#22 ENAMEL WIRE
L2	420–450 450–500	4½ 3½		
L3	420–450 450–500	5½ 3½		
L4	ALL	3 TURNS ¼″ LONG	#27 DRILL (0.144″ DIA) SPACE TURNS	MADE WITH #22 TINNED COPPER
L5	ALL	4 TURNS ¼″ LONG		
L7	ALL	1½ TURNS 1/16″ LONG		
L8	ALL	2½ TURNS ⅛″ LONG		
L6, L9, L11, L13	ALL	PER FIG 4	NONE (PC BOARD)	
L10, L12	ALL	PER FIG 4	FERRITE BEAD	#32 ENAMEL WIRE
L14	4.5 MHZ (NTSC SOUND SUBCARRIER)	8 TURNS #22″ ENAM	TOROID	#22 ENAMEL WIRE

NOTE—DUE TO INDIVIDUAL WINDING TECHNIQUE AND NORMAL CIRCUIT TOLERANCES, L1, L2, L3, AND L14 MAY REQUIRE ONE TURN MORE OR LESS THAN SHOWN IN TABLE. L4, 5, 7 AND L8 MAY HAVE TO BE SQUEEZED OR SPREAD LENGTHWISE. ALL DIMENSIONS ARE TAKEN FROM AVERAGE OF SEVERAL WORKING UNITS. INDIVIDUAL UNITS MAY VARY SOMEWHAT FROM GIVEN DIMENSIONS DUE TO TOLERANCES AND WINDING TECHNIQUES, AND INSTALLATION.

Fig. 3-5. Coil data for video transmitter.

UHF circuitry can be easily dealt with. As long as the design is exactly duplicated there is no reason to encounter "nightmare, off the wall, weirdo" problems. The coils are easy to wind and the largest ones have only eight or nine turns of wire. In fact, several are only loops or pieces of wire, since inductors required at 420 – 500 MHz are usually in the .01 to 0.1 microhenry range. However, the PC board is compact, and parts are very small. A small iron with a pointed tip is recommended, especially for the soldering in the nine chip capacitors. A discussion of precautions at VHF-UHF also appears in our February 1986 Radio Electronics article, *Build this Wireless Video Link*, and the reader interested in building this project might also find this article very informative.

Referring to Fig. 3-2 a detailed circuit description will now be given. Xtal oscillator Q1, a 2N3563 VHF NPN transistor, is biased to initially 10 volts and 5 milliamperes Q point by resistor R1, R2, and R3. Crystal CR1 acts as a series resonant "bypass" to ground only at the crystal series resonant frequency (52.5 to 62.5 MHz). At this frequency Q1 acts as a common-base amplifier. Tuned circuit L1 and C2 in series with C5, together with stray capacitance of about 1 to 2 pF, form a load for the collector of Q1. C3 and C4 bypass the "cold" end of L1 solidly to ground for ac signals. Internal feedback from collector to emitter occurs in Q1 via the intrinsic collector to emitter capacitance of Q1 about 2 picofarads. C1, a 56 pF capacitor, forms a voltage divider to feed back a portion of collector signal to the emitter. Note that C1 is *not* an emitter bypass but a part of the feedback network of oscillator Q1.

At the series resonant (crystal) frequency of CR1, Q1 acts therefore as a grounded base oscillator. An rf signal is generated at this frequency. Dc is supplied to Q1 stage through decoupling resistor R4. Collector current, once Q1 starts to oscillate, is dependent on the tuning of L1, but typically is 5 – 10 milliamperes.

A portion of this voltage (about 1.5 volts rms) across L1 is fed to Q2 by the voltage division between C2 and C5. C5, an 82 pF capacitor, has a low impedance at twice the oscillator frequency. Q2 is biased initially the same as Q1 via R5, R6 and R7. C6 is a bypass capacitor, as are C7 and C8. Q2 acts as a frequency doubler by the fact that a large drive signal from Q1 causes rectification in the emitter base junction of Q2. This produces appreciable harmonic generation. Keeping the impedance low in the E-B circuit of Q2 by using a large value (82 pF) for C5 results in efficient harmonic generation.

C7 and L2 are tuned to normally twice the crystal frequency. R9 supplies dc to Q2 stage. Tuning is accomplished via a slug in L2. C10 couples rf energy at 2X Xtal frequency to a second tuned circuit L3-C11-C12, also tuned to twice crystal frequency. Use of two tuned circuits assumes good selectivity and improved rejection of unwanted frequencies. This is important for a clean signal from the transmitter. R8, in the collector of Q2 suppresses a tendency to unwanted UHF parasitic oscillation.

Fig. 3-6. PC board of video transmitter installed in a housing.

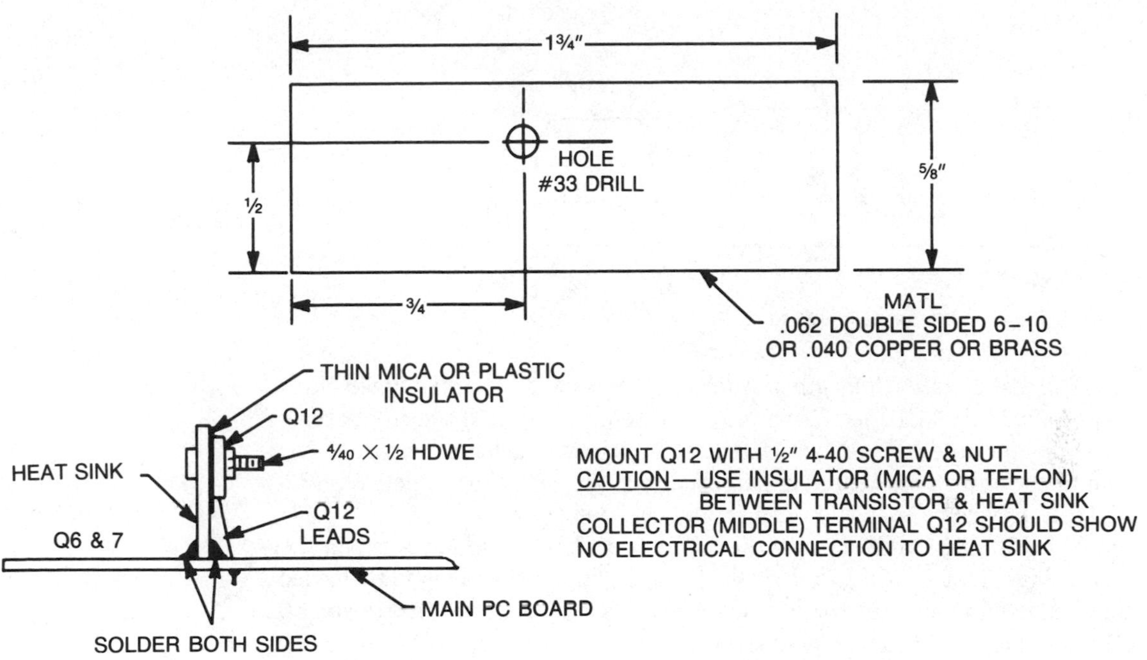

Fig. 3-7. Q12 heatsink and mounting.

Q3 is fed energy at 105 to 125 MHz from the junction of C11 and C12. R10, R11, and R12 bias Q3. Since the rf level at the base of Q3 is high, the rf level affects the bias. Typically Q3 runs at 10 to 15 milliamps collector current. Q3 is an MPS 3866, a 400 MHz medium power (1 watt dissipation) plastic transistor. It offers supperior performance at 250 MHz to the 2N3563 used at Q1 and Q2. Except for frequency, operation is similar to the Q2 stage. Q3 doubles the input frequency to 210 to 250 MHz. R13 suppresses UHF (> 300 MHz) possible parasitic oscillation C15 and L4 are tuned to twice the input frequency. C14 is a bypass cap, 470 pF. The .01μF used in Q1 and Q2 is ineffective at 250 MHz and not used here, the 470 pF being sufficient. R14 feeds dc to Q3 stage. Note that now the output tuned circuit is tuned by variable capacitor C15 and L4 is fixed. This is because slug tuning is no longer practical, the coil L4 having too few turns C16 couples energy to tuned circuit C17, L5, and C18. This forms a double tuned circuit at 210 – 250 MHz. C17 is the tuning capacitor. C18 is a variable capacitor to optimize matching into Q4, the last (third) doubler. R15, a 330 ohm resistor, with a ferrite bead to act as an rf impedance (see Fig. 3-8) in series with it, completes the base circuit dc path for Q4. Bias now is supplied entirely by the drive signal. No extra dc bias is applied. The emitter of Q4 is directly grounded, since bypassing of emitter circuits at 420 – 500 MHz is difficult without some loss of rf gain. However, a low value of R15 keeps dc stability adequate.

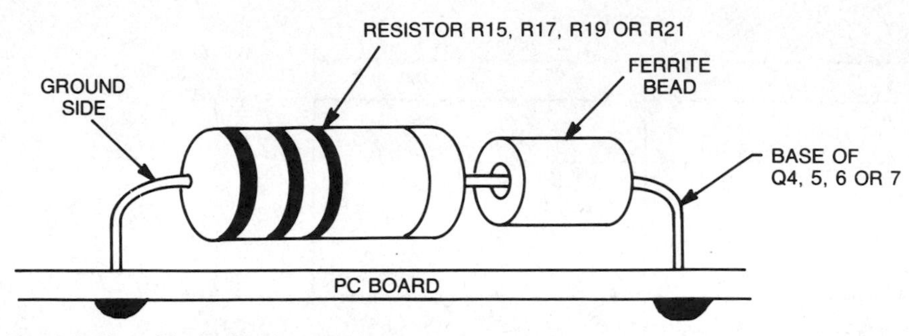

Fig. 3-8. Use of ferrite beads on resistors.

C19 and L6 (a short length of wire is all that is needed) form a tuned circuit at 420 – 500 MHz. C19 and C20 provide low frequency (video) and rf bypassing. C19 provides little bypassing at UHF. Its purpose is to kill stray low frequency gain of the Q4 stage. C20 is a chip capacitor, the only type effective at 420 MHz. It provides a solid rf ground for the cold end of L6.

The 420 – 500 MHz signal at the collector of Q4 is fed to tuned circuit C21 and L7 via C32. C21 and L7 match the low base impedance of Q5 to the collector circuit of Q4 and form a double tuned UHF circuit together with C19 and L6. R17 and a ferrite bead provide a low dc impedance but a high rf impedance to the base of amplifier Q5.

Q5 amplifies the UHF signal to about 30 milliwatts. L8 acts as an rf choke. C22 and C23 perform bypassing for video and UHF respectively. Note that if Q5 is to be video modulated (low power version) C22 *must* be deleted, since it would cause loss of high frequency video components. R18 supplies dc to Q5. If Q5 is to be modulated, the dc must be modulated with video, and R18 must be returned to the modulator output (emitter of Q12, shown as point A in the diagram of a PC board) rather than the 12V supply.

C24, a 470 pF chip, couples rf output but blocks dc (and video, if applicable) from tuned circuit C25 and L9.

In the low power version L9 acts as the output tuned indicator. A link of wire (see Fig. 3-3) in the close proximity to L9 (this link is called L9A) picks up rf from L9 and transfers it to output jack J2A. J2A and L9A are not used in the 2-watt version. Output is 1 to 30 milliwatts, depending on the proximity of the link L9A to inductor L9.

In the 2-watt version, L9 acts as a matching inductor to the base of driver transistor Q6. Q5 is fed straight unmodulated +12 volts dc. Q6 therefore gets the full 30 milliwatts drive from Q5. R19 and the ferrite bead on one lead of it provide high rf impedance and low dc resistance to the base of Q6. Since a ferrite bead looks more like a high resistance rather than a reactance at high frequencies, the effective Q is very low. This prevents the possibility of parasitic oscillations that could occur if a conventional type solenoid wound rf choke were to be used. C27, L11, and chip capacitors C28 and C29 match the collector impedance of Q6 to Q7. L10 is a ferrite bead choke made with three turns of wire wound through a ferrite bead, in toroidal fashion (see Fig. 3-4). This results in a very low Q, resistive, about 1,000 ohms, impedance and again avoids possible parasitics. C26 is a coupling capacitor, a chip to minimize stray inductance. C30 is a bypass capacitor chosen to provide a short circuit to UHF while looking like a high impedance at 10 MHz or lower, so the video signal component of this power supply voltage (modulating) is relatively unaffected. R20 is a decoupling resistor. Q6 draws about 130 milliamperes current at modulation peaks (sync tips). It is an MRF559, MRF627 or equivalent.

Q7, an MRF630 is similar to Q6 in operation. Q6 supplies about 300 to 500 milliwatts drive to Q7. R21 and its ferrite bead on one lead (see Fig. 3-8) allow low dc base circuit resistance with high rf impedance as for Q6. L12 is an rf choke exactly the same as L10. L13 and C33 form the collector matching circuit, together with mica capacitor C34, to match the 50 ohm load impedance to the optimum collector load impedance needed by Q7. *Note that a 50 ohm load must always be present at J2.* Otherwise, Q7 may be damaged a tolerance of ±50% (25 to 100 ohm) is permissible here. However, optimum performance is obtained with a 50 ohm load. Suitable 50 ohm coax must be connected from C34 (on the PC board) and J2, with short connections (1/4" or so). Any length of coax can be used, but for best results, keep the coax short. We used RG174/U PVC type but a type such as Teflon coax (RG188/U) would be better. From J2, standard coax (RG8U, RG58/U, etc.) will do. Remember, feedline loss can be very high at 420 MHz and up.

C31 is an rf bypass chosen for the same reasons as C30. Q6 and Q7 are fed video-modulated 12 volts from Q12 which will be discussed later.

Input video from J3 (standard 1V p-p negative sync, etc.) is fed through C43 to clamp diode D3. Note that C43 is apparently incorrectly polarized. This is to allow for certain video equipment that may have a dc component of up to 16 volts present on the video output. If you do not expect to encounter this, you can reverse the polarity of C43 if you wish. The low reverse voltage (0.6 volt) appearing across it doesn't seem to do any harm. D3 clamps the maximum negative input level to −0.7 volts and avoids serious overmodulation at sync tip level. If you wish, you can dc couple from J3 directly into R32, the video gain control, if your interfaces permit. Also, note optional 82 ohm termination resistor R32A (not on PC board). This resistor can be used if you want the transmitter video input to be a line termination (the usual case). Use it unless you are in a situation where loop-through (several other video loads in parallel) is required. It was not placed on the PC board so this possibility would not be compromised. The resistor R32A can be soldered across J2, in the usual case.

R32 acts as a video gain control. Video from R32 is fed to the base of video amplifier Q11. The collector of Q10 is fed by current source transistor Q10. Q10 is biased to about 50 milliamperes collector current by R34, R35, and R36.

RESISTORS 1/8W or 1/10W 5%	
R1,R5	3.9k
R2,R6, R31,R11	15k
R3,R7,R15	330 ohm
R4,R9,R12, R16 - R19, R35	100 ohm
R8,R13	10 ohm
R10	680 ohm
R20	10 ohm 1/4W
R21	22 ohm
R22	100k pot, PT10V
R23	22k
R24,R29	100k
R25	33K
R26	4.7k
R27	10k ohm
R28	470 ohm 1/4W
R30	2.2k
R32,R33	1k pot, PT10V
R34	15 ohm
R36	1k ohm
R37	3.3k

CAPACITORS	
C1	56pF NPO
C2,C12	33pF NPO
C3,C7,C19, C22,C38	.01 disc
C4,C6,C8, C13,C14	470 disc
C5	82pF
C9,C11	15pF
C10	2.2pF
C15,C17, C19,C21,C25, C27,C33	2 - 10 trimmer
C16,C32	1pF NPO
C18	2 - 22pF
C20,C23, C24,C45	470pF chip
C26,C30, C31	100pF chip
C28,C29	22pF chip
C34	5pF Mica
C35,C36, C37	1μF/35 or 50V elec
C39	10μF/16V elec
C40	3 - 40 trimmer
C41	220pF NPO
C42	470pF NPO
C43	220μF/16V elec
C44	10μF 15V chip Tantalum
C46	100pF NPO

Fig. 3-9. Parts list—Video Transmitter, Project 4.

This permits the collector of Q11 to supply plenty of drive to modulator Q12, and eliminates the need for a low value resistor from the collector of Q11 to the power supply rail (+12V). This enables the base of Q12 to nearly approach the +12V supply level and allows a higher positive swing of the emitter of Q12 than a resistor from Q11 to +12V would permit, due to the base drive needs of Q12. Q12 an MJE180 is configured as an emitter follower. It must supply all the current to Q6, Q7, (or Q5) and must provide a very low supply impedance and very high slew rate. The low impedance is necessary for both full rf power output and control of parasitic oscillation tendencies in the Q6 and Q7 power amplifier. Also, the load is capacitive, due to the bypassing from C30, C31, and somewhat from C26. The Q12 circuit, in tests, can supply nearly 12 volts of video into a 10 ohm load. This is 1.2 amperes of current. Q12 must be heat sinked. R37 provides feedback around the modulator to establish both Q point, video gain, and bandwidth. R33 sets the exact Q point (voltage seen at point A, the emitter of Q12), under zero drive conditions about 5 to 6 volts dc to Q6 and Q7. R33 is

SEMICONDUCTORS

Q1,Q2	2N3563
Q3,Q4,Q5	MPS3866
Q6	MRF559 or MRF627
Q7	MRF630
Q8	2N3565
Q9	MPF102
Q10	2N3906
Q11	2N3904
Q12	MJE180
D1	1N757A
D2	MV2112
D3	1N914
D4	1N4007

CRYSTAL

1	XTAL

COILS

2′	#22E
2′	#22 tinned
2′	#32E
1	Toroid 76T188
7	Beads, Ferrite
3	Blue slugs, Cambion

MISCELLANEOUS

1	PC board
1	TO 220 insulator
1	8/32 x 3/4 screw

HARDWARE

1	8/32 screw
1	4/40 screw & nut
1	TO 220 insulator

Fig. 3-10. Parts list—Video Transmitter, Project 5.

adjusted for maximum undistorted symmetrical video at point A, while R32 controls video drive to Q11. Supply bypassing must be effective at the collector of Q12 due to the high current, fast waveforms handled. A 10μF 15V chip tantalum was used, conventional electrolytics are found to be somewhat less effective, at C44, the main supply bypass capacitor.

Dc power is fed to the transmitter at J4. A diode D4, a 1N4007, is provided to serve as reverse polarity protection. It is cheap insurance against inadvertent damage to Q6, Q7, Q10, Q11, and Q12 due to reversed supply. It is not mounted on the PC board. Connect it directly across J4.

Audio is fed to gain control R22 from jack J3. Input level is between 10 millivolts to 1 volt at high impedance, allowing direct interfacing with most microphones or other audio sources. Audio from R22 is fed through coupling capacitor C35 to Q8, a 2N3565. Q8 is biased from R25, R24, and R23. C36 is a bypass capacitor to prevent audio degenerative feedback and loss of gain. Collector load resistor R26 supplies dc to Q8. C37 couples audio to R27, and blocks dc. Note that no pre-emphasis (the providing of high frequency boost) has been used. If you want to use it, for better high frequency audio response, charge C37 to .001 microfarad and set the gain control R22 up higher to compensate for loss. We found we did not need it in our application, the audio being adequate. A ferrite bead on base lead of Q8 and C45 reduces rf pickup by Q8, which cause buzz on the audio.

R27 couples audio to varactor diode D2, an MV2112. R29 provides dc bias of +9V to the varactor. The varactor diode varies the capacitance of D2 (56 pF at 4 volts, about 33 pF at 9 volts bias) at an audio rate. The capacitance of D2 appears across 4.5 MHz oscillator coil, L14. Q9, an MPF102 FET, together with capacitors C41, C42, C40, and L14 form a Colpitts type rf oscillator operating at 4.5 MHz. C40 is a trimmer (variable) capacitor to set the frequency to exactly 4.5 MHz. L14 is a toroidal coil to minimize both size and stray magnetic field generation. C46 provides rf grounding for D2 while blocking audio. When D2 charges capacitance, the oscillator frequency shifts. Therefore, an audio voltage component on dc causes a frequency modulation (FM) of the 4.5 MHz signal generated in Q9 circuit. R30 provides operating bias for Q9. Resistor R31, a 15k ohm resistor, couples this sound subcarrier (4.5 MHz FM signal) into the video amplifier, which modulates it onto the rf signal along with the video.

R28, zener diode D1, and bypass capacitors C38 and C39 supply a regulated 9 volt nominal dc voltage to Q9 and varactor D1. The regulation prevents oscillator drift if the supply voltage were to vary. A frequency counter can be connected to point A to set C40 to exactly the value needed for 4.5 MHz sound subcarrier frequency.

Construction of this transmitter should pose no special problems as long as care is taken to duplicate the prototype as closely as possible. Some precautions to take are:

1. Use only G10 .062″ thick epoxy fiberglass board material. Other materials and thicknesses could be used, but may result in different tuning conditions and stray capacitances. However, do not use paper base phenolic material—it is too lossy at the high frequencies present in this transmitter.
2. Q12 must be heat sinked. A possible 3 watts dissipation makes some form of heat sinking mandatory. The method of Fig. 3-7 has proven adequate if at least one oz. copper is used. If possible use .040 copper or brass, but this is not absolutely necessary. Q7 is adequately heat sinked if the metal case is soldered to the PC board ground plane.
3. Solder as many component leads as possible, on top of the board, that pass through the ground plane to both top and bottom of the board. In particular, the ground lugs on all trimmer capacitors should be soldered on both sides, and also most of the resistors that have one side connected to ground. The idea is to ground as much of the ground plane (shield top side) to the ground foil on the component side, in as many places as possible. This is especially important around Q4 through Q7.
4. Use chip capacitors where specified. Do not substitute ordinary leaded capacitors. Use only those components specified, no substitution.
5. Keep all components as close to the PC board as possible, leads as short as possible.
6. Take care to make coils as accurately as possible. While some error can be tolerated, accurate work will make tune-up easier and ensure duplication of results.

7. Never operate this device without a load of ≥2:1 VSWR connected to the rf output.

Construction is started by first installing all resistors and then D1 and D3. (Do not forget ferrite beads on R15, R17, R19 and R21.) Next, install all disc ceramics .01 and 470 pF. The NPO capacitors are installed next. After the NPO capacitors are installed, install potentiometers R22, R32, and R33. Solder grounded sides of R22 and R33 to both sides of the PC board. Next install all trimmer capacitors. Note that C18 and C40 are different from the rest. Solder ground tabs of all trimmers to both top and bottom of PC board. Next, install Q1 through Q5 and Q8 through Q11. Do not install Q6, Q7 or Q12 as of yet. Now, wind and install L1 through L9, and L14. If you are building the low power version, leave out any components associated with Q6 and Q7, except L9. Last, install chip capacitors C22, C24, C44, and C20. If you are building the low power version make modifications shown in Fig. 3-3 and make sure to omit C22. Check all of the PC board for shorts, solder bridges, and trim away any excess foil with a sharp knife (X-ACTO type or equal). Make sure excess foil on the top side is not touching any component leads not intended to be grounded. Slight misregistration of the top foil may cause this. Simply trim excess foil away with the X-ACTO knife. Even if you are building the high power full version, Q6, Q7, and associated circuitry is best installed after the remainder of the board is completed and tested. Now install Q12 and heat sink per Fig. 3-7 and also Fig. 3-3. Note that the heat sink also serves as an rf shield for the Q6 and Q7 power amplifier, if used. Be sure to solder the heat sink to the top foil side (ground plane), on each side of the heat sink where it butts against the PC board. Note that the Q12 case should be insulated from the heat sink. Use a TO-220 insulator (cut to size) or a scrap of mica, Mylar, polyethylene, or Teflon tape used in plumbing work. You are now ready to test the main part of the board. If you are constructing the 2-watt version, Q6, Q7, and associated components will be installed after testing the rest of the PC board.

After checking for shorts, opens, and solder bridges, measure the dc resistance between B+ and ground. It should measure greater than 200 ohms. If lower check for the cause before proceeding further.

Next, install the slugs in L1, L2, and L3 if you have not already done so. The slugs should be initially set fully inside the coils. Set R22, R32 and R33 about halfway between extremes of rotation. Set C40 halfway meshed. Set all other trimmers to half mesh. Final settings will depend on operating frequency, coil construction technique, and application.

Next, apply +12 volts to the B+ line after connecting the negative supply lead to the ground plane of the PC board. Immediately observe power supply current. If over about 130 mA there may be a problem. If anything smokes or gets hot immediately remove power and find the problem before proceeding.

If all seems OK, connect a VOM (analog meters are easier to use for this test, but a DVM is OK) across R3 and then R7. You should read about 2 volts dc

(1.5 to 3Vdc OK). Next, connect the VOM to the emitter of Q3. You should read 1 volt or less. Now, connect the VOM to point A (emitter of Q12) and ground. Verify that adjusting R33 through its full range can vary the voltage at point A between less than 5 volts to greater than 11 volts. Set R3 for full voltage (11V) at point A for now. Next, measure voltage at the collector of Q8, about 4 to 7 volts is OK. Next, measure voltage across D1 (1N757A). It should be between 8 and 10 volts dc, more or less indicates a problem in Q8, Q9, or associated circuitry. Check for 8 to 10 volts across D2. If it reads 1 volt, D2 is either installed backwards or is shorted.

If all is OK install crystal CR1. Connect the VOM across R7. Apply 12 volts. Now slowly back the slug of L1 out of the winding. You will find that the voltage across R7 will suddenly increase then slowly decrease as the slug in L1 is tuned. Adjust the slug for maximum voltage (3 to 5 volts) and then back out the slug for about a 10% drop. This will ensure stable oscillation. As a check, a frequency counter connected to the junction of C2 and C5 should indicate the crystal frequency. An unstable reading indicates possible problems in that the crystal is not controlling the frequency. If this is the case, try readjusting L1.

Next, connect the VOM across R12. Adjust L2 and L3 for maximum voltage. This will be 1 to 2 volts. If the slugs in either L2 or L3 have no definite peak, add or subtract a turn from that coil as required, after first checking C9, C10, C11, and C12 for correct value.

Connect an rf probe to the junction of L9 and R19 (or to the junction of C25 and L9 if building the low power version). Adjust C15, C17, C18, C19, C21, and C25 for maximum reading. You should be able to obtain at least 1.5 volts of rf energy at the junction of R19 and L9. For the low power version, about 2 volts at the junction of C25 and L9. If OK this checks out Q1 through Q5 stages. In the case of the low power version, connect a 47 ohm resistor now to J2A. Adjust C25 and the position of L9A with respect to L9 for desired output. Do not couple L9A too close to L9—just enough for no more than about a volt across the 47 ohm resistor.

If you are building the 2 watt version now is the time to install Q6 and Q7. See parts placement diagram Fig. 3-11. If you are building the low power section, skip these steps to follow concerning the PA.

After installing Q6 and Q7, install L10, L11, L12, and L13. Last, install chip capacitors C26, C28, C29, C30, and C31. See Fig. 3-11 for correct chip placement. Do not overheat the chips. Make sure the PC board is tinned in the areas where chips are installed. The best ways to install them is to first tack solder to one side of the chip capacitor. Then solder the other side. Now, resolder the first (tack soldered) side. See Fig. 3-12. Do not overheat. Use a 25 watt iron with a pointed tip. Small fine point needle nose pliers or tweezers should be used to manipulate the chip capacitors. C44 is somewhat large, but C28 and C29 are tiny. Finally, install C34 and a suitable length of small 50 ohm coax cable to J2. Check all joints for solder bridges. Make sure that the metal case of Q7 is soldered to the ground plane (top side). Q7 has a reverse pin out—the emitter

CAPACITORS	TYPE	QTY
C1	56 pF NPO	1
C2, C12	33 pF NPO	2
C3, C7, C19, C22,C38	.01 DISC	5
C4, C6, C8, C13, C14	470 DISC	4
C5	82 pF	1
C9, C11	15 pF	2
C10	2.2 pF	1
C15, C17, C19, C21 C25, C27, C33	2-10 TRIMMER	7
C16, C32	1 pF NPO	2
C18	2-18 or 2-20 TRIMMER	1
C20, C23, C24, C45	470 pF CHIP	4
C26, C30, C31	100 pF CHIP	3
C28, C29	22 pF CHIP	2
C34	5 pF MICA	1

Fig. 3-11. Parts list—Video Transmitter, Project 6.

CAPACITORS	TYPE	QTY
C35, C36, C37	1 μF/35 or 50V ELEC	3
C39	10 μF/16V ELEC	1
C40	3-40 TRIMMER	1
C41	220 pF NPO	1
C42	470 pF NPO	1
C43	220 μF/16V ELEC	1
C44	10 μF 15V CHIP TANTALUM	1
C46	100 pF NPO	1

Fig. 3-12. Parts list—Video Transmitter, Project 7.

SEMIS		QTY
Q1, Q2	2N3563	2
Q3, Q4, Q5	MPS3866	3
Q6	MRF559 OR MRF627	1
Q7	MRF630	1
Q8	2N3565	1
Q9	MPF102	1
Q10	2N3906	1
Q11	2N3904	1
Q12	MJE180	1
D1	IN757A	1
D2	MV2112	1
D3	IN914	1
D4	IN4007	1

COILS	HARDWARE
2′ #22E	1– 8/32 SCREW
2′ #22 TINNED	1– 4/40 SCREW & NUT, LOCKWASH
2′ #32E	1– TO 220 INSULATOR
1– TOROID 76T188	
7– BEADS, FERRITE	
3– BLUE SLUGS, CAMBION	

MISC	CRYSTAL
1– PC BOARD	1—XTAL
1– TO 220 INSULATOR	
1– 8/32 × 3/4 SCREW	

Fig. 3-13. Parts list—Video Transmitter, Project 8.

is internally connected to the case. Install Q7 and connect leads to the underside of the PC board using as little lead length as possible.

Apply power and quickly adjust C25, C27, and C33 for maximum power into a 50 ohm load connected to J2. An rf power meter is nice to have here, but you can use a 47 ohm 2-watt *carbon* resistor or a dummy load made up as in Fig. 3-19, using ten carbon resistors to make up a 47 or 51 ohm 2-watt load. The rf probe of Fig. 3-19 can be connected to the hot side of the resistors (center connection of connector) to read the rf voltage.

You should get at least 1.5 watts (about 8.5 volts rms) into the 50 ohm load. It will become warm due to heating. Power supply current will be about 500 mA to 700 mA. Now adjust R33 for an output voltage of about half, or 1/4 power as read on the power meter if used. Leave the rf load connected.

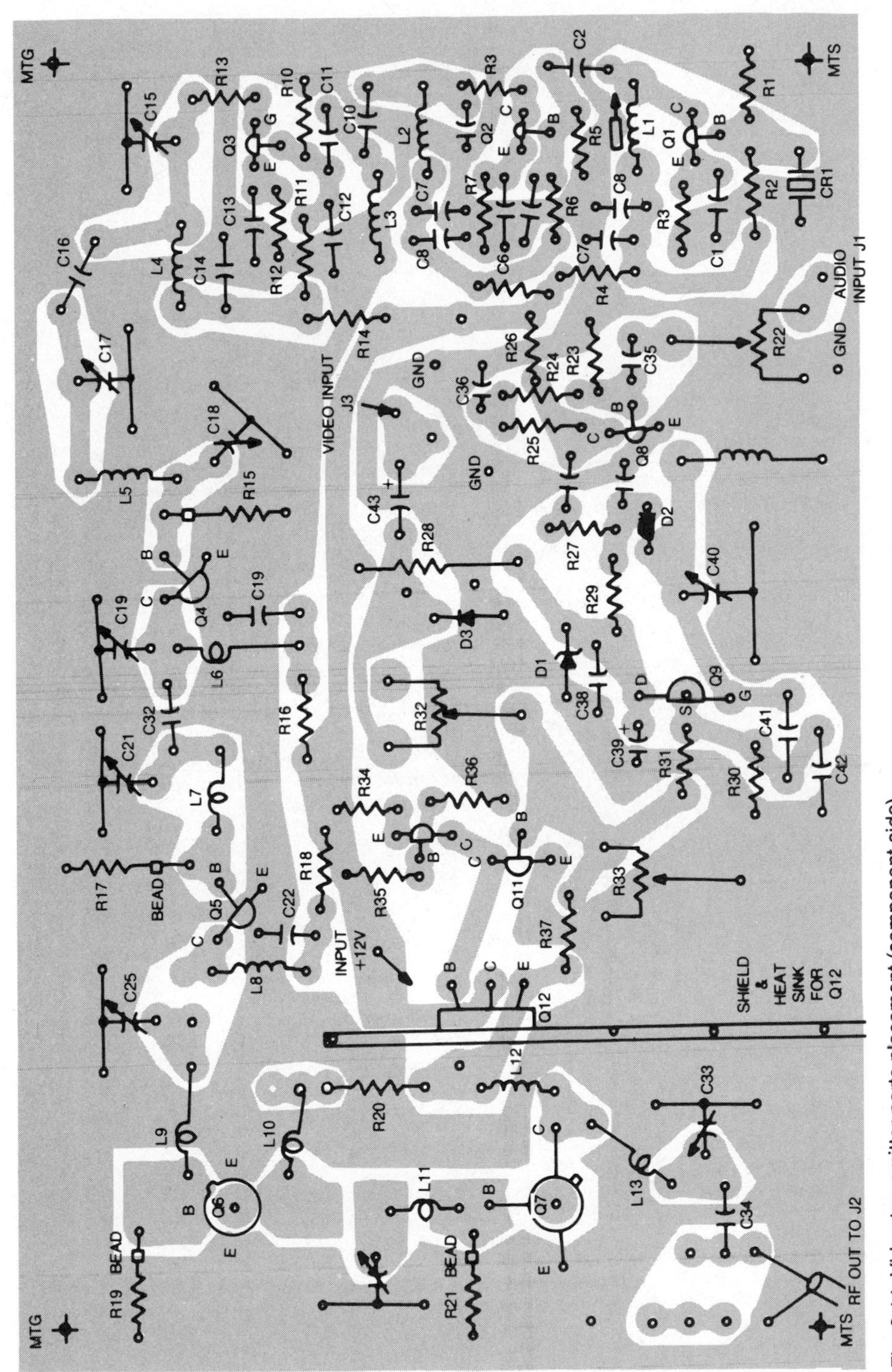

Fig. 3-14. Video transmitter parts placement (component side).

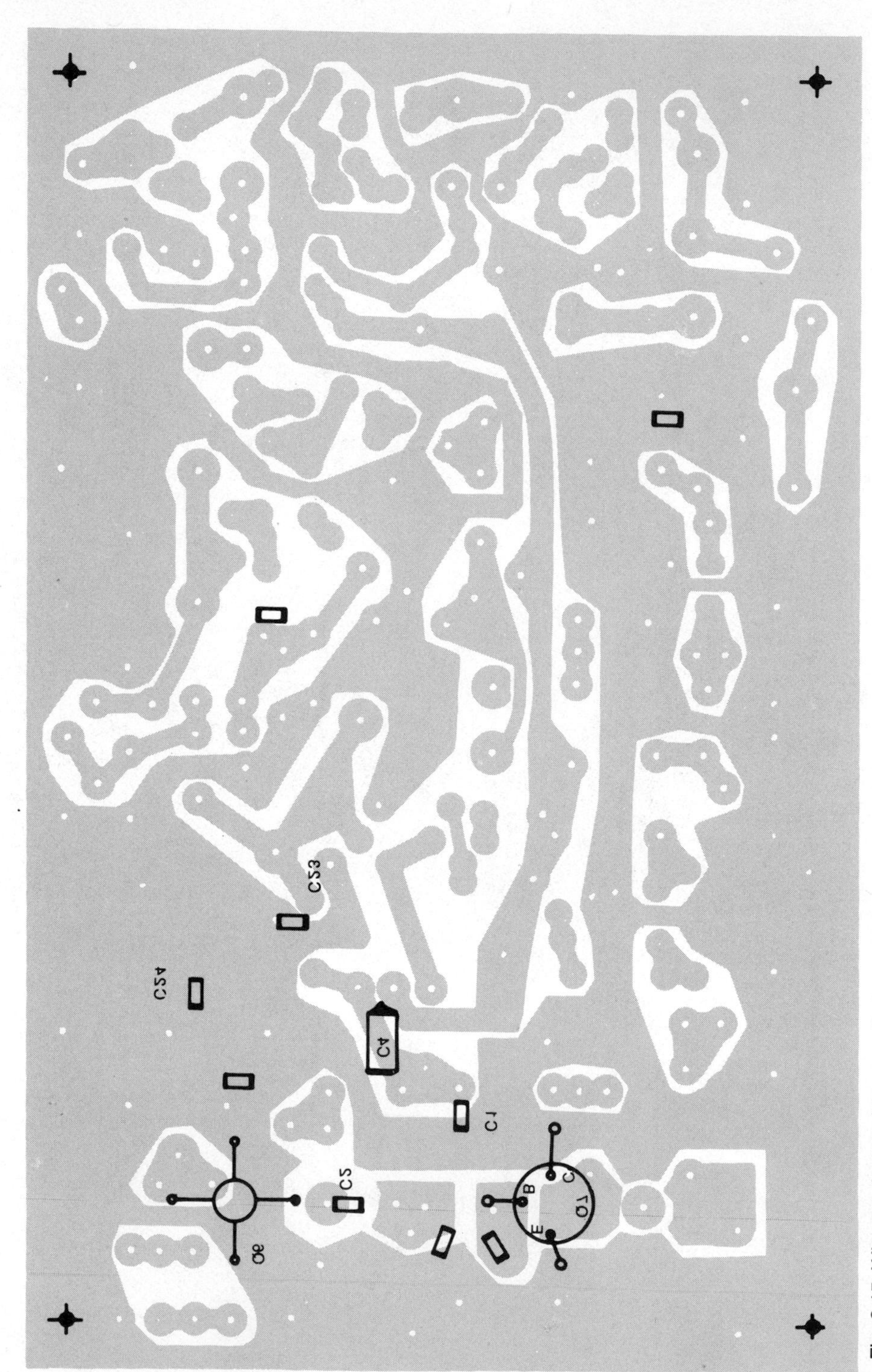

Fig. 3-15. Where to install chip capacitors.

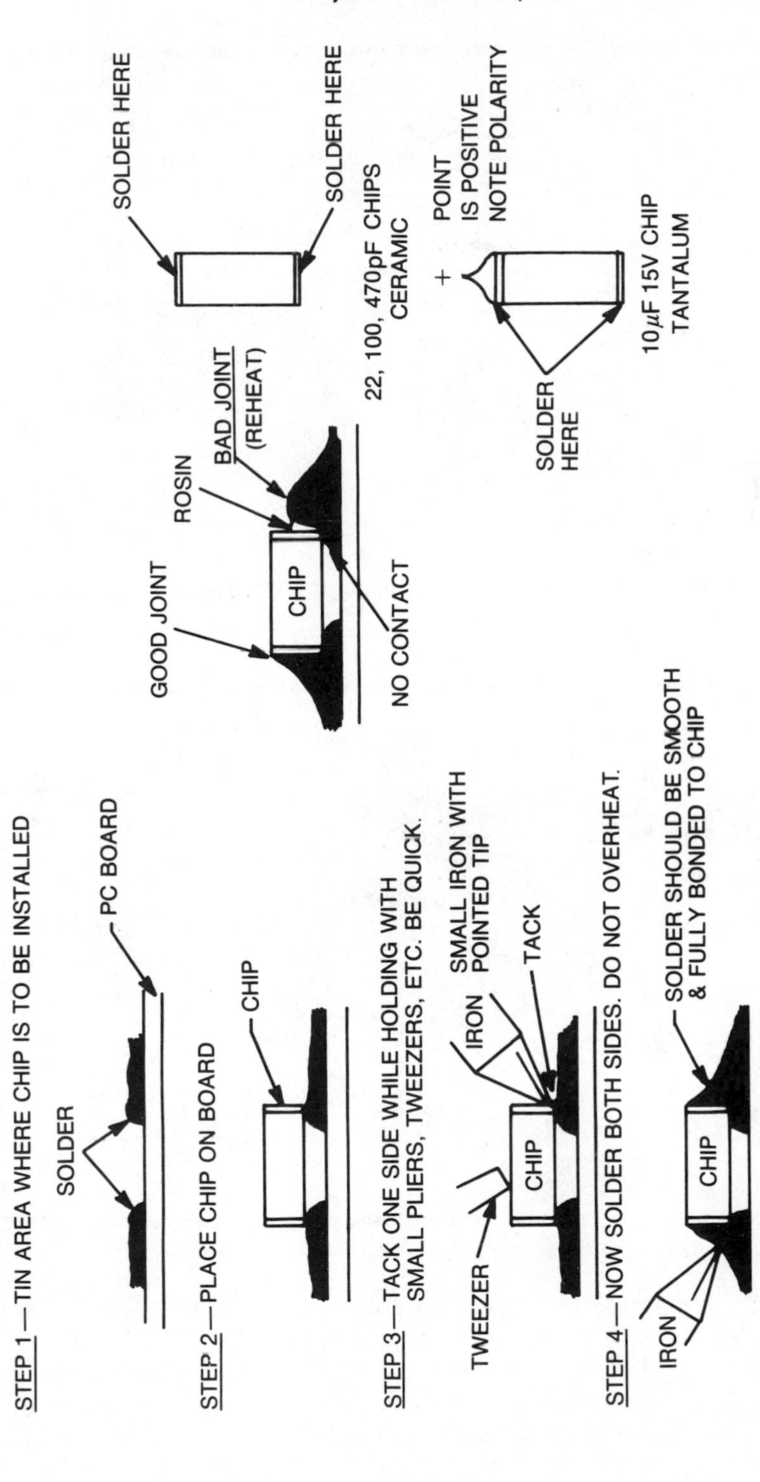

Fig. 3-16. How to install chip capacitors.

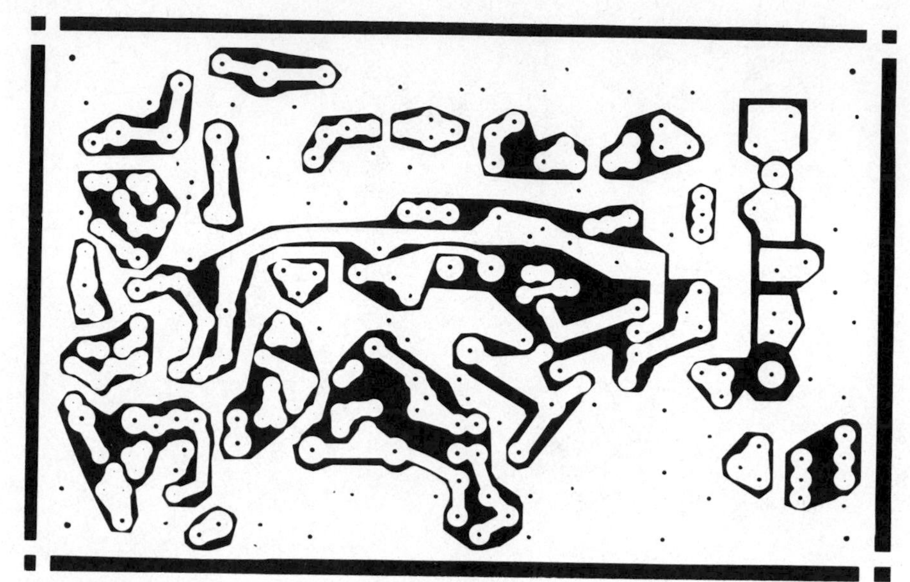

Fig. 3-17. Artwork for video transmitter (foil side).

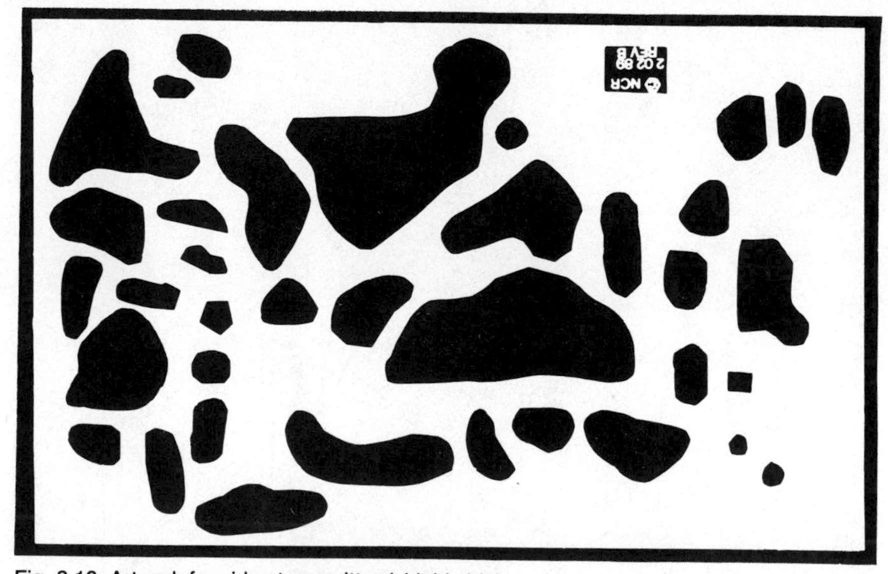

Fig. 3-18. Artwork for video transmitter (shield side).

Next, for either the low or high power unit, adjust R33 for about +6 volts at point A (emitter of Q12). Connect a frequency counter to point A. Adjust C40 for exactly 4.500 MHz. Now apply video and audio signals to J3 and J1 respectively. Using a TV receiver tuned to the transmitter frequency, adjust video gain R32 for best picture contrast and stability. Now adjust the audio level with R22.

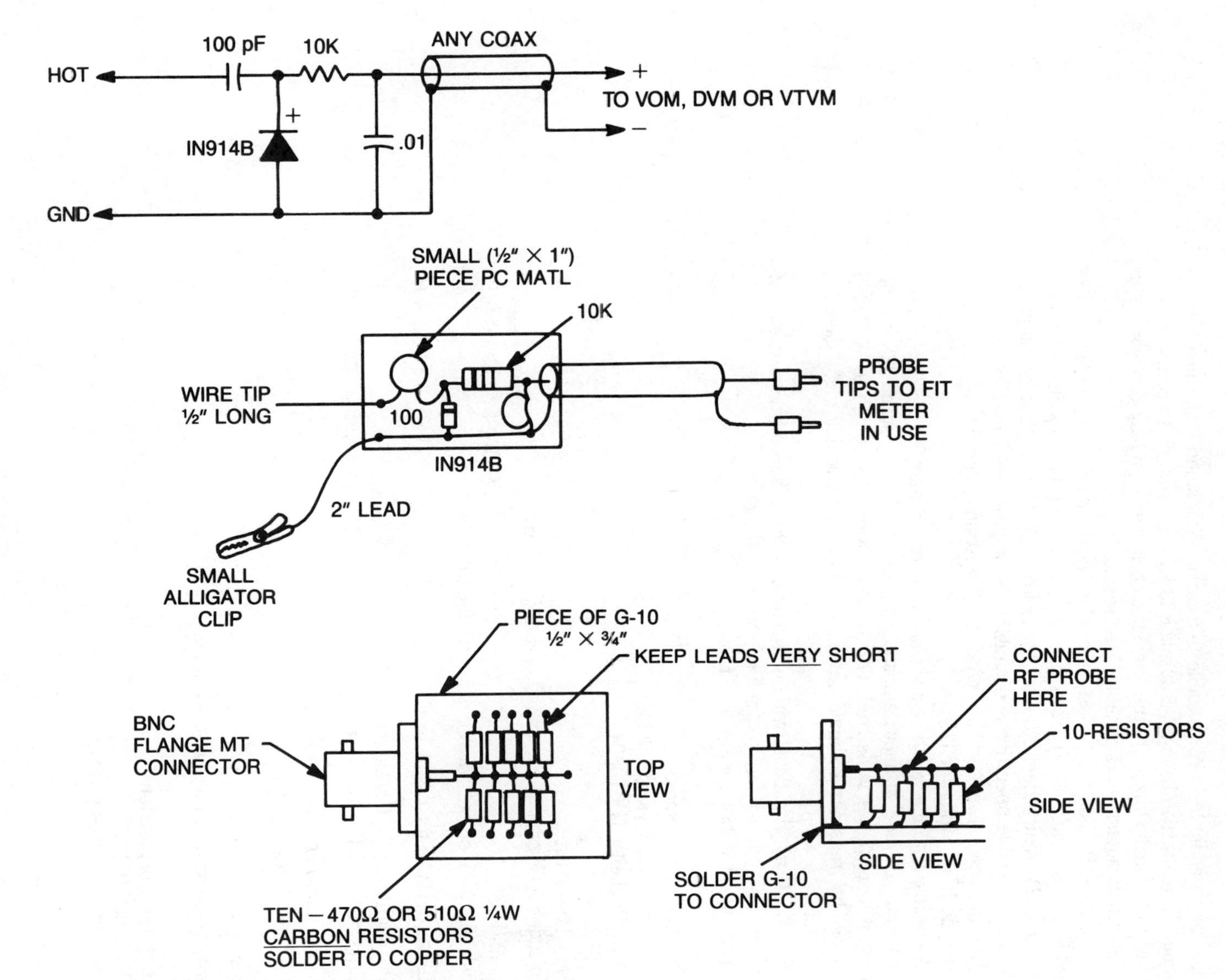

Fig. 3-19. Dummy load and rf probe for video transmitter.

Adjust for an audio level comparable to a commercial station on a neighboring channel. Now alternately adjust R32 and R33 for maximum video contrast without instability, audio buzz, or other evidence of clipping. You may also wish to go over all tuning adjustments again for best results.

Mount the PC board in a case, preferably metal, and connect leads from the board to suitable jacks for J1, J2 or J2A, and J3. Also provide a suitable connector for the 12V input if desired. Also, the case can contain either an ac supply or batteries if portable operation is desirable. Use appropriate batteries to handle the 100 mA (low power) or 500 mA drain (2-watt unit) Nicads are suggested. Use a type BNC fitting for J2.

A suitable antenna would be a six inch whip or a center fed dipole 12″ long. For amateur TV use, a linear amplifier may be installed between J2 and the antenna if more power is desired. For the low power version use the 6″ whip antenna. The transmitter should be mounted in a well shielded metal case for best results.

Finally, *do not* broadcast without a suitable license beyond the confines of your home or you may find yourself in trouble with the law.

The following kits are available from:

North Country Radio
P.O. Box 53, Wykagyl Station
New Rochelle, NY 10804

— Low Power Kit includes ATV crystal for operation at 439.25 MHz	$79.95 + $2.50 S/H
— 2-Watt Kit includes ATV crystal for operation at 439.25 MHz	$104.95 + $2.50 S/H
— Extra Crystals for CH14 or CH15	$6.50 + $1.50 S/H
— PC board only + Cores and Chip caps and D2 (partial kit)	$49.95 + $2.50 S/H
— Video Link February 1986 RE + Reprint of Article	$69.95 + $2.50 S/H

New York residents must include sales tax.

Crystals can also be purchased from:

Crystek Corporation
P.O. Box 06135
Fort Myers, Florida 33906

PROJECT 5: CARRIER CURRENT TRANSMITTER

A method of wireless transmission and reception of audio, data, or even video signals that uses ac power lines as a transmission medium is called carrier current. It uses rf carriers in the low frequency (100 – 500 kHz) range that can be suitably modulated with the information to be transmitted. Simple AM, FM, or related modulation methods can be used to place the information on the carrier.

Carrier current techniques are useful for coverage of a wide area, such as a complex of buildings, or a large building with one or two way communications, and since rf radiation can be minimized, only those areas served by power lines are covered. Applications are many, but a few useful in the home are:

a. Wireless extension speakers—mono or stereo;
b. Wireless intercom or paging system;
c. Remote control of appliances and other devices;
d. Headphone applications for both privacy and as aids to those with hearing difficulties;
e. Monitoring and security/surveillance applications, such as audible monitoring of doorways, baby rooms, etc.

If the power level is kept to only that needed for the applications intended, there is little danger of RFI and unwanted radiation of these low frequency signals.

The ac power system in the average home, while commonly a three-wire 120/240V 60 Hz system, with a grounded neutral varies widely in construction. The size of the house number of circuits, appliances, wiring methods, and layout pretty much make each individual house a unique situation. Conceivably even a small house that is all electric, with electric heat, stove, well pump, water heater, and the usual run of appliances can place a load of over one hundred amperes on each side of the 120/240V system. See Fig. 3-21. At 120 volts, 100 amperes represents a 1.2 ohm load across the line to neutral. This load can vary widely. For example, usually only a few appliances may be operating at night, such as lights and a TV or stereo system. On a cold winter day, there is the load from heating appliances. The laundry may be simultaneously in operation with an electric dryer going. The well pump may be operating intermittently. At the same time, the electric stove may be in operation. On a summer evening, only a few lights may be in operation. Any combination of weather factors, time of day, day of the week, etc., all impose a widely varying load on the power system. Therefore, as an rf transmission medium, the impedance may be indefinite. Complicating this factor is the fact that certain loads may have a very high rf impedance or be a near short circuit to rf, especially if those loads have built-in rf bypassing such as TV sets, stereos, computers, etc. However, the line cords on these devices have some inductance. A further complication is

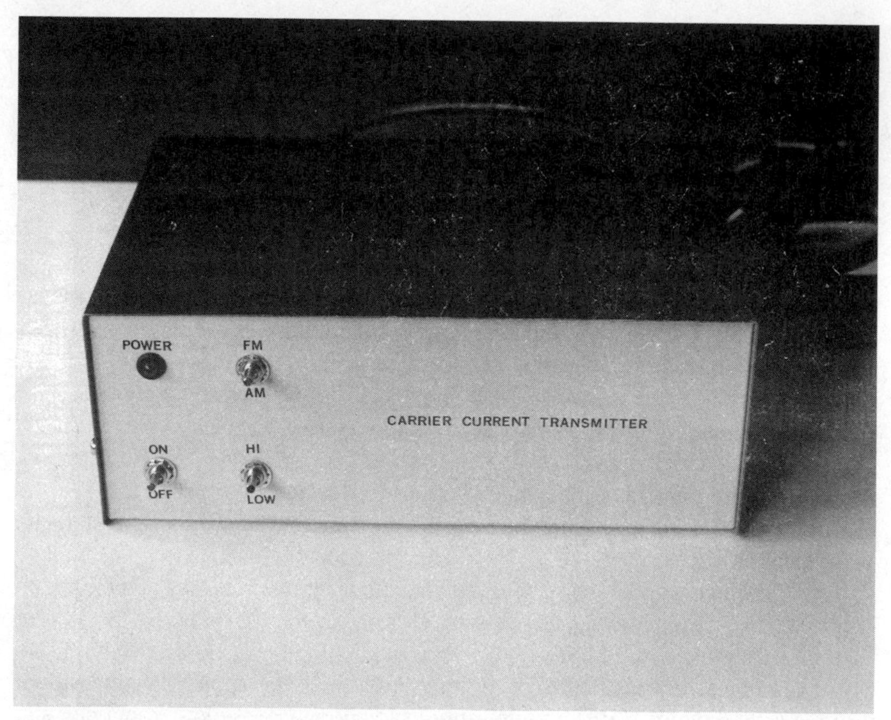

Fig. 3-20. External view of carrier current transmitter.

the fact that all newer homes have a three wire system in the 120V circuits, where there is an extra grounding wire in addition to the normally grounded neutral and the "hot" lead. This extra ground wire is separately run to the ground at the service entrance and is very important for safety considerations being used to ground all metallic appliances, fixture housings, and metal cabinets and boxes, to prevent their becoming "hot" if there is any internal short between the hot side of the 120V circuit and the appliance or fixture housing. This ground can act as an rf shield in the case of BX or conduit style wiring.

In addition to the unpredictability of the rf properties of the average home wiring system, another problem is the presence of power line noise. Noise voltages are generated by many appliances connected to power lines, and unfortunately these noise voltages are high in the frequency range (100 – 500 kHz) useful for carrier current transmission. Offenders are motors, fluorescent lamps, neon signs, contacts or relays, and induced line noise from the inevitable corona discharges and leakage of insulators on the primary side (usually 7200 volts) of the power distribution system. In addition, some of the worst noise is due to modern solid-state triac and SCR devices, rectifier diodes, computers, and sweep radiation at 15.7 kHz from TV sets. The ac power line in the typical modern household is not only an rf transmission medium of considerable unpredictability, it is a hotbed of noise and interference. If you will sit down and write a list of every electrical device in your home, down to the last light bulb, you will be surprised at how many motors, fluorescent lamps, light dimmers, etc., and other noise generators are present (undoubtedly you will come up with at least twenty, and possibly over a hundred noise sources).

The situation is not hopeless as one would at first think. It is easy to generate considerable signal power levels at these low frequencies. Solid-state devices of the inexpensive variety, such as the NE566 voltage controlled oscillator, NE565 phase-locked loop, and low cost audio power transistors work well at frequencies in the 100 – 500 kHz range. It is no problem to generate several volts of rf signal, couple it into the power line, and to receive this signal, even if over 100dB of losses in the lines are experienced, with good signal to noise ratio. The use of a phase-locked loop detector with FM signals allows this degree of performance. Even a simple AM system is useable for certain applications, if the line noise is not too severe. This article will describe a carrier current transmitter and a simple receiver useful for many applications. Subsequent articles will describe equipment for transmission of stereo audio, data, and even video signals. Schematics, PC layouts, and details of construction will be furnished, and also sources of ready made PC boards and parts kits will be provided.

Since a transmitter is needed for any system, while any one of several receiver approaches may be used, from simple to complex, the transmitter will be described first. See Fig. 3-22, Fig. 3-23, and Fig. 3-24 for the following discussion.

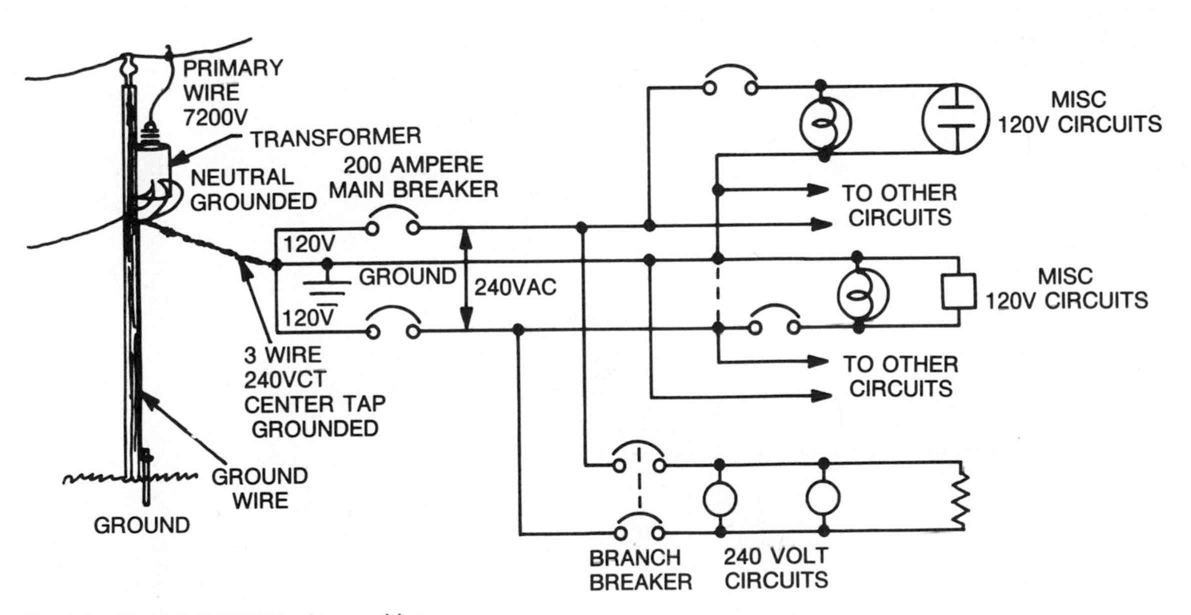

Fig. 3-21. Typical 120/240Vac home wiring.

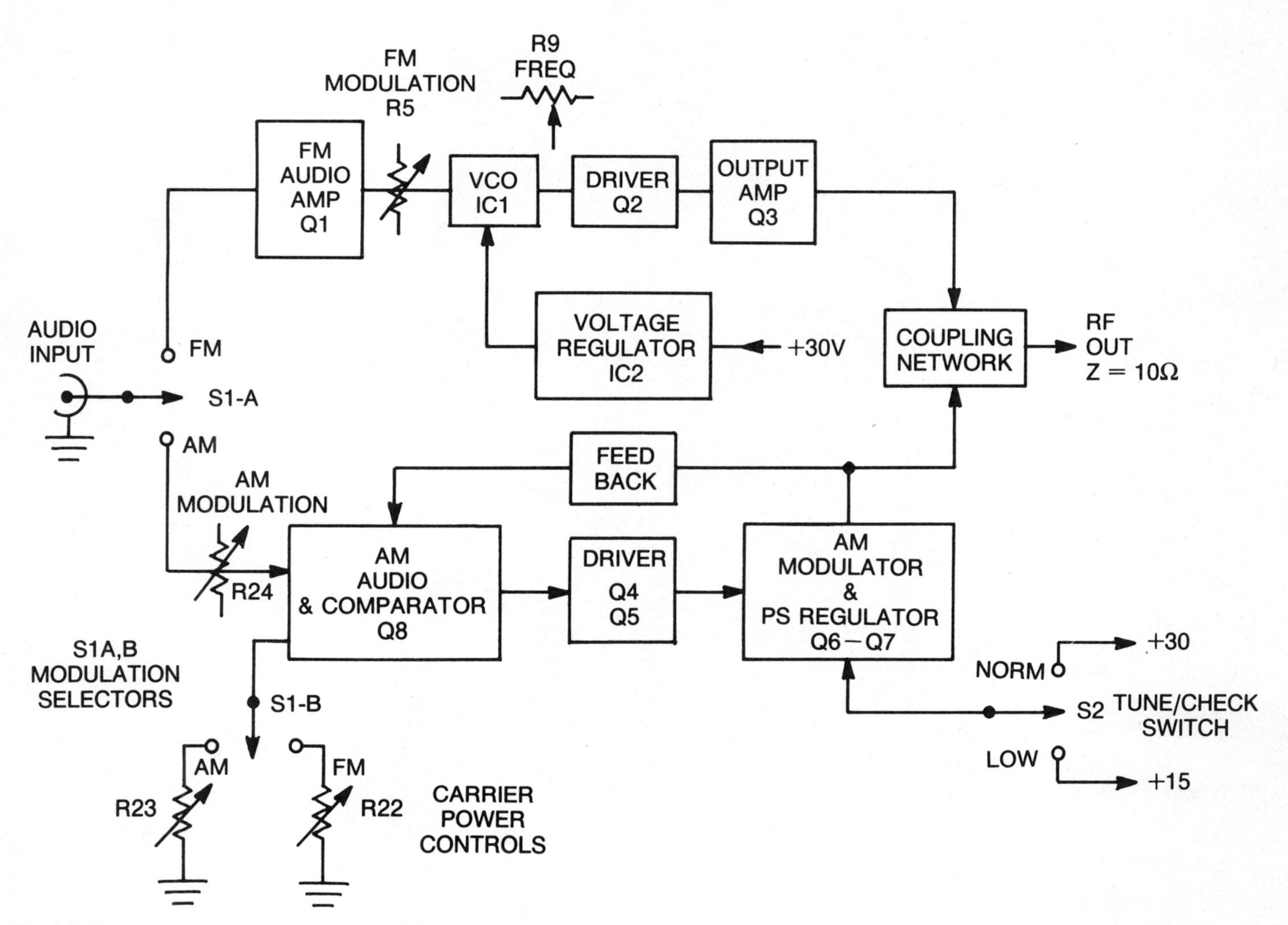

Fig. 3-22. Carrier current transmitter block diagram.

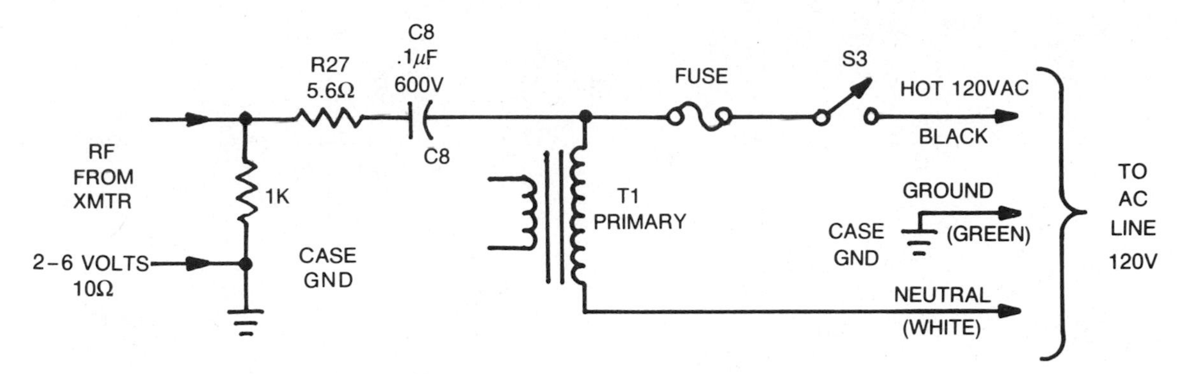

Fig. 3-23. Method of coupling rf signals to power line.

The transmitter to be described is built in a small Radio Shack Case, P/N 270-274A. It uses a 3″×4″ single sided PC board. Three of the transistors are heat sinked to a piece of 1/8 or 3/32″ aluminum mounted flat against the rear of the case. The case is 8 1/4×6 1/8×3″. While a smaller case could be used, the Radio Shack case is readily available and easy to work using hand tools. A power transformer Radio Shack, P/N 273-1512A is used to power the unit. It runs cool and is easy to obtain. The transmitter circuitry is contained on the PC board (except the power supply, switches, and ac line coupling components). Parts are readily available from either mail order suppliers, local parts houses, or as a kit from the source mentioned at the end of this article.

A block diagram of the transmitter is shown in Fig. 3-22, and schematics are shown in Fig. 3-23 and Fig. 3-24. The transmitter will deliver up to 5 or 6 volts of rf signal into a 10 ohm resistive load. This is peak power for AM or continuous power for FM. Frequency coverage is nominally 250 to 300 kHz for standard (narrowband) AM. The limits are set by the output coupling network, which could be changed as required.

The decision as to whether to use simple AM or either narrowband (less than 15 kHz deviation) FM or wideband FM (greater than 30 kHz) depends on the application. For transmission of music or data, FM may be better since it has greater noise immunity. For speech or noncritical applications, AM may be perfectly satisfactory. In fact, the frequency range of 150 to 300 kHz is used for AM broadcasting in Europe and Asia, with high power transmitters and low sensitivity, simple receivers, usually only a simple AM portable with a low frequency tuning range. There are many small portable radios available, which tune this range and if the reader has one of these, only this transmitter would be needed to make up a simple wireless speaker or headphone system. If noise is a problem, either a low sensitivity AM receiver with relatively high transmitter power can be used, or better, the FM approach with a PLL FM receiver. The transmitter to be described permits either mode with no modifications, simply by switch selection.

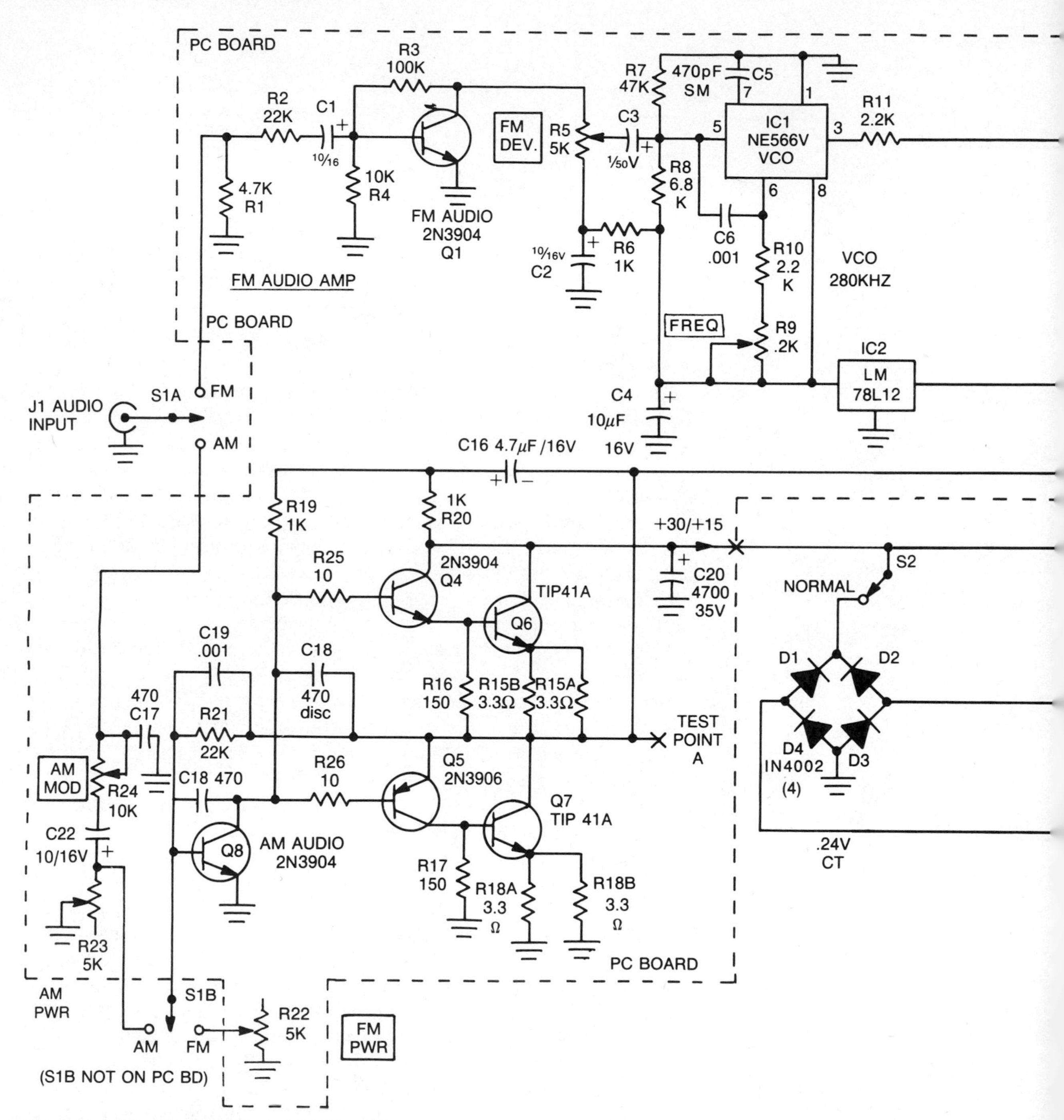

Fig. 3-24. Schematic of carrier current transmitter.

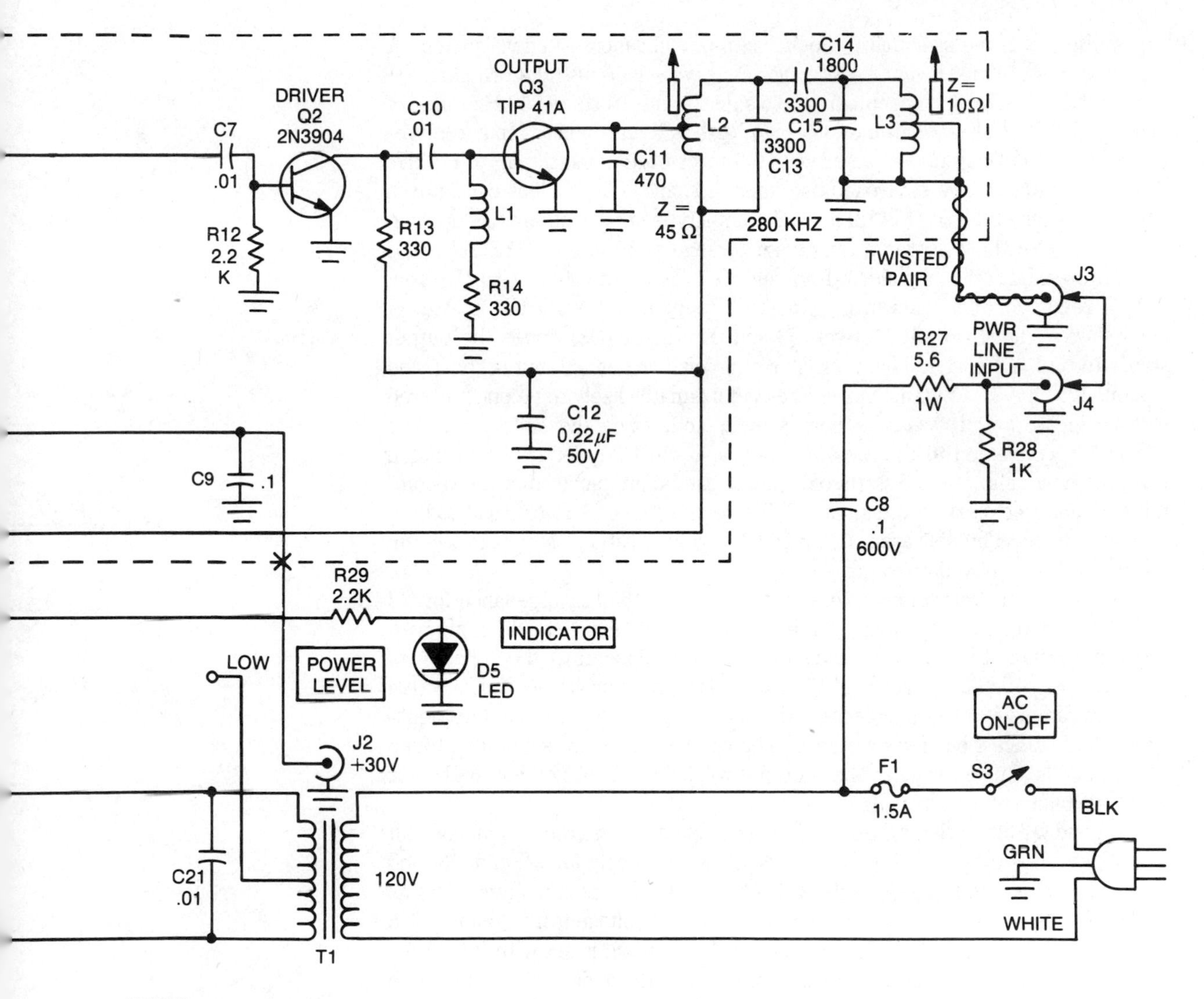

NOTES

1) COMPONENTS WITHIN DOTTED LINE ARE ON PC BOARD. ALL OTHERS ARE NOT.
2) RES ALL ¼W ± 10% OR BETTER EXCEPT AS NOTED
3) Q3, Q6, Q7 MUST BE HEAT SINKED

In Fig. 3-22, the transmitter block diagram, audio input is fed via switch 1A to either Q1 or Q8 for FM and AM audio. First we will discuss the FM part. Q1 is an audio amplifier that accepts an audio signal in the 10 Hz to 80 kHz range of about 0.5 volt peak-to-peak into a 5 ohm load. (Q8 has similar input requirements). The audio gain is adjusted with R5 to provide nominally up to 60 kHz deviation of the voltage controlled oscillator IC1, and NE566. This oscillator is set to nominally 280 kHz via R9 and C5. Stability is very good for this application (with ±1 kHz) over room temperature range of 60 to 80°F (15 to 27°C). The NE566 and audio amplifier Q1 are supplied regulated voltage (+12V) from IC1, a regulator. This aids in keeping the frequency of VCO IC1 stable. A square wave signal from IC1 driver Q2 and this driver (Q2) drives the output transistor Q3, a TIP41A. This transistor is larger than needed, but is cheap and has enough "beef" to withstand severe load mismatch likely to be encountered in this application. It *must* be heat sinked, and, since the collector of the TIP41A is connected to the mounting tab, also must be electrically insulated from the heat sink. Up to 5 watts of rf power can be produced, (but *not* recommended due to excessive radiation) by this device at 280 kHz. Normally, about one watt of rf will be delivered. (With 100% AM modulation, the peaks will run up to 4 watts at maximum modulation.)

A coupling network is used to provide a nominal 45 ohm impedance for Q3 and a 10 ohm output impedance to the power line coupling network, along with the suppression of harmonics. Since a square wave appears at the collector of Q3, primarily odd harmonics (3rd, 5th, etc.) are predominate. A check with a communications receiver plugged into the same ac power line revealed no significant AM broadcast band interference. The coupling network is a double tuned circuit, and is simple to align. The rf output will light a small 12V automobile or flashlight bulb to full brilliance.

Q6 and Q7 are series connected pass transistors across the nominal 30 volt dc supply. They provide a 5 to 25 volt adjustable dc supply for Q2 and Q3, and act as regulators. In the AM mode, Q4, Q5, and Q8 act as audio amplifiers as well as dc control transistors. R22 sets the dc output voltage from Q6 and Q7 in the FM mode. In the AM mode, a nominal dc level output is set with R3. In the AM mode, audio is coupled to Q8 via R24, C22 and the audio wave form (same level 0.5V peak-to-peak into 5k ohm) is amplified by Q8, Q4, and Q5 and therefore is superimposed on the output of Q6 and Q7, which is set by R23 in the AM mode. Therefore, the voltage fed to Q2 and Q3 has its amplitude modulated by the audio fed to Q8. This produces an amplitude modulated output from the transmitter. By proper setting of R24 (audio level) and R23 (carrier level), excellent modulation linearity and 100% modulation can easily be secured. Frequency response is limited to the 10 kHz and below region (although if needed, can be extended to over 50 kHz), since AM will primarily be used for speech purposes in most instances.

Power (+30 V) is supplied by T1, a 24V CT transformer, and D1 thru D4, a bridge of 1N4002 diodes or any other similar types suitable for the job. Transmitter output is supplied to J3, for test purposes. A 10 ohm 2-watt carbon resistor can be used as a load. J4 connects to the ac line coupling network to enable coupling transmitter signal into the ac line. Normally, J3 and J4 are jumpered. J2 is used as a +30 V output to power up an external accessory, such as a stereo generator.

If stereo audio is needed, a suitable stereo encoder must be used to generate a stereo signal, which can then be connected to J1. A suitable generator is the next project in this book. The FM oscillator at 88 – 108 MHz can be omitted from the circuit, simply using the encoder portion. This will be discussed later in detail. Also, data signals in the audio range can be handled by this transmitter, as can video (medium or slow scan) signals. Standard NTSC video will not work since only about 60 – 70 kHz bandwidth is available. More on this later.

Referring to Fig. 3-24, a detailed circuit description will follow: Audio input of normally 0.5V peak-to-peak, 5k ohm impedance, is plugged into J1. S1A selects either FM or AM modulation. For FM modulation, audio appears across R1, a 4.7k resistor to serve as a termination for the 5k ohm audio source. Audio is then coupled through gain set resistor R2, C1, into the base of audio amplifier Q1, a 2N3904. R3 and R4 bias Q1, and the ratio of R3 to R2 determine the gain of Q1 stage, at about four. Output is developed across collector load resistor, a pot, designated R5. Audio from the wiper of R5 is taken by C3. C2 and R6 decouple audio and stray VCO signal from the B+ voltage. About 0.5 to 1.0 volt peak-to-peak audio appears at pin 3 of IC1, at an output frequency of 200 to 350 kHz, depending on the setting of R9. C7, and R11 couple the ac component to the base of driver Q2. R12 provides a dc path to ground for the base of Q2 and allows Q2 to generate its own base bias. A square wave of about 8V peak-to-peak appears at the collector of Q2, amplitude modulated if AM is used. C10 couples this waveform to the base of Q3. C10, L1, and R34 provide waveshaping and a dc bias return for the base circuit of the power amplifier Q3. IC2 supplies a regulated 12V to IC1 and Q8. C4 and C9 are bypass capacitors to ensure stability of IC2.

Q3 provides power amplification of the 280 kHz signal (nominal) from Q2. C1N acts as a harmonic bypass at the collector of Q3 and helps protect Q3 from spikes or parasitic oscillation. The collector connects to a tap on L2 at about a 45 ohm impedance level. C12 provides a bypass and rf return for the B+ supply to Q2 and Q3. If AM modulation is employed, this supply voltage is modulated and instantaneously varies from nearly zero to over 27 volts. L2, C13, C14, C15, and L3 form a bandpass filter for the range 200 to 350 kHz, and also match the impedance of the collector circuit of Q3 to a 10 ohm (nominal) load impedance. Q3 *must* be heat sinked, and the collector (tab) also insulated from ground. A mica washer is used for this purpose, with a light coating of silicone grease to aid in heat transfer. *Do not* operate the transmitter without a heat sink on Q3, Q6, and Q7 or they will overheat and be destroyed. (If an LM7812 is used for IC2 instead of a 78L12, the larger device LM7812 requires no heat sink.)

In the FM mode Q6 and Q7 act as a pair of pass transistors, supplying dc to Q3. Dc from across filter capacitor C20 (4700μF 35V) of about +30V is applied to the collector of Q6 and driver Q4. R20 and R19 supply dc bias to the bases of drivers Q4 and Q5 through oscillation suppressors R25 and R26 (10 ohm resistors). This bias would tend to fully turn on Q4 and cut off Q6, except for dc feedback through R21. R16 and R17 provide dc base bias for Q6 and Q7. R15 A and B, and R18 A and B are 3.3 ohm resistors to provide emitter stabilization for Q6 and Q7. (Note that Q5 is a PNP device.) Q8 is connected as a common emitter amplifier with its collector to R19, R25, R26, and the bases of Q4 and Q5. Q8 receives its bias from R21, which connects to the emitter of Q6 and collector of Q7, and is also the dc supply voltage for Q2 and Q3. If this voltage (at TPA, see Fig. 3-24) rises, it will tend to turn on Q8 harder. Q8 will draw more current through R20 and R19 dropping the drive to A4 and increasing drive to Q5. This makes Q6 conduct less and Q7 conduct more, lowering the voltage at TPA, cancelling the tendency for it to rise. A similar but opposite effect occurs if the voltage at TPA tries to fall. In this case Q8 tends to conduct less, Q4 and Q6 more, and Q5 and Q7 less, raising the voltage at TPA. The exact voltage at TPA depends on the ratio of R21 to either R23 (for AM) or R24 (for FM) and the base-emitter turn on voltage of Q8 (about 0.6 volt). Therefore, R23 and R24 can set the dc level at TPA. This is nominally about 10 to 20 volts for FM and about 12–14 volts for AM, depending on the power level needed. C19 provides extra feedback at frequencies above about 7,000 Hz when AM modulation is used. C18 a 470pF capacitor, is connected from the collector to the base of Q8 to control the bandwidth of that stage. Without it, oscillation tends to occur at certain bias settings of R23 and R24.

When AM modulation is required, audio is fed to R24, a level control, and rf bypass C17. Audio from R24 is coupled to the base of Q8 through C22 and S1B. R23 determines the quiescent point of Q6 and Q7 and the no-signal resting (static) voltage to Q2 and Q3. Now, Q8, Q4, Q5, Q6, and Q7 act as an audio amplifier, producing a modulated dc voltage at TPA. In fact, this circuit is exactly the same as a simple audio power amplifier, the exception being the dc load current passing through Q3. For applications as a loudspeaker amplifier, a large capacitor is used in series with the speaker. In our case, since there is a dc component through Q2 and Q3, no capacitor is used. Q6 must be heat sinked, as well as Q7. As with Q2, the transistors must also be electrically insulated from the heat sink. Q6 dissipates more power than Q7 since it carries the dc current through Q2 and Q3. *Do not* operate without a heat sink. Destruction of Q6, Q7, Q3, or all of them will occur due to overheating.

C16 couples an ac signal to the junction of R20 and R19. This maintains a constant currect in R19 and improves the available positive voltage swing from Q6. It is called the "boot strap" capacitor and is also used in audio amplifiers.

There are also components not on the PC board. R29 and D4, a LED, are used as a power indicator and are at the discretion of the constructor. They are not essential and may be omitted if desired. Also, a small incandescent lamp rated for 30 or 36 volts could be used.

Transmitter output is connected to J3 via a twisted pair (we used two lengths of No. 24 AWG hookup wire of two different colors, anything handy will do) to prevent excess radiation. At 280 kHz, they are not at all critical—the length will be about 6″ or so and anything will work.

J4 is connected to R28, R27, and C8 to the hot side of the ac power line, *After* the fuse F1 rated at 1.5 or 2.0 amps. Rf from J3 is fed into J4 via a short jumper. R28 limits ac voltage developed through coupling to the line via C8 to about five volts. Otherwise a mild but uncomfortable shock can be gotten from J4 if the center pin is touched. R27, a 5.6 ohm 1 watt resistor provides a stabilizing effect on the impedance seen by the transmitter. It also limits ac line current to a reasonable value in case C8 shorts. F1 will blow instantly in this case, since J4 connects to the top on L3 which is at 60 Hz ground potential. C8 is a 0.1μF coupling capacitor. This has a reactance of about six ohms at the transmitter frequency. Since the power line may have only a few ohms impedance, this capacitor is needed. It *must be 600Vdc or better,* or else rated for *250Vac service minimum.*

NOTE: If this transmitter is used in an application where no ac line coupling will be used, C8, R27, R28 and J4 can be deleted. Do not operate this unit for long periods of time with J4 opened. Short J4 when not used or during test, in case C8 should short, so F1 will blow. *Do not change the ac input circuit to an ungrounded (2-wire system). Make sure the case of the transmitter is grounded to either earth ground, cold water pipe, or grounded to an electrical system ground (conduit, metal boxes, etc.). This is taken care of if a 3-wire grounding type cord and properly wired outlet is used.*

RF from C8 is coupled through F1 to the ac line. S3 is used as an on-off switch. Use switches rated for 125Vac operation in this application. T1 primary acts as a choke and has a very high impedance. See Fig. 3-23 for this coupling detail.

S2 is used to select either the full transformer voltage or half of it. Normally, the full voltage is used. During tests use the low positions. This reduces the chances of blowing something during initial setup. It may be also used (in the FM mode only) to reduce transmitter output if desired (depending where R22 is set). J2 provides +30V to an accessory, such as a stereo MPX generator. D1 through D4 form a rectifier bridge. C21 helps to suppress rectifier switching noise and reduce the possibility of spurious 60 Hz transmitter modulation.

Construction and Alignment Hints

Construction is straightforward. Nothing is particularly critical, but use good construction practices. Use only rosin core solder. Keep all leads as short

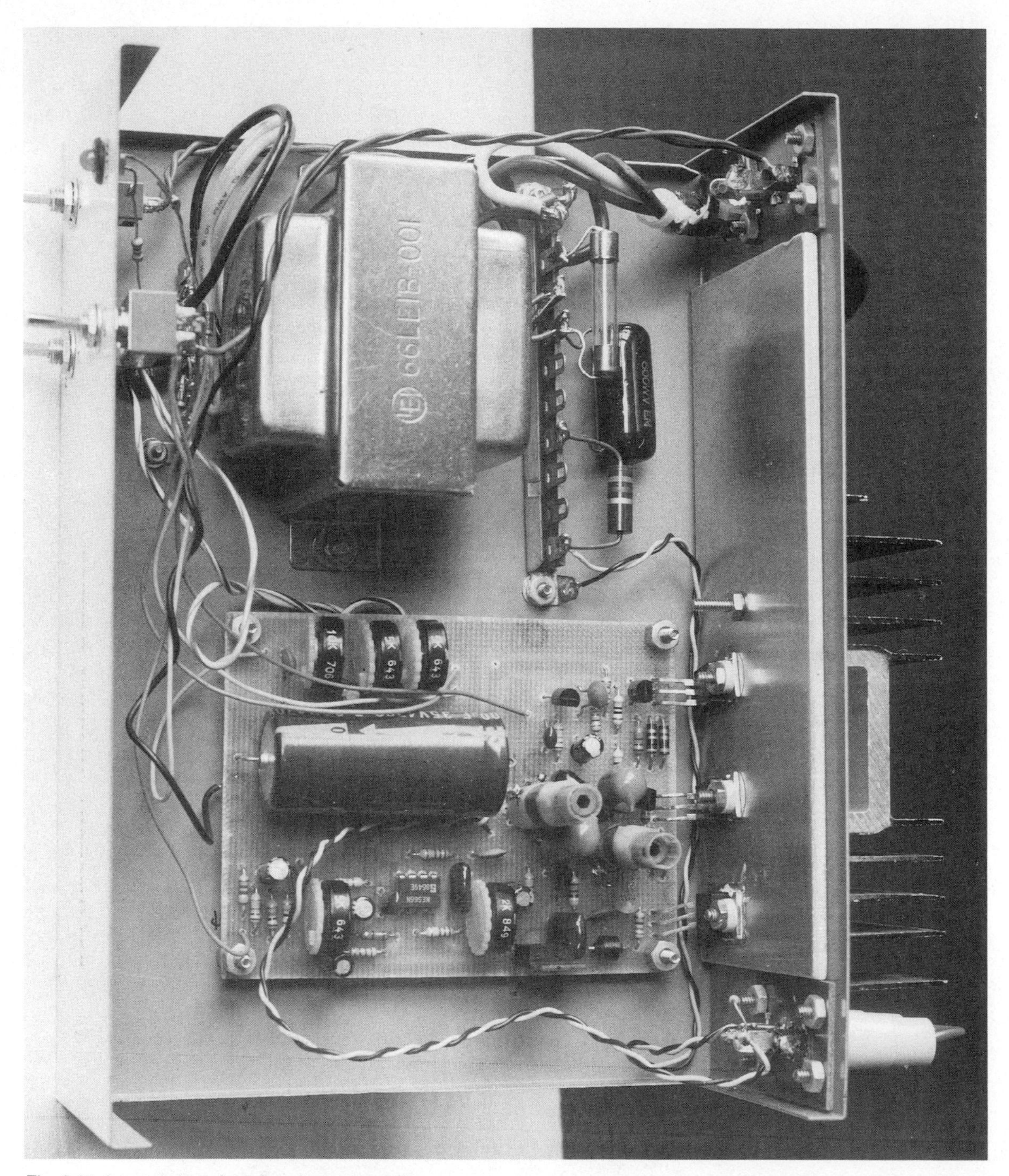

Fig. 3-25. Internal view of carrier current transmitter.

as possible. The leads to and from J1, J3, J4 should be either twisted pairs (4 or more twists per inch) or else shielded cable or miniature coax such as RG 174/U, etc. Mount all components not on the PC board on suitable tie strips or standoff insulators. Twist all ac leads and keep them away from leads carrying audio (J1 wiring, S1 wiring). Use adequately rated components, *especially for C8*. Only use a 3-wire ac grounding type cord, and suitable outlet. If you live in an older home with two prong outlets, make sure the chassis is grounded either to the outlet box or a cold water pipe. Do *not* omit F1. It can be a 3AG 1.5 or 2A type, with either wire leads or a fuseholder mounting, as desired. Also, Q3, Q6, Q7 *must* be heat sinked and electrically insulated from the chassis. Use TO-220 type mounting hardware, or sheet mica cut to fit, and fiber or plastic bushings to insulate mounting screws from the transistor tab. Nylon screws can be used but metal screws with insulator washers are also satisfactory. See Fig. 3-26 for details. It is best to use the PC layout in Fig. 3-27 and placement of

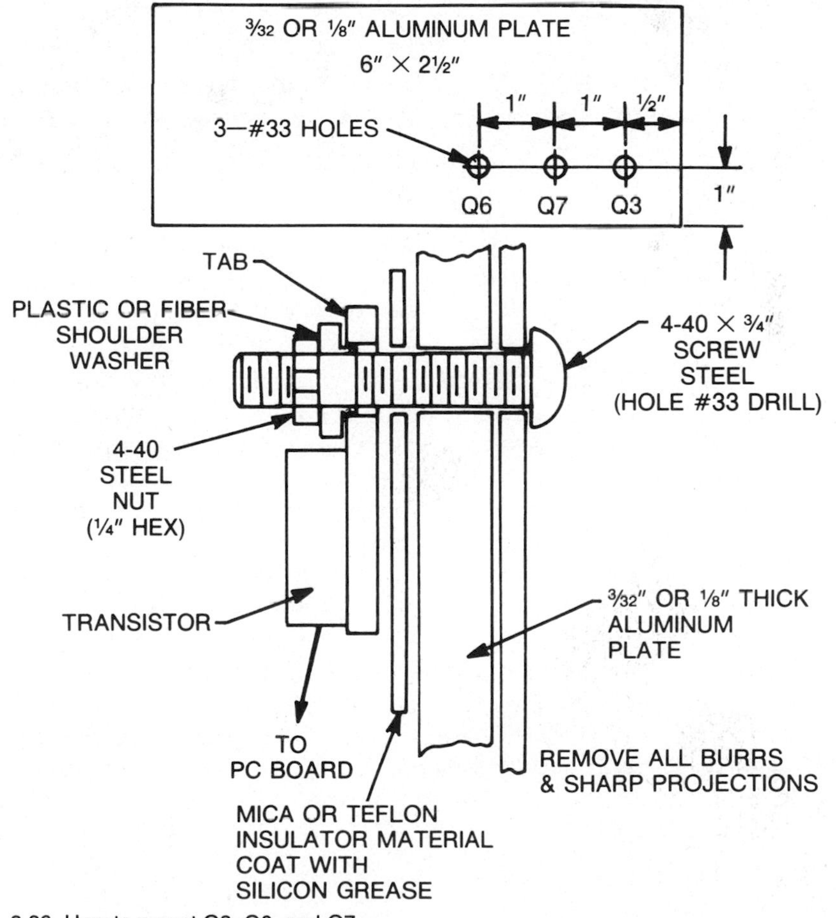

Fig. 3-26. How to mount Q3, Q6, and Q7.

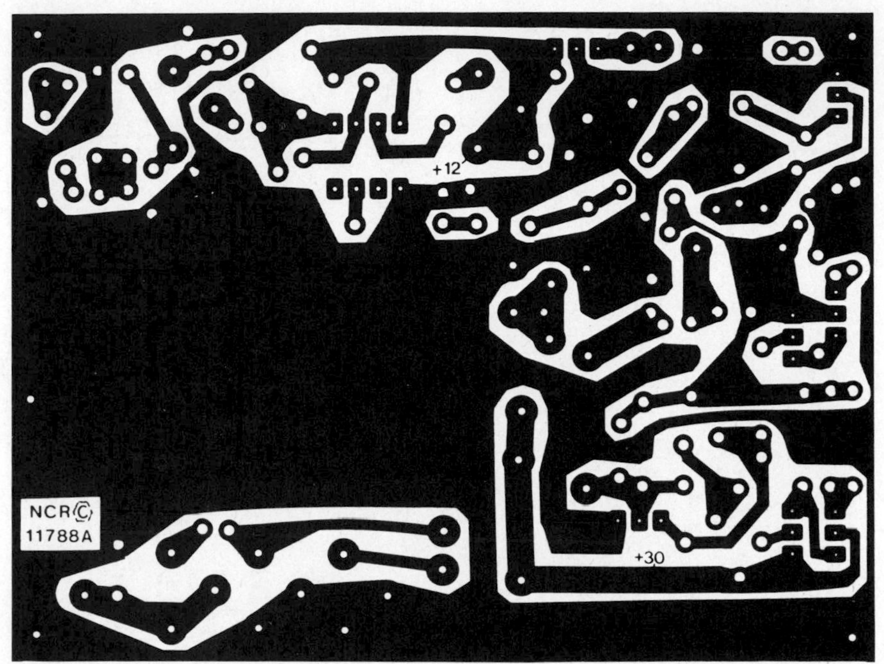

Fig. 3-27. PC board artwork carrier current transmitter.

parts shown in Figs. 3-28 and 3-29. The PC board can be homemade using .050 or .062 material or can be purchased (see parts list), as can be L2 and L3 (see Fig. 3-31). Some transistor substitutions can be made at the constructors risk. Several substitutions are listed, using Sylvania/Philips ECG devices. They should work well, although there may be slight variations in performance. Also, other equivalent devices would probably be okay as well. Again, do not skimp on C8—a shock hazard could result if C8 shorts, as well as some nasty surprises.

Refer to Fig. 3-30 for waveforms. After construction, check for shorts, poor connections and solder bridges. Make sure Q1 through Q8 are properly inserted and that Q3, Q6, and Q7 tabs do not short to the case or heat sink otherwise Q7 may be destroyed or D1 through D4 may blow. The B+ line should read at least 200 ohms to ground with no power applied. Reverse the ohmmeter probes if less than this is read and take the highest of the readings. Also, allow several seconds for C20 to charge. Many VOMs and VTVMs have the plus (red) lead negative when ohms are read.

After everything looks OK, place S1 in the FM position, S2 in LOW, and S3 off. Now apply 120Vac to the transmitter. Connect a dc voltmeter to the junction of D1 and D2. Quickly turn on S3—you should see 25Vdc. Quickly turn off S3 if not, and troubleshoot. If OK, next check for 15 volts across C20. If OK, turn off S3 and connect the dc voltmeter to TPA. Set R22, R23, R24 to maximum resistance. Set R5 and R9 to center positions. Connect a flashlight bulb (use a 6V or higher bulb) to J3. Set the slugs in L2 and L3 half way into the windings.

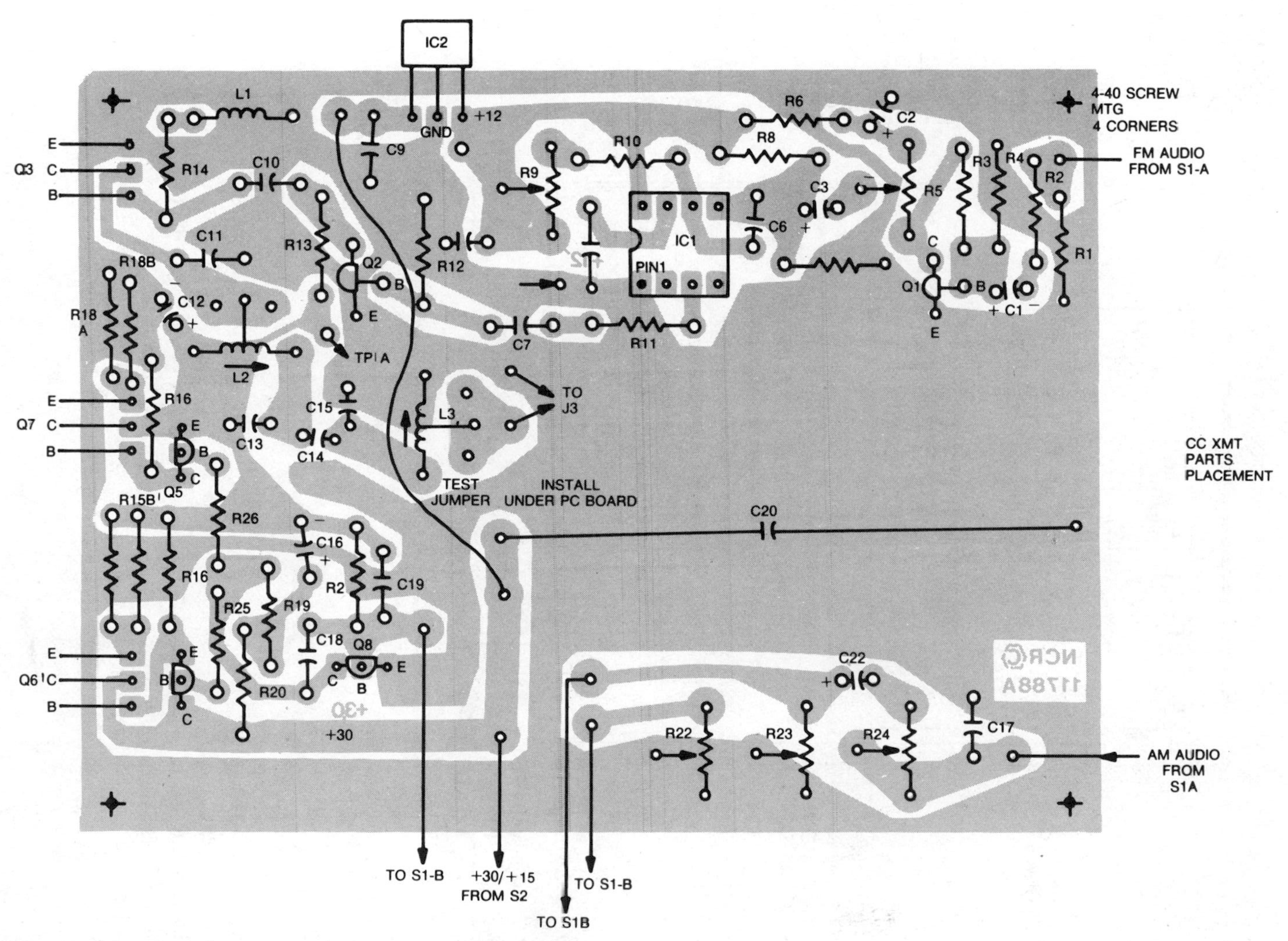

Fig. 3-28. Component placement carrier current transmitter.

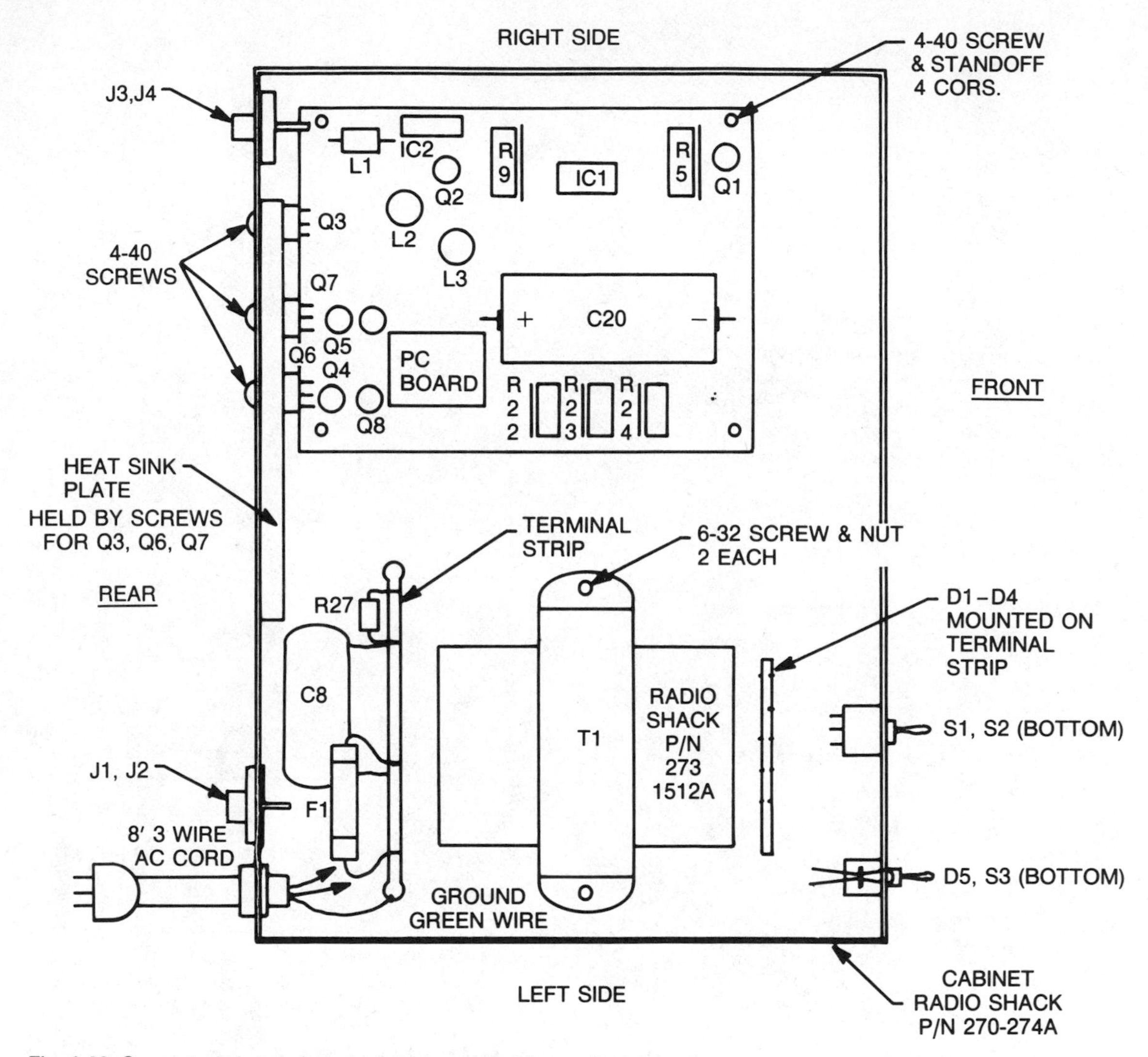

Fig. 3-29. Suggested chassis layout of carrier current transmitter.

Use only the correct tuning tool, otherwise you may damage L2 and L3. A plastic TV alignment tool is best to use in this application. Short out the J4 center conductor to ground. Now apply power. The voltage at TPA should be less than 5 volts. Now adjust R22 (make sure S1 is in FM position) for about 8 volts at TPA. R22 should smoothly control this voltage, from less than five up to about 13 volts or more. Now check voltage at pin 8 of IC1 for +12 volts. Then make

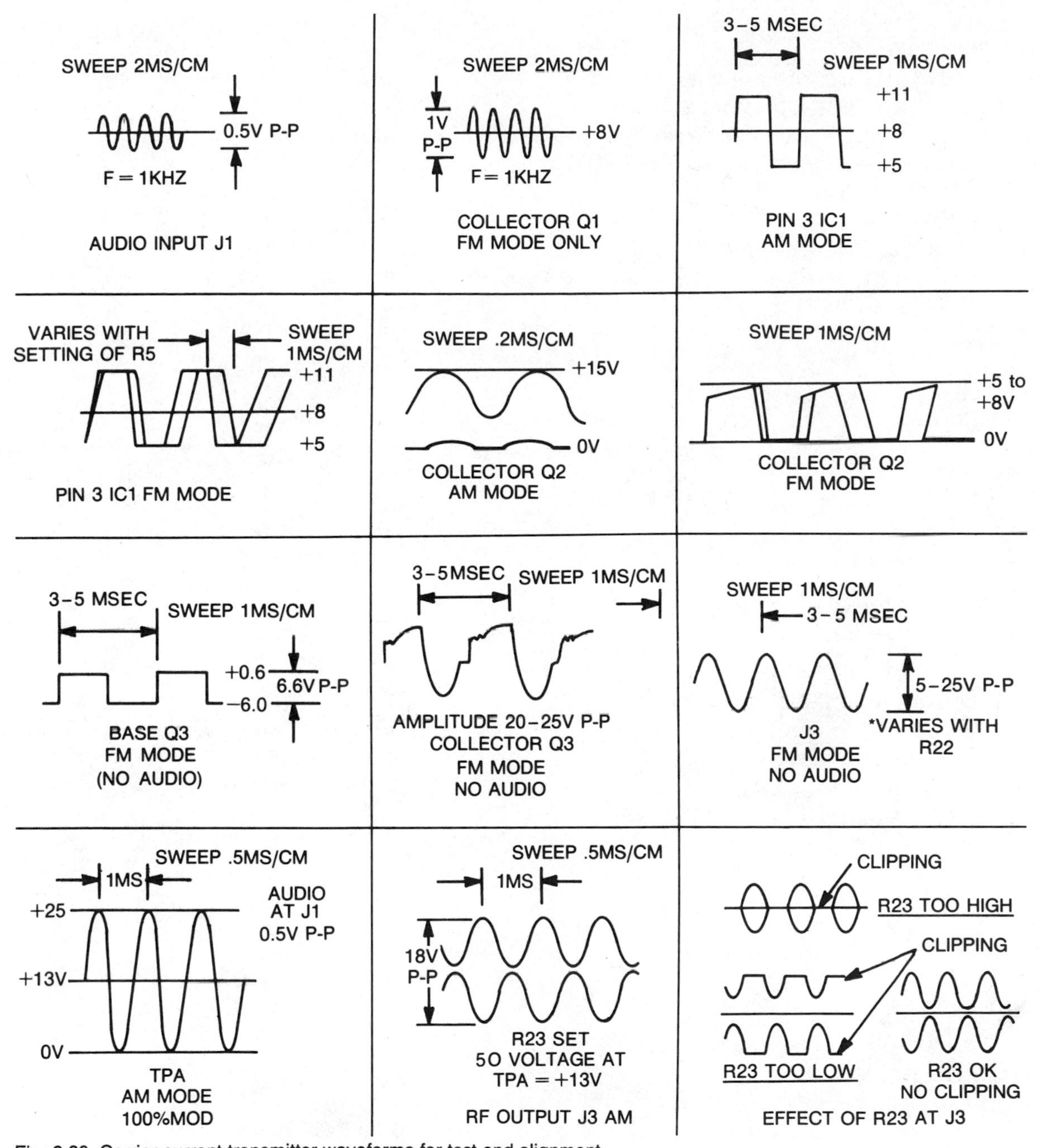

Fig. 3-30. Carrier current transmitter waveforms for test and alignment.

the following checks:

1. Collector Q1—about + 8 volts.
2. Collector Q2—4 to 10 volts.
3. Collector Q3 same voltage as TPA, about 8 volts. Use a meter that is not affected by rf fields, such as a VOM or properly shielded VTVM or DVM.
4. D5, if used, should light.
5. Collector Q5 about 0 to 0.5 volts.
6. Collector Q4 about 1.0 to 1.5 volts more positive than the voltage at TPA.

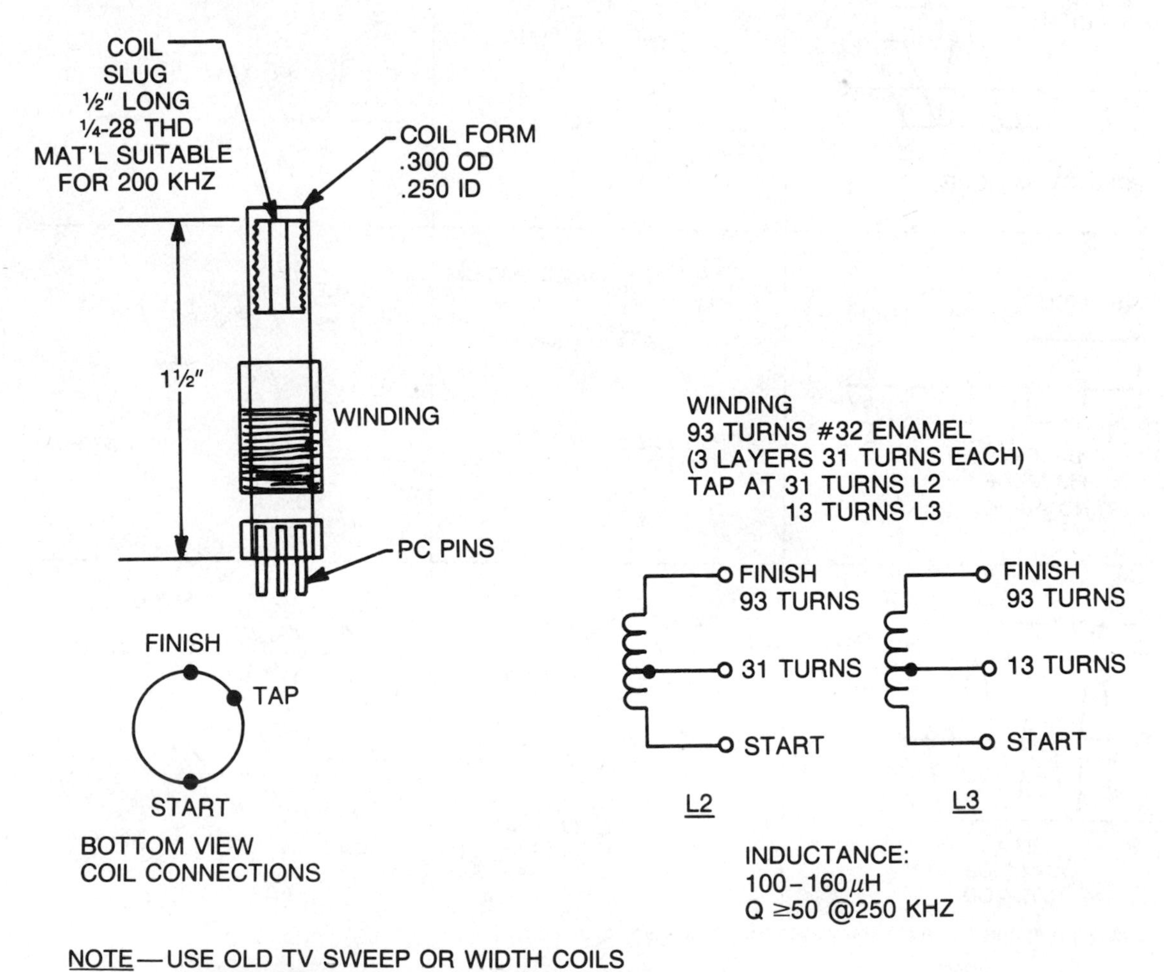

Fig. 3-31. Coil details—carrier current transmitter.

If anything is wrong, remove power and troubleshoot. If all OK, next connect a frequency counter to the collector of Q2 and verify that R9 adjusts the frequency from 200 to 350 kHz or thereabouts. Set R9 for 280 kHz. See Fig. 3-30 for waveforms at collector Q2 and at other points. If no frequency counter, you can use a scope to verify a waveform period from 3 to 5 microseconds. Set R9 for a period of 3.57 microseconds. Next connect a scope to J3. Adjust L2 and L3 for maximum output. Next, vary R9 to produce frequencies from 200 to 350 kHz. It should be possible to obtain nearly constant ouput over a range of 220 to 340 kHz ($\pm$60 kHz from 280 kHz). Adjust L2 and L3 to achieve this. A 6V bulb connected to J3 (use a No. 44 pilot lamp if available) should show a dim glow and can be used as an output indicator if no scope is available. A 10 ohm 2-watt resistor is an excellent load if a scope is available to observe the rf output. If not, use the bulb.

Next, place S2 in the normal position and adjust R22 for 15 volts at TPA. The lamp should glow brightly. Now place S1B in the AM position. Adjust R23 for R14 volts at TPA. The lamp should glow brightly.

Apply a 0.5 volt peak-to-peak 1 kHz sinewave to J1. Adjust R24 until 100% modulation is obtained (see Fig. 3-30). The bulb will brighten with modulation. Adjust R23 for best modulation symmetry if a scope is available. Check for waveforms in Fig. 3-30.

Now switch S1A back to FM. Check for the waveform in Fig. 3-30. Adjust R5 as required. Finally, run the transmitter into either the light bulb or the 10 ohm resistor for an hour or so to check for overheating. Q3, Q6, and Q7 should not get so hot that you cannot hold your finger on them.

This completes construction, alignment, and testing of the transmitter. Install the transmitter in an area where there is some air circulation, as some heat is generated.

For reception of carrier current signals several receiver systems are applicable. The main objective is to recover the transmitted signals with minimum noise and distortion.

One simple application is a cordless headphone system. A simple loop antenna possibly tuned to the carrier frequency picks up the signal from a nearby power line, Ac cord, or outlet. The signal is amplified, detected and used to drive a small pair of earphones. The receiver is designed such that the total gain is relatively low (60dB or less) and a relatively strong signal on the power line overrides most noise. This simple system is useful for noncritical applications, such as TV audio for the hearing impaired, etc. A more general application would use a higher gain AM receiver with AGC and would have both audio and data demodulators, and possibly a subcarrier detector for control applications. The AM approach uses less bandwidth than the wideband FM approach, but is less immune to power line noise.

Another approach is a wideband FM system. This would have a much better noise immunity but more bandwidth is required. A rule of thumb for FM systems is that the necessary bandwidth is twice the deviation plus the highest modulating frequency. For FM stereo, we need about 53 kHz audio bandwidth. This means that transmitter deviation would be limited to 30 to 50 kHz, since the maximum practical bandwidth at these low carrier frequencies lies in the 75 kHz to 100 kHz range. The power line noise becomes very severe below 150 kHz, and at 100 kHz a powerful Loran C signal can cause problems. Above about 350 kHz excessive power line radiation may become a problem. The receiver can be used to handle modulation having components up to about 60-75 kHz. This would include FM stereo, data at better than 10 band rates, and medium scan television video (not NTSC) or facsimile signals, and subcarriers for control purposes. A receiver of this type and a simpler AM receiver will be described elsewhere in this book. See Fig. 3-32.

A kit of parts consisting of the printed circuit board and all parts that mount on the board is available from:

North Country Radio
P.O. Box 53, Wykagyl Station
New Rochelle, NY 10804

Price:	Carrier-Current Transmitter	$54.50	plus $2.50 for
	Carrier-Current AM Receiver	28.50	postage and
	Carrier-Current FM Receiver	38.50	handling.

New York residents must include sales tax.

PROJECT 6: WIRELESS STEREO FM AUDIO LINK

In many cases some form of speaker or headphone extension is desirable for a hi-fi setup. Some examples would be for individual listening, or listening at a distance, either in the same room, another room, or even outside the house. Audio sources might be a CD player, turntable, tape deck, etc. Simple, inexpensive "walkman" type AM-FM-cassette decks nicely solve the problem, if headphone listening is acceptable and either prerecorded cassette tape or FM program material is satisfactory. However, there is no readily available solution if the desired program material is in the form of CDs, reel-to-reel tape, records, or TV stereo audio. Infrared cordless headphones are available, but usually their range is limited to one or two rooms, and a separate receiver is necessary for each listener if independent mobility is required. In addition in certain applications, more range might be desirable.

One solution to these problems would be a device that can broadcast stereo audio over the FM broadcast band, on a vacant channel around 88 MHz. Cheap "walkman" receivers can be used as headphone drivers, since only FM stereo capability is needed. The authors have seen these receivers, complete with headphones, for less than $10 (or even $6) at discount outlets. This makes extra receivers feasible at very low cost. (Almost everyone seems to own one or more of these receivers nowadays anyway.)

1/4W RESISTORS

R1	4.7k ohm
R2	22k ohm
R3	100k ohm
R4	10k ohm
R5	5k pot
R6	1k ohm
R7	47k ohm
R8	6.8k ohm
R9	2k pot
R10	2.2k ohm
R11	2.2k ohm
R12	2.2k ohm
R13	330 ohm
R14	330 ohm
R15 A&B	3.3 ohm
R16	150 ohm
R17	150 ohm
R18 A&B	3.3 ohm
R19	1k ohm
R20	1k ohm
R21	22k
R22	5k pot
R23	5k pot
R24	10k pot
R25	10 ohm
R26	10 ohm
*R27	5.6 ohm 1W
*R28	1k ohm
*R29	2.2k ohm

CAPACITORS

C1	10μF/16V
C2	10μF/16V
C3	1μF/50V
C4	10μF/16V
C5	470pF SM 5%
C6	.001 Mylar
C7	.01 disc 50V 250Vac
C8	0.1μF 600Vdc
C9	0.1/50V Mylar
C10	.01 disc 50V
C11	470pF disc
C12	.22μF/50V electrolytic
C13	.0033 10% 250V Mylar
C14	.0018 10% 250V Mylar
C15	.0033 10% 250V Mylar
C16	47μF/16V
C17	470pF disc
C18	470pF disc
C19	.001 Mylar or disc
C20	4700μF/35V elec
C21	.01 disc 50V
C22	10μF/16V

COILS

L1	470μH choke
L2	100 – 160μH TAP 33%
L3	100 – 160μH TAP 14%

SEMICONDUCTORS

IC1	NE566 V
IC2	LM7812 or LM78L12
Q1	2N3904, ECG 123
Q2	2N3904, ECG 123
Q3	TIP41A, ECG 152
Q4	2N3904, ECG 123
Q5	2N3906, ECG 159
Q6	TIP41A, ECG 152
Q7	TIP41A, ECG 152
Q8	2N3904, ECG 123
*D1	1N4002
*D2	1N4002 or equal
*D3	1N4002
*D4	1N4002
*D5	LED, Red, Yellow, Green etc.

****MISCELLANEOUS PARTS**

J1,J2	Dual stereo jack
J3,J4	RS P/N 274 – 332
F1	1.5 or 2A fast blow fuse
S1	RS P/N 275 – 1546
S2	RS P/N 275 – 326
S3	RS P/N 275 – 326
T1	24VCT, 1.5A (120V Primary) RS P/N 273 – 1512
1	Ac line cord 3-wire
1	PC board
1	Cabinet
2	Terminal strips
	Hardware as required

**Not included in list

Fig. 3-32. Parts list, Project 5.

We will describe a very low power wireless FM stereo transmitter that is easily constructed, reasonable in cost, and where several receivers are needed, probably the least costly of all approaches. It generates a complete multiplex stereo signal, has good fidelity, good channel separation, and has a range of 50 – 100 feet in a typical residential environment. It is tuned to an unused FM channel and is received in exactly the same manner, as if it was simply another FM station.

Fig. 3-33. Wireless stereo FM audio link transmitter.

The idea of a flea-powered radio transmitter for use with a home audio system is not new. Those readers who were into electronics in the 1950-1960 era will remember AM broadcast band "phono oscillators" used for this application, to "broadcast" records through nearby AM radio. However today this application requires much better audio fidelity, and stereo capability, which the old AM approach cannot deliver. Further, the need for stereo necessitates generation of an FM Multiplex composite signal, which is costly, using vacuum tube techniques. Modern IC devices make Multiplex generation much simpler and reasonably accomplished, without too much complexity and "tweaking." Allowable circuit tolerances, with regard to pilot frequency accuracy, can be relaxed since most modern FM stereo receivers using PLL stereo demodulators can handle ± 100 Hz or more tolerance in the nominal 19 kHz pilot frequency. This eliminates the need for an expensive low frequency crystal or a divider chain. Operational amplifiers make accurate matrixing easy to achieve, and IC doubly-balanced modulators eliminate the need for critical balanced coil or transformer specifications, and expensive filter networks, using special, hard to get parts. Only three coils are used in the multiplex generator, none of them critical. Two small hand wound coils are used in the transmitter made of No. 22 gauge magnet wire and threaded ferrite cores. These coils, a PC board, and all necessary parts to fabricate the PC board are available from the source listed at the end of this article.

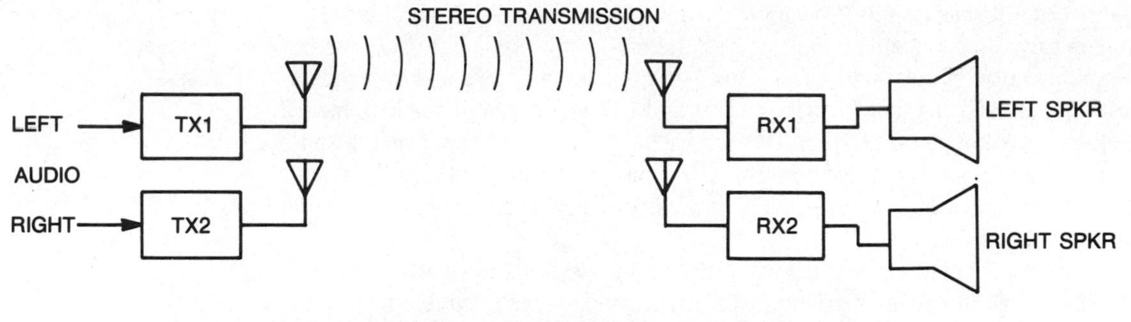

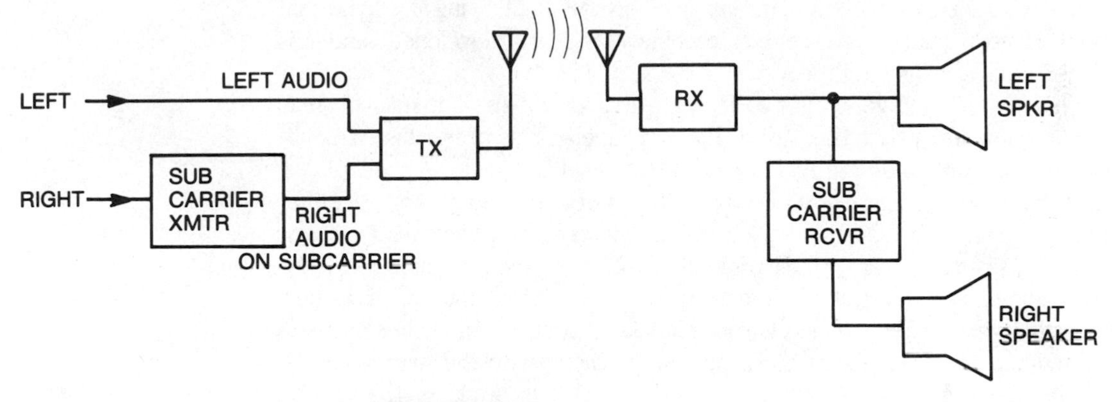

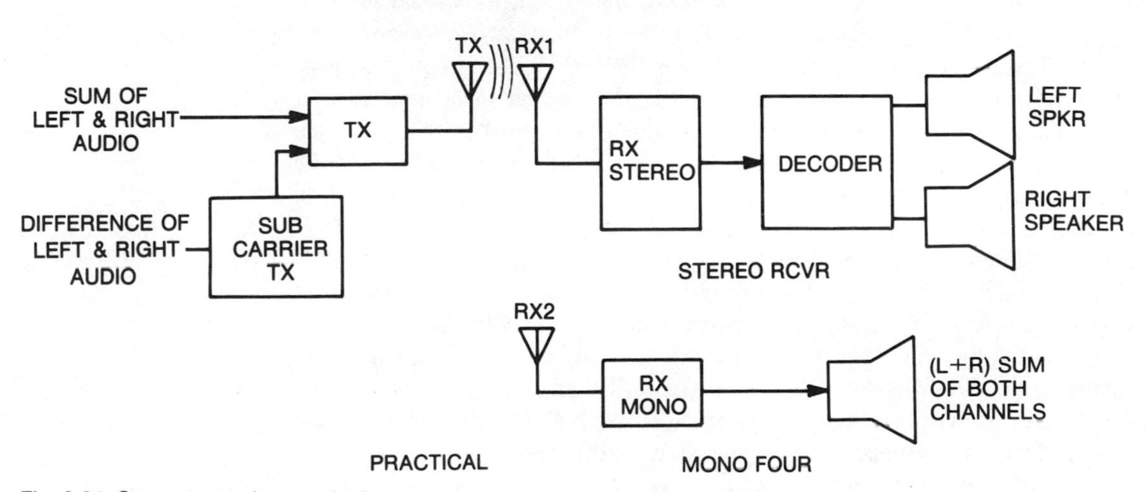

Fig. 3-34. Stereo transmitter methods.

A brief description of FM multiplex will follow. Basically the problem is to transmit two audio channels, one for left, the other for right audio. A simple but impractical way is obviously to use two separate transmitters and two receivers—one of each channel. However, since an FM channel can be made to have a frequency response up to 75 or 100 kHz or so, we could transmit one channel on a subcarrier, say at around 50 – 100 kHz. This works fine, but is not compatible with existing mono receivers. These receivers would only get one of the two channels (left or right).

Whenever anything new is introduced into a mass market with millions or billions of dollars worth of existing equipment, it is always desirable not to disturb the status quo. When stereo was introduced, existing monophonic systems had to receive the same program quality as before, without losing any significant amount of performance. A scheme to achieve this was developed, and this scheme is universally used today.

What is done is to generate two separate audio channels. The main channel (L + R) is the sum of both the Left + Right channels. This channel is transmitted as standard audio on the main (0 – 15 kHz) baseband audio. A monophonic receiver will receive this and everyone will be perfectly happy. Also, another channel is generated. This channel is the *difference* between left and right, or (L – R). This is done simply by algebraically subtracting the two channels. If the audio channels are identical in content, as in a monophonic source, this difference signal will be zero. This channel modulates a subcarrier, chosen as 38 kHz, and the 38 kHz subcarrier is transmitted along with the main audio. A monophonic receiver will simply ignore this channel. In practice, the 38 kHz carrier is suppressed in order to allow the information-carrying sidebands from 38 to 53 and 38 to 23 kHz to account for more transmitter deviation. Since the 38 kHz carrier is necessary for detection of the subcarrier, it must be supplied in some way at the FM stereo receiver. In order to ensure synchronism of the 38 kHz subcarrier, it is reconstructed by using the 19 kHz (half frequency) pilot carrier sent along by the transmitter. This 19 kHz pilot carrier is generated at the transmitter from the 38 kHz suppressed subcarrier. It is better to do it this way anyhow, since the 19 kHz subcarrier is out of both the 0 – 15 kHz (L + R) frequency range and the 23 – 53 kHz subcarrier range. This makes for less crosstalk, and also makes it easier to recover the subcarrier without interference from the program audio. The 19 kHz pilot carrier usually is 10% of the full audio amplitude, and the (L + R) and (L – R) channels each use 40% of the available audio amplitude for transmitter modulation. The 19 kHz pilot carrier can also be used to turn on the receiver stereo decoder to control a stereo program indicator. An additional subcarrier can be placed higher, at 67 kHz, for the purpose of carrying another "hidden" channel. This is called SCA an abbreviation for Secondary (or Subsidiary) Communications authorization.

When an FM stereo signal is produced, it has the frequency spectrum of Fig. 3-35. If we wish to generate a stereo FM MPX signal, the signal must have the spectrum of Fig. 3-35 so it can be received on an ordinary FM stereo

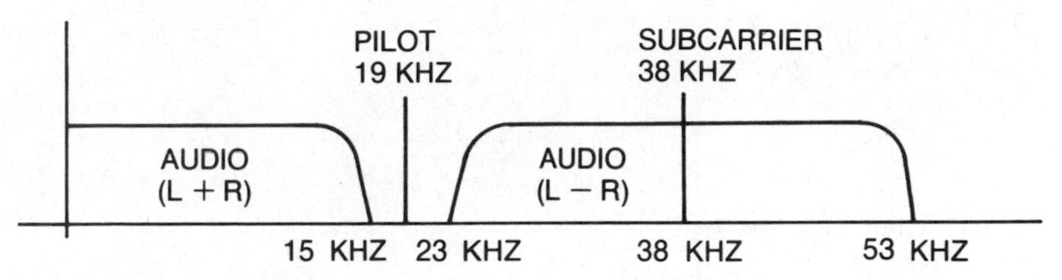

Fig. 3-35. Stereo signal—frequency spectrum.

receiver. A block diagram of such a system is shown in Fig. 3-36. Refer to Fig. 3-36 for the following discussion. Two audio inputs, labeled L (left) and R (right) are fed into a matrix consisting of two halves of an LM1458 dual operational amplifier. One half of the IC uses a simple summing configuration to algebraically add the two audio waveforms. It has a gain of −1 for each input so each input contributes equally to the output of the op amp. The output is the algebraic sum of the L and R channels. It is like a mixer in an audio setup, simply adding the two inputs. The other half of the LM1458 IC op amp is set up to subtract the two inputs algebraically, so that the Right input is subtracted from the Left input to form (L − R). The output of this op amp consists of the *difference* of two inputs. If the inputs are the same, the output from the (L − R) amp would be zero.

The difference (L − R) signal is fed to a balanced modulator, MC1496 IC device. This modulator produces sum and difference frequencies of the audio and subcarrier input. Both the audio input and the subcarrier input are suppressed by the IC, only the sum and differences appearing at the output. A balance control is used to set the correct dc voltages on the IC to achieve maximum carrier suppression. By the way, the subcarrier frequency is 38 kHz. If a 1 kHz audio signal (L − R) and a 38 kHz subcarrier are mixed in this stage, only a signal having 37 kHz and 39 kHz components appears at the (L − R) subcarrier output. If the audio were 10 kHz, we would have 28 and 48 kHz components in the output.

The 38 kHz signal is derived from a 76 kHz oscillator and a ÷ 2 flip-flop. The 76 kHz oscillator and buffer amplifier consist of 2N3565 transistors and an oscillator coil, capacitors, and resistors. Frequency division is done using a CD4027 Dual D flip-flop. One flip-flop divides the frequency by 2, producing a 38 kHz square wave. The other flip-flop divides this 38 kHz square wave again by 2 to obtain a 19 kHz square wave. Both 38 kHz and 19 kHz square waves are converted to sinewaves by filtering out only the fundamental frequency, using a tuned circuit. This method assures a constant phase relationship between the 19 and 38 kHz signals, and also isolates the L-C oscillator from any interaction or accidental modulation.

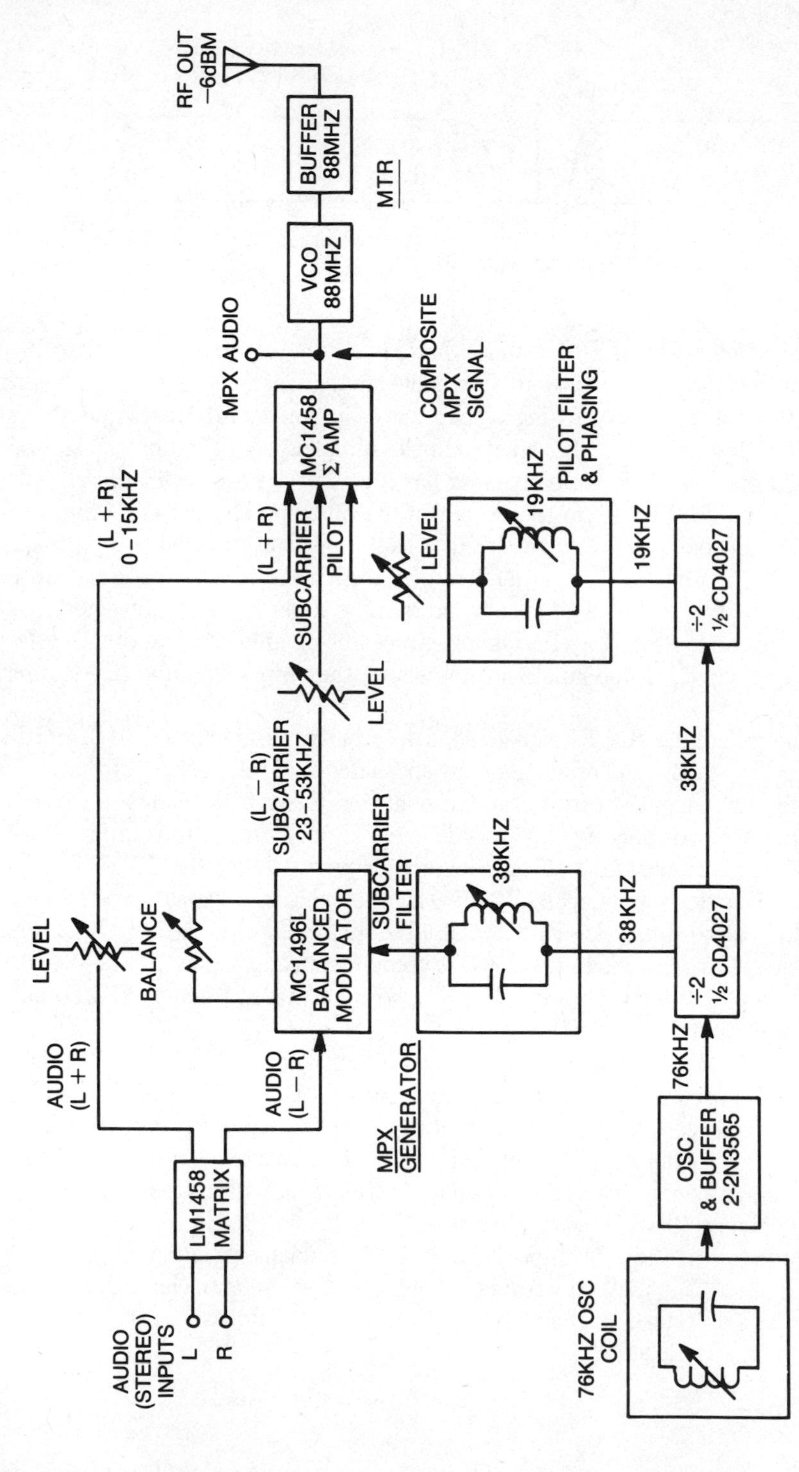

Fig. 3-36. Block diagram of FM stereo transmitter (MPX generator portion).

The L + R signal, 38 kHz (nominal) L − R signal, and 19 kHz pilot signal are summed in another MC1458 op amp section, using an ordinary summing amplifier configuration, with a gain of 2. The output of this summing amplifier is a composite multiplex signal. This signal is fed to a voltage controlled oscillator operating at the low end of the FM broadcast band. Level controls are provided to set the deviation caused by each portion of the MPX signal to correct levels.

The VCO runs at a low level, from a zener diode regulated supply to ensure frequency stability. It is fed to a buffer amplifier to isolate the VCO from the antenna output connector, further increasing stability. Output is kept low. Output is about 1/4 milliwatt. Range should be limited to only that necessary. About 100 feet will be obtained with a typical ''walkman'' FM stereo and an antenna on the transmitter of about 12″ in length. *Do not* make any changes to increase range. The circuit is capable with some modifications, of over 20 milliwatts output, which would produce excessive signal and could carry up to a mile or more. Therefore, *do not* use any more antenna than you need or the FCC might decide to get you. Do not be a broadcast station—you have been warned.

A schematic of the stereo transmitter is shown in Fig. 3-37. It is designed to handle the audio range from 100 Hz or lower to up to 15 kHz. Examination of the schematic will reveal many apparently large value electrolytic capacitors. The reason for this is to minimize phase shifts at low audio frequencies. Remember that we are adding and subtracting signals and that we have to preserve levels, ratios, etc. For example, if we wish to cancel two signals, as in the L − R channel, this means that exactly the same signal on each input gives zero output. If there is a 1% difference in level this would only cancel 99% of the signal, leaving 1% or −40dB down. There would be a 40dB down residual signal. This may not be too bad but remember that phase shift also contributes. If we had a phase error of say 4° (which we would encounter by passing a 1 volt 100 Hz signal through a simple R-C circuit designed to cut off at 1/10 of this or 10 Hz, the output amplitude would be 99.7% of the input. This is good for −50dB cancellation (.003 volt). However an out of phase component of 0.07 volt would remain from phase shift effects. This would allow only −23dB cancellation. (However, we are also interested in differential phase between the two channels as well.) It should not take much thought to see that unwanted very small phaseshifts that could normally be neglected in other circuits could degrade phase response so that the stereo generator would have degraded stereo channel separation. Enough said. By using very large capacitors for coupling we can eliminate these perplexing headaches, and maintain good separation at lower audio frequencies.

Audio inputs for L and R are connected across R1 and R5 respectively. Coupling capacitors C1 and C2 couple audio to R2 and R4, the input resistors for the L + R channel, and R11 and R12, the input resistors for the L − R channel. R3 is connected between R2 and R4. Ideally, if R2 equals R4 then R3 is set at the midpoint. R3 is used to set the ratio of gains for L and R produced by IC1A, an LM1458, to exactly 1:1. R6 sets the stage gain to about unity for each

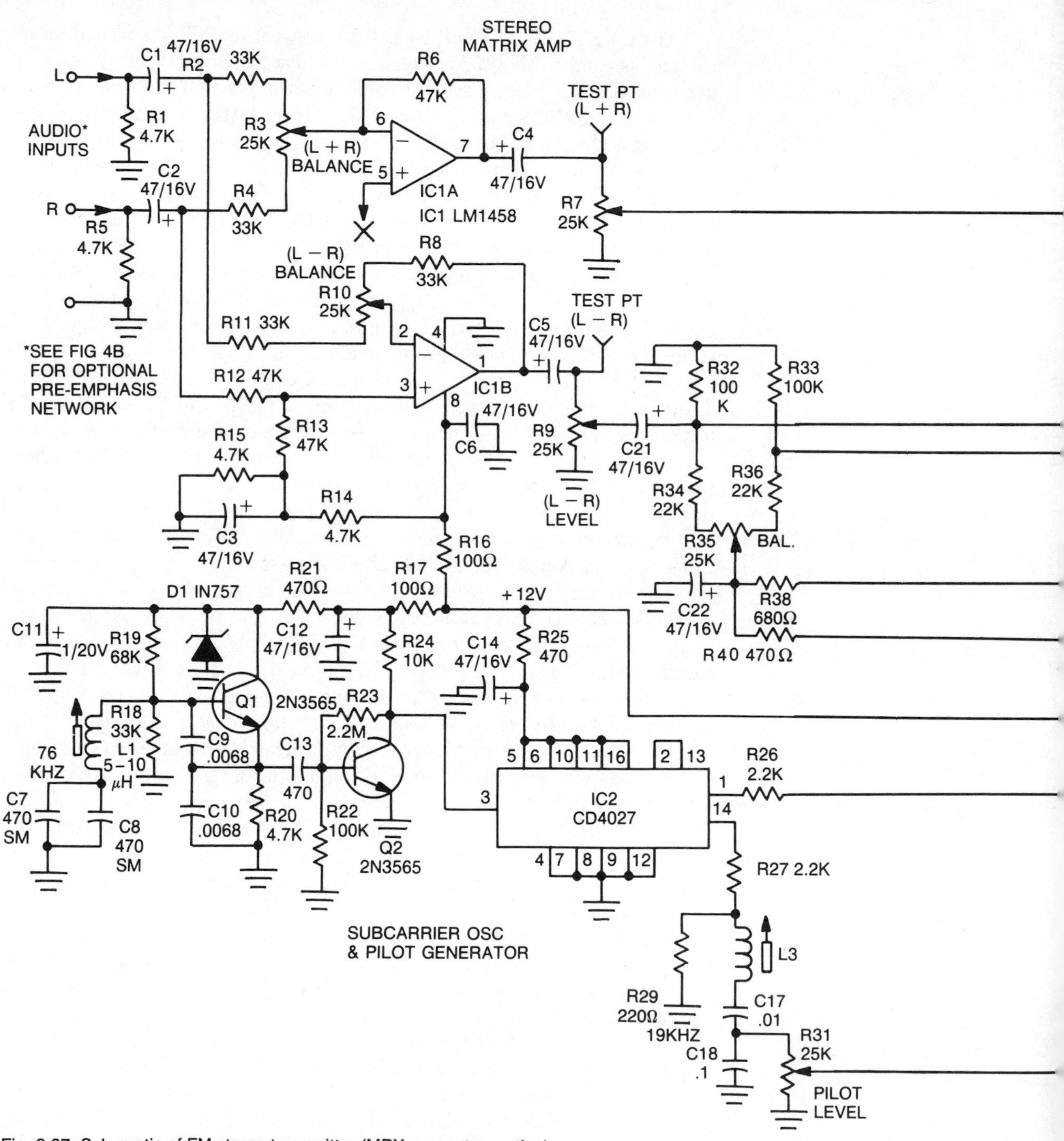

Fig. 3-37. Schematic of FM stereo transmitter (MPX generator portion).

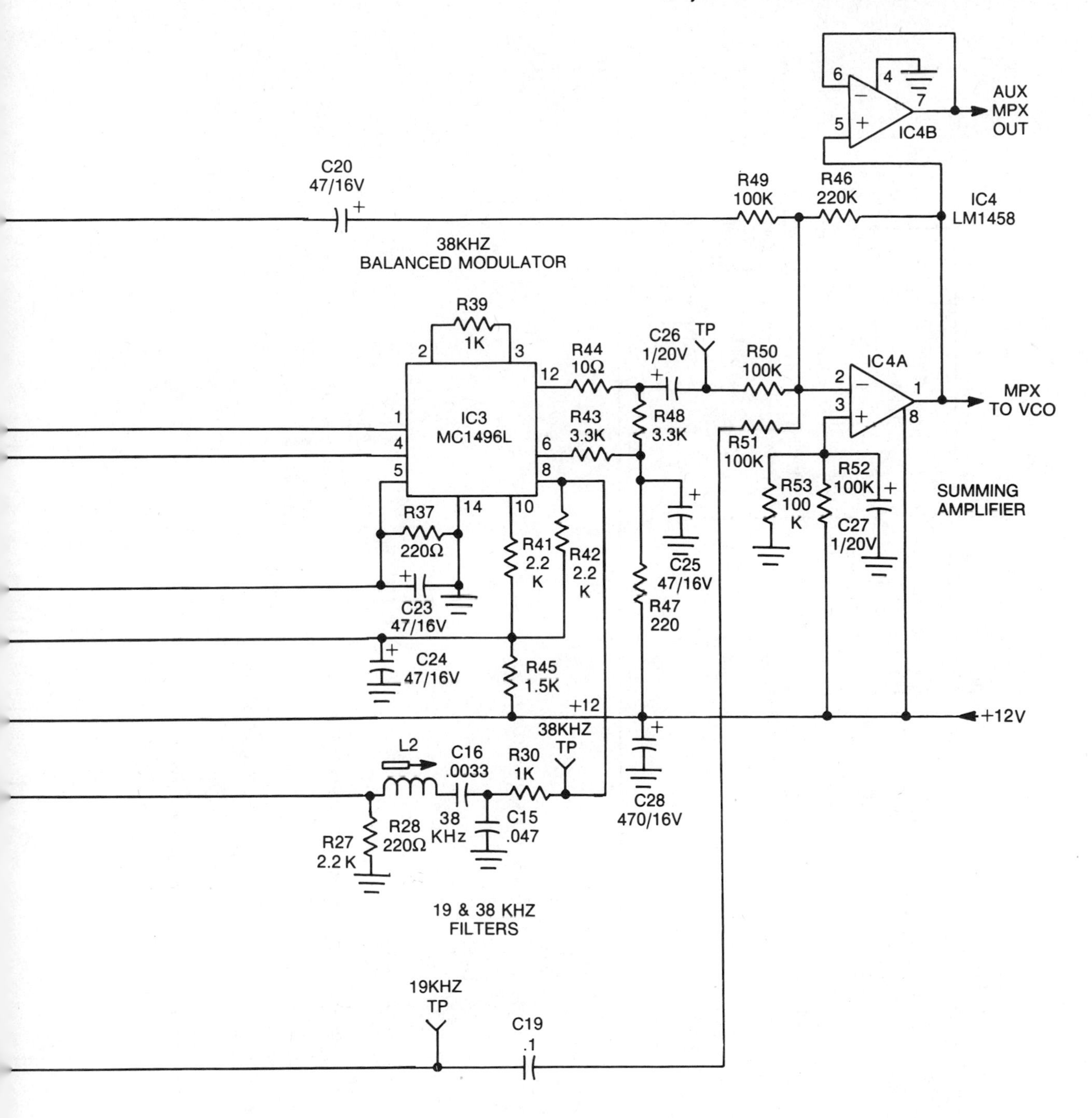
AUX
MPX
OUT
IC4B
C20
47/16V
R49
100K
R46
220K
IC4
LM1458
38KHZ
BALANCED MODULATOR
R39
1K
R44
10Ω
C26
1/20V
TP
R50
100K
IC4A
IC3
MC1496L
R43
3.3K
R48
3.3K
R51
100K
MPX
TO VCO
R52
100K
R53
100
K
C27
1/20V
SUMMING
AMPLIFIER
R37
220Ω
R41
2.2
K
R42
2.2
K
C25
47/16V
R47
220
C23
47/16V
C24
47/16V
R45
1.5K
+12
+12V
38KHZ
TP
L2
C16
.0033
R30
1K
C28
470/16V
R27
2.2 K
R28
220Ω
38
KHz
C15
.047
19 & 38 KHZ
FILTERS
19KHZ
TP
C19
.1

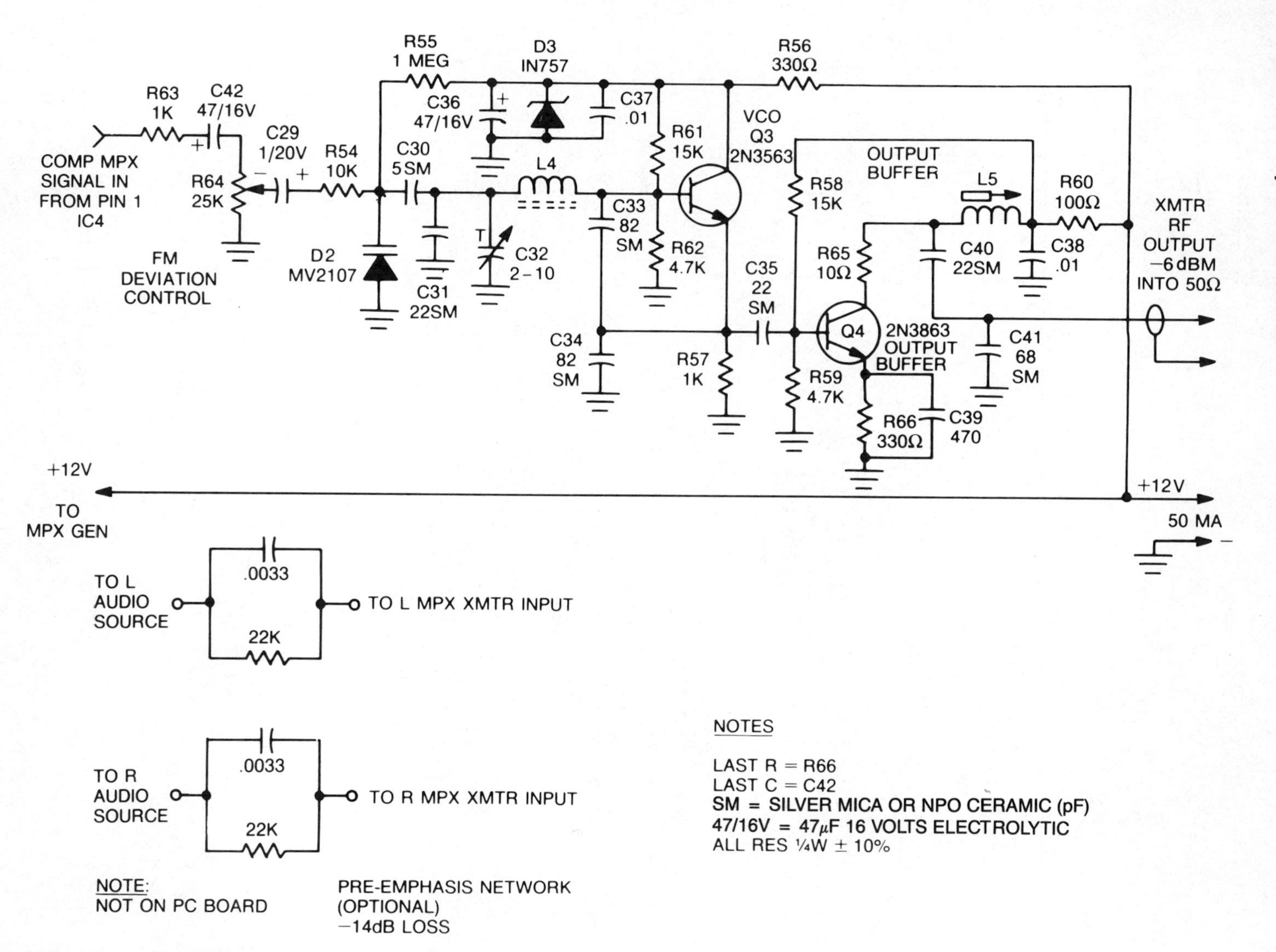

Fig. 3-38. Schematic of FM stereo transmitter (rf section).

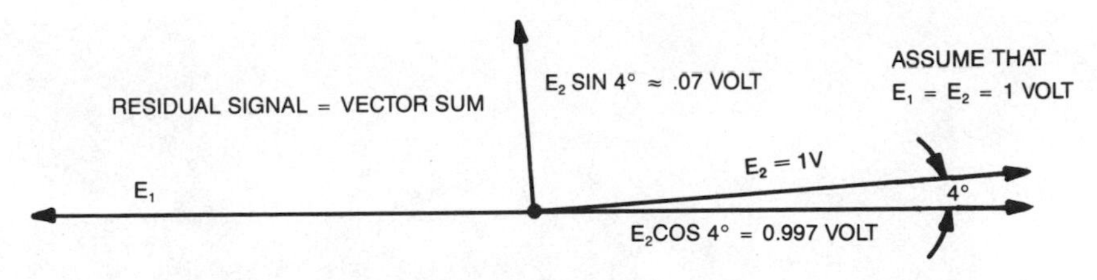

Fig. 3-39. Effect of phase shift on signal cancellation.

channel. R3 is adjusted for exactly equal gain than either L or R channels. It is more important that the gains be equal than exactly unity. C4 couples the L + R sum channel output to level control R7.

IC1B is the (L − R) difference amplifier. R8, R10, and R11 form an adjustable network similar in function to R2, R3, and R4. R12 and R13 couple signal from the R channel. R14 and R15 and C3 form a bias network to bias both op amp sections to a Q point equal to half the supply (+12V) voltage. Since, viewed from the noninverting (pin 3) input, the gain of IC1B would be 2 if R11 equals R8, and R10 is at midpoint. R12 and R13 divide the R signal nominally by 2. R10 is set such that, with L and R inputs exactly equal (shorted together with a clip lead) the output from IC1B seen across R9 (coupled from pin 1 through C5) is exactly zero. This fulfills the necessary condition that (L − R) = 0 if both inputs are identical.

A pilot signal at 19 kHz and a 38 kHz subcarrier signal is also necessary. Q1 is an oscillator running at 76 kHz. R18, R19, R20, and R21 provide dc bias to the Q1 stage (2N3565). Zener D1 maintains the dc bias voltages constant. C11 provides bypassing. C7, C8, L1, C9, and C10 form the frequency determining network. L1 is adjustable and is used to set the oscillator frequency to exactly 76 kHz. C13 couples oscillator signal to buffer Q2, also a 2N3565. R24, R23, and R22 bias Q2. C12 and R17 provide filtering to remove 76 kHz components from the B+ line. The collector of Q2 provides a negative going pulse train to IC2, a dual JK flip-flop. This flip-flop divides the 76 kHz signal by 2 to get 38 kHz, and by 2 again to get 19 kHz. R25 and C14 provide B+ supply decoupling for IC2. R26 and R27 provide 38 kHz and 19 kHz signals to R28 and R29, about 1 volt peak-to-peak in each case. These signals are square waves. We need sinewaves, so two filters, at 38 kHz (L2, C16, and C15) and at 19 kHz (L3, C17, and C18), provide harmonic filtering to pass only the fundamentals. R30 couples the 38 kHz sinewave (about 1 volt p-p) to modulator IC3 and R31, a level control, couples a variable level 19 kHz signal through C19 to the output summing circuitry. Phase adjustment of the subcarrier and pilot is achieved by adjusting L2 and L3, respectively.

The subcarrier (L − R) component of the MPX signal is produced by IC3 and associated components. IC3 is an LM1496 balanced modulator. It produces an output that is the sum and difference of the two input frequencies at pin 1 (L − R audio) and pin 8 (38 kHz), without feed through of either input. This is exactly what is needed.

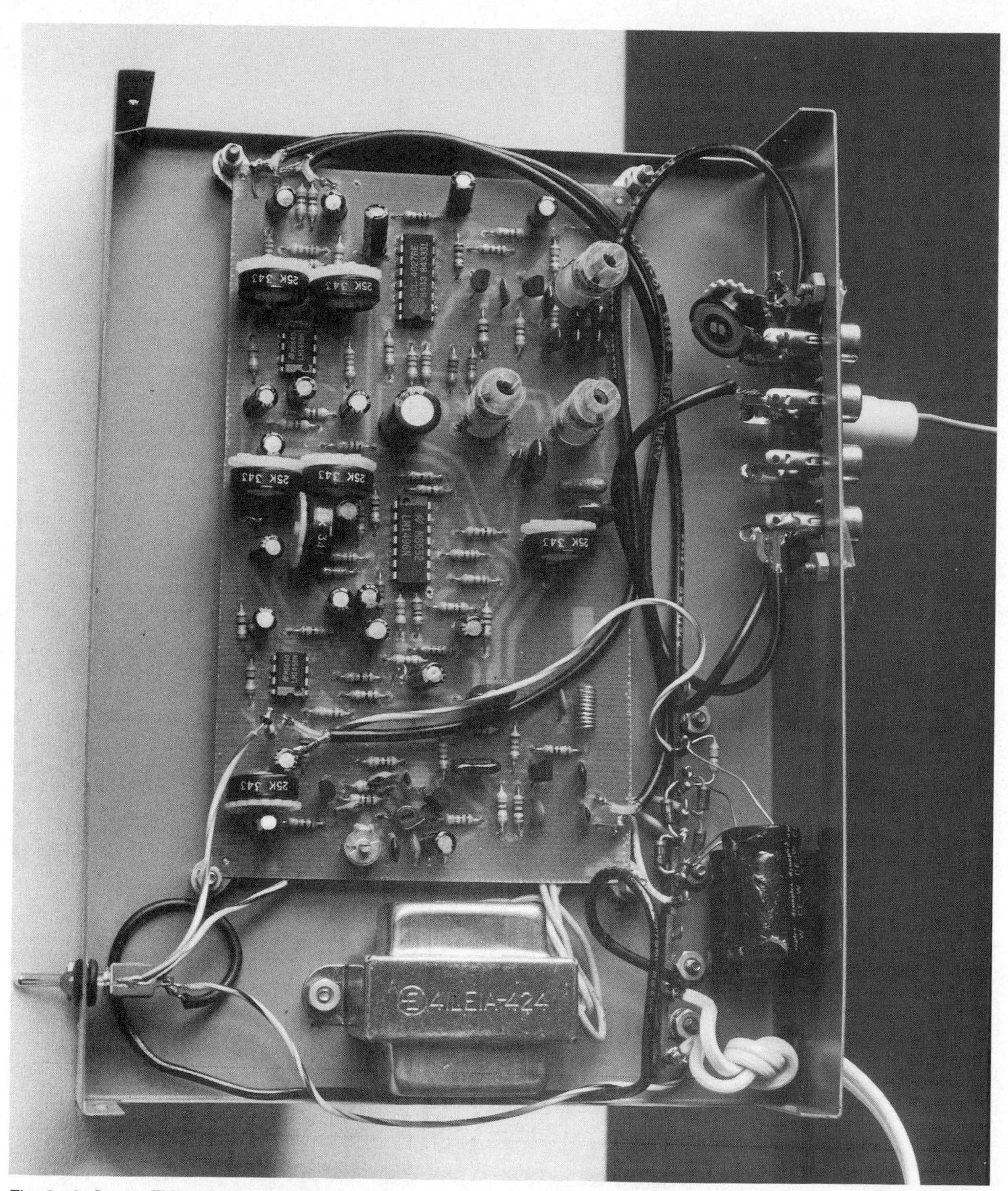

Fig. 3-40. Stereo FM audio link mounted in a case.

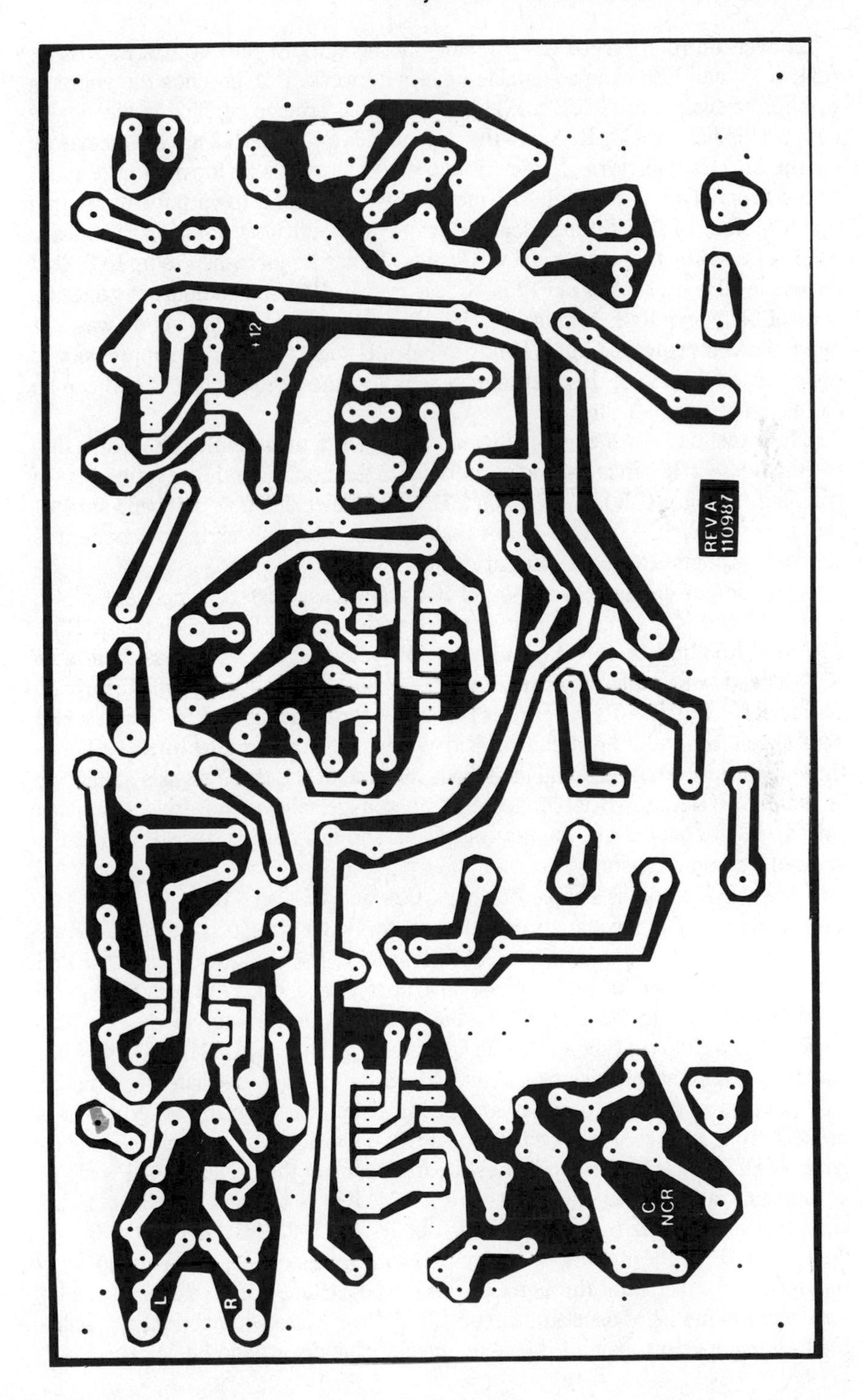
+12
REV A
110987
C
NCR
L
R

Level control R9 feeds (L – R) audio through C21 to pin 1 of IC3. R32, R33, R34, R35, and R36 is an adjustable dc bias network. C22 grounds the wiper of R35 for ac signals. R37, C23, R38, R40 and C24 provide correct dc bias levels and decoupling for IC3. R39 sets the gain of IC3. R41 and R42 are bias resistors for the 38 kHz (subcarrier) input circuitry. R45 provides dc for the entire input circuit bias network. R47 and C25 provide decoupled bias to output circuits (pin 6 and pin 12) of IC3 through load resistors R43 and R48. R44 is an auxiliary level set resistor that is nominally 10 ohms. It may be increased to up to 2.2k if excess level is present at pin 12 of IC3. However, R43 should equal in value the sum of R44 and R48, within ±10%. This level setting adjustment was not needed in our prototype but is handy to have. If imperfect carrier suppression is experienced with IC3, this feature can help, and the level to pin 1 of IC3 can be increased to compensate.

In actual use, with zero (L – R) input (R9 set a minimum), R35 is adjusted to null the 38 kHz carrier at the output of IC3 (test point on the negative lead of coupling capacitor C26). Audio from C21 will produce a double sideband subcarrier at pin 12 of IC3. Gain is about three times. This subcarrier component is coupled through C26 to the output circuit.

The output circuit, comprised of IC4 and its associated components R46, R49, R50 and R51 is a summing amplifier with a gain of 2, at each input. R52, C27, and R53 bias the op amp at half supply voltage Q point, as was done with IC1. The (L + R) audio from the wiper of level control R7 is coupled through C20 to R49. The (L – R) subcarrier is coupled through C26 to R50. The 19 kHz pilot signal from pilot level control R31 is coupled through C19 to R51. These three signals are summed and amplified, the sum of the three signals appearing at pin 1 of IC4A. IC4B is connected as a voltage follower to drive an output (MPX OUT) connection, for test purposes and if a composite MPX signal is needed for some reason.

The MPX signal is coupled through C29 and R54 to a varactor diode D2. This diode acts like a voltage variable capacitor. Q3 and associated components make up an oscillator operating at 88 MHz. R61, R62, and R57 bias Q3. R56 and D3, a zener diode, and bypass capacitors C36 and C37 supply voltage of nominally 9 volts to Q3 and R55, a bias resistor for varactor diode D2. The oscillator is a Colpitts type (as in the Q1 circuit), with C31, C32, C30, L4, C33, and C34 comprising the tank circuit, which determines oscillator frequency. C35 couples rf energy to buffer stage Q4. When a composite MPX signal is present from R64, it also appears across D2. This signal modulates the capacitance of D2, which in turn modulates the frequency of the oscillator. MPX signal is supplied via R63, C42, and level control R64. R64 is set for a drive level to D2 such that a ±75 kHz peak deviation is obtained at full audio input level (about 0.5V p-p at L and R inputs). Varactor D2 forms a capacitance and is coupled to the oscillator tank circuit through C30. As the capacitance of D2 varies, it modulates the frequency of oscillator circuit Q3 and associated components. Oscillator frequency is set, using C32, to an unused channel around 88 MHz.

Q4 is a buffer stage. It has low gain and its chief function is to isolate oscillator (Q3) stage from the effects of variable loading caused by ''antenna'' effect or other output load. R58, R59, R66 and R60 bias Q4. R65 suppresses a tendency toward unwanted UHF oscillations. C39 bypasses the emitter of Q4 and C38 is the collector circuit bypass capacitor. L5, C40, and C41 provide a matching circuit for a 50 ohm output load or antenna. Output level is 100 – 120 millivolts rms into 50 ohms, about 1/4 milliwatt.

C28 provides bypassing of the supply line. About 50 milliamperes at 12 volts is required to run the MPX transmitter. It should be very well filtered, with less than 10 millivolts p-p of ripple. Either a small power supply using an LM7812 IC regulator, or a set of 8-AA penlight cells can be used as a power source. See Fig. 3-41 for a suitable ac power supply circuit using easily available Radio Shack components.

We did not use any audio pre-emphasis, as the MPX transmitter sounded good without it. The ''walkman'' stereo used with this unit has excessive treble anyway.

Construction of the MPX Generator is not very critical. We suggest using the PC layout in this article. It has been made single sided for ease of duplication and plated through holes are not required. Double sided PC boards with plated through holes are in our opinion, not suitable for easy home or experimenter fabrication. Most hobbyists do not have access to photographic equipment that would be needed, and producing the plated through holes requires special techniques. For those who would rather purchase a drilled, ready made PC board, a board is available from the source listed at the end of this article. L1, L2, L3 are also available, and we would recommend the use of these coils so that correct performance will be obtained. L4 and L5 are only a few turns of wire and must be hand wound on an 8-32 screw thread, using an 8-32 screw. A small ferrite slug taken from an old CB radio rf transformer or a TV set will do for a core, as the core can be threaded into L4 and L5 to obtain the required inductance, and glued in place with a small dab of cement such as Q-dope, crazy glue, or even hot glue from a glue gun. By the way, the hot glue seems to be very good at high frequencies, from a loss standpoint. All resistors are 1/4W ±10% tolerance types. Be sure to use the type of capacitors specified, especially in the VHF circuits involving Q3 and Q4, and in the circuits involving L1, L2 and L3 as component Q and stability are important in these circuits. The ICs are not critical as to manufacturer, and although we used 2N3565s and 2N3563 transistors, any suitable low noise audio and VHF rf types, respectively, would probably work okay. A complete kit of parts that mount on the main PC board is available from the same source as the PC boards, if desired.

Construction should begin with the mounting of all fixed resistors and diodes on the PC board. Next, install all capacitors. Be sure to orient electrolytic capacitors as specified. Next, install the potentiometers, transistors, and IC devices. Last, mount the coils L1, L2, L3, and the hand wound coils L4 and L5 (see Fig. 3-42). Be careful with L1, L2, and L3. They are wound with fine

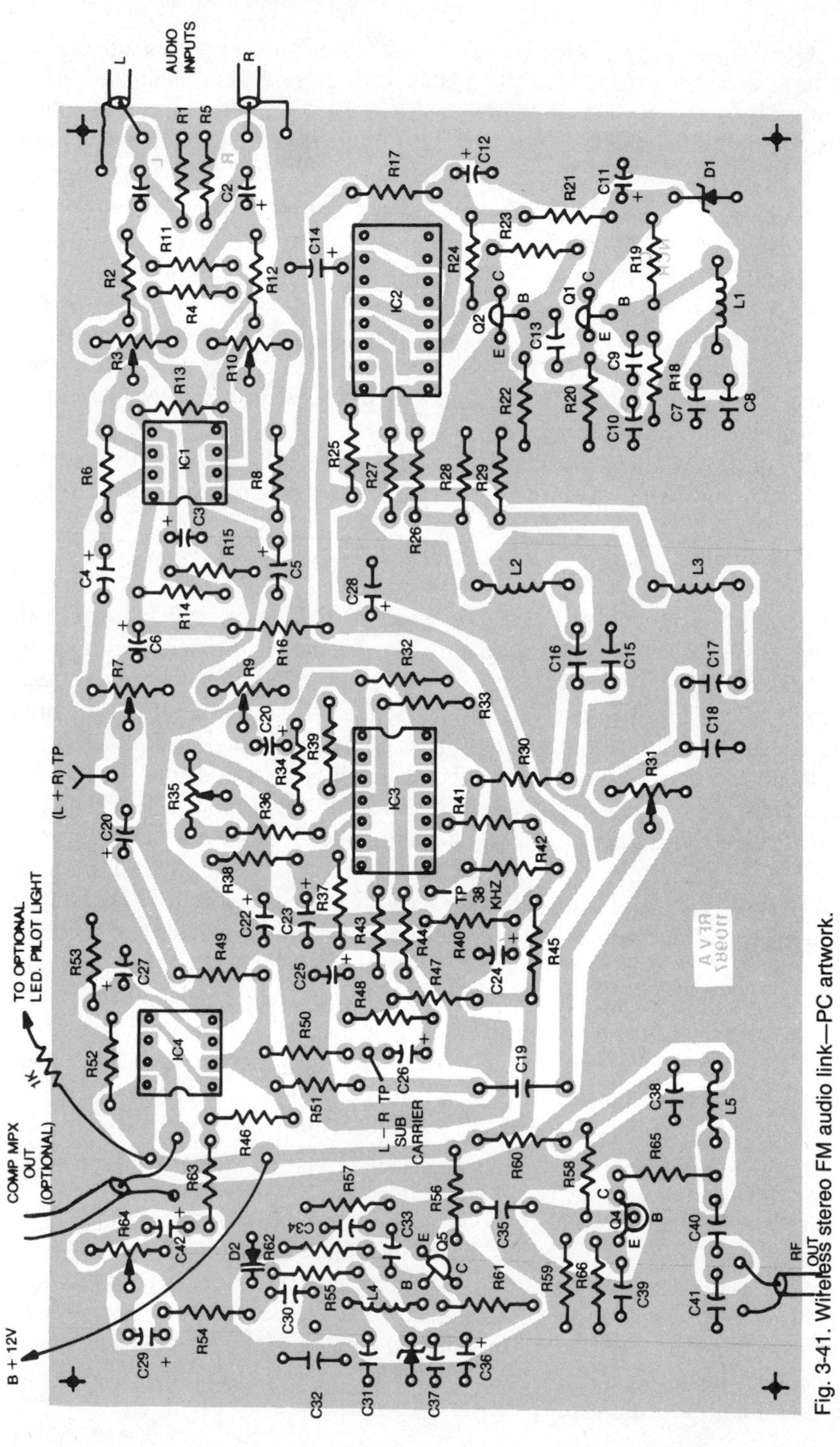

Fig. 3-41. Wireless stereo FM audio link—PC artwork.

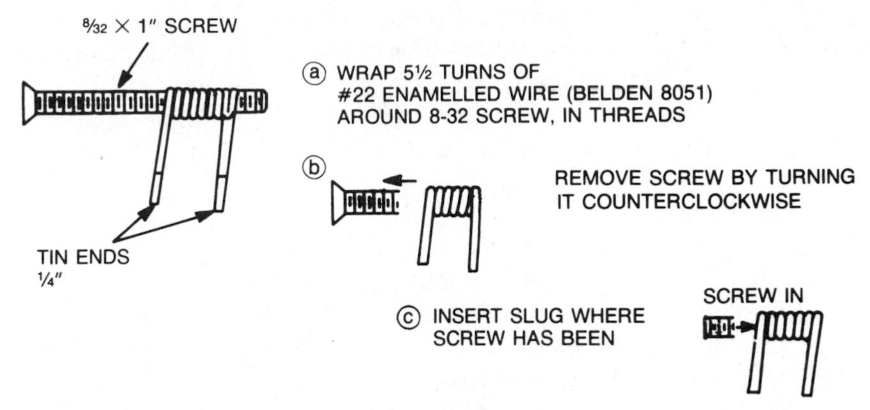

Fig. 3-42. Stereo FM audio link coil construction.

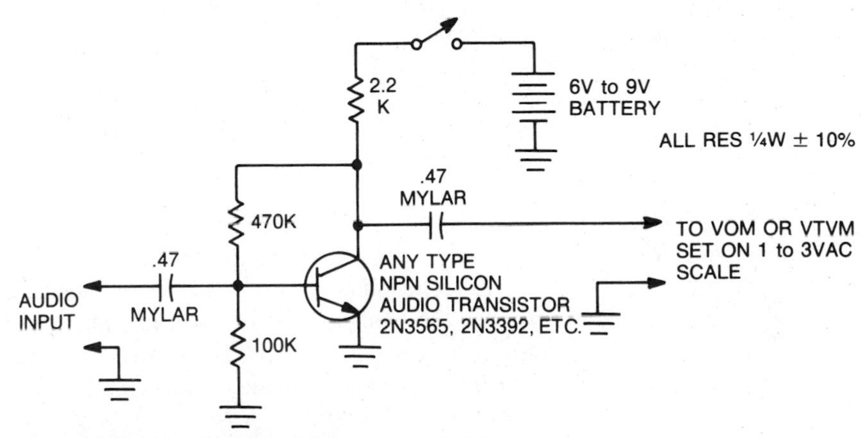

THIS AMPLIFIER WILL GIVE SUFFICIENT GAIN TO DETECT MILLIVOLT LEVEL AUDIO SIGNALS & ENABLE ADJUSTMENTS OF R3, R10, R7, R9, R35, R31 AND L2, L3 TO BE MADE

NO CALIBRATION IS NECESSARY AS MOST ADJUSTMENTS ARE FOR NULLS, PEAKS, OR ARE DONE BY EAR AS WELL AS BY METER

Fig. 3-43. Oscilloscope or VOM amplifier for alignment of MPX circuitry in transmitter.

wire and can be somewhat delicate. Use only rosin core solder, nothing else. When all components are installed, carefully check your work for shorts, opens, poor solder joints, and correct component orientations. In particular, check all ICs and transistors. Wrong orientation may cause irreversible damage to the IC devices and/or no operation of the respective circuitry. Refer to Fig. 3-41 for parts placement and orientation.

After you are *sure* that everything is okay, connect +12V to the B+ twice on the PC board. *Quickly* make the following voltage checks. Use a 10 megohm VTVM (or FET VOM) if possible.

Pin 8 of IC4	=	+12 volts	
Pin 1 & 7 of IC1	=	+6 volts (5–7 OK)	
Pin 1 & 7 of IC4	=	+6 volts (5–7 OK)	
Pin 16 of IC2	=	+11 volts	
Pin 12 of IC3	=	+8 to +10 volts	Refer to
Pin 8 of IC3	=	+4 to +6 volts	Fig. 3-41
Pin 4 of IC3	=	+3 to +4 volts	PARTS PLACEMENTS
Pin 5 of IC3	=	+0.5 to +1.5 volts	
Collector Q1	=	+8 to +9 volts	
Emitter Q1	=	+2 volts (1.5–3 OK)	
Collector Q2	=	+6 to +10 volts	
Collector Q3	=	+8 to +9 volts	
Emitter Q3	=	+2 volts (1.5–3 OK)	
Emitter Q4	=	+1.5 to +2.5 volts	
JCT R54 and D2	=	+6 to +9 volts (Use VTVM or FET VOM) (or +3 to +9 volts if 20k ohm/v VOM is used)	

If all these voltages are present things are probably okay. Nothing should get very warm or hot to the touch. If it does, something is wrong. When all these checks are okay, set up the test setup shown in Fig. 3-44. An oscilloscope is desirable but not necessary. An audio amplifier can be used in conjunction with a VOM set on its ac voltage range if a scope is not available. An audio oscillator is necessary, or if not available, a stereo audio source, such as the line output terminals of a tape deck, CD player, etc.

A steady tone of 1,000 Hz or so is necessary. A suitable source, if one does not have an audio oscillator, is a recording made from an on the air tone, such as a TV test pattern. Most TV stations broadcast a test pattern at some hour of the day (early AM or just before sign off at night). You can record the tone on cassette tape and use the audio. About 15 to 30 seconds duration is enough for the tests to be made. A test LP record is also a possible tone source. Figure 3-45 shows waveforms.

1. Connect audio of about 0.5 volts peak-to-peak, 1 kHz frequency, to the L input and ground. Set R7, R9, R3, and R10 to middle positions (half resistance). Connect scope or audio level meter or other indicating device to junction of C4 and R7. Note the audio level. It should be nearly equal to the input. Now reconnect the input audio to the R input. The level should be equal to that obtained when the L input was used. If not, adjust R3 until the L and R inputs yield *exactly* the same level. This should occur near the middle of the adjustment range of R3.

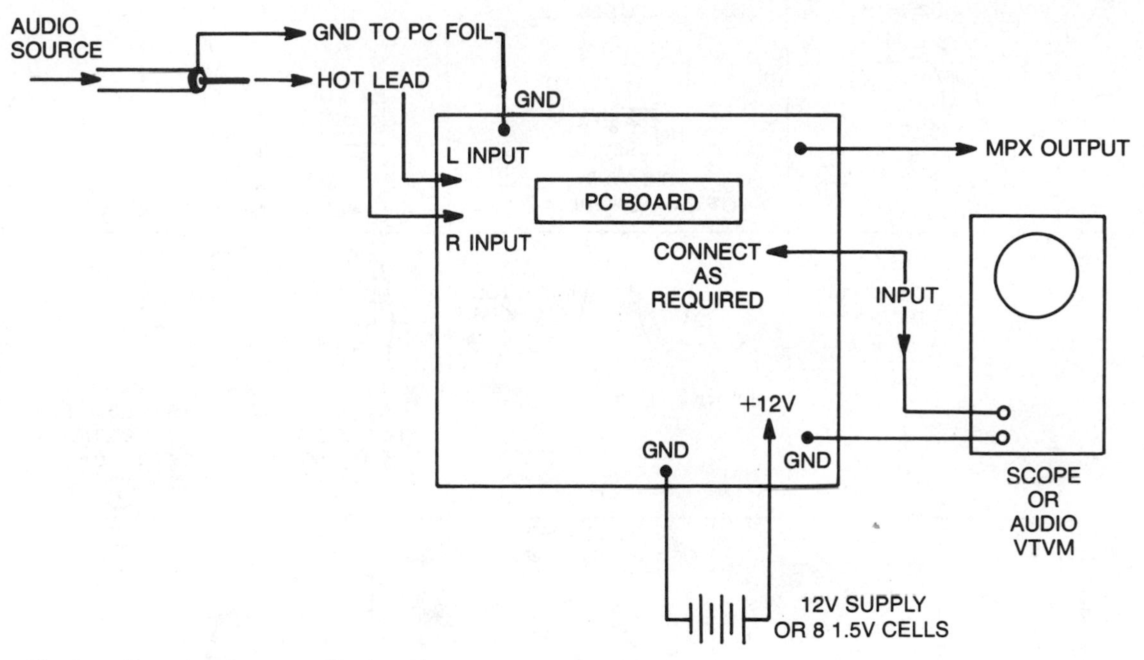

Fig. 3-44. Test setup for transmitter board.

2. When Step 1 is satisfactorily completed, this time connect the scope or audio level meter across R9 (JCT C5 and R9 and ground). Connect, as in Step 1, 0.5V of audio to L input. The level seen on the scope or audio level meter should be approximately the same as the input level. Now transfer the audio input to the R channel. The level should now still be about equal to the input. Connect a jumper between the L and R inputs (clip lead) so that the same audio input is on both L and R. *The audio across R9 as seen on the scope or level meter should now drop to zero or nearly so*. Adjust R10 for *lowest* output, ideally zero. This and Step 1 check out the audio matrixing circuit consisting of IC1 and associated components.
3. Connect a frequency counter, if you have one, to the collector of Q2. Adjust L1 for 76.000 kHz. If you do not have access to a frequency counter you can use an AM radio receiver. Run a wire from the collector of Q2 to the antenna of the receiver or simply lay the wire near the receiver. Adjust L1 for a whistle as heard on the station audio, then adjust L1 for lowest audio tone (pitch). Now tune to 1140 kHz. A station on this frequency should simultaneously be experiencing the audio interference. Rock the slug of L1 back and forth slightly to verify this. If not, go back to 760 kHz and readjust L1. This will guarantee that you

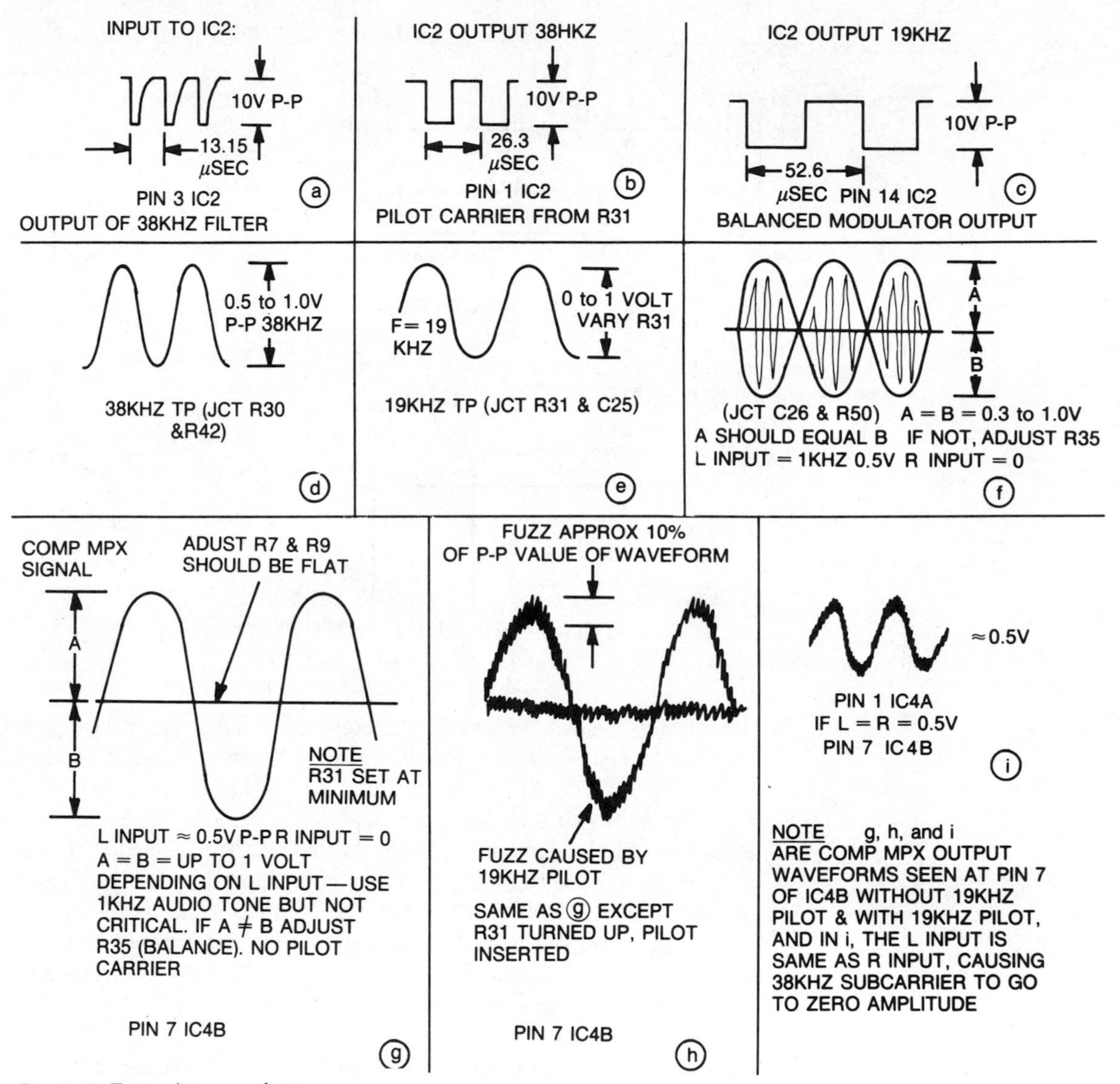

Fig. 3-45. Transmitter waveforms.

are receiving the 10th and 15th harmonics of 76.000 kHz. If the oscillator is set to, say 84.4 kHz you will get a beat or 760 but not simultaneously on 1140.

4. Now adjust R31 to halfway open. Connect the scope or audio indicator (Fig. 3-43) to the junction of R31 and C19. Adjust the slug of L3 for maximum output (19 kHz). This should occur with the slug well into the coil. If you must back out the slug almost all the way, you are probably not tuned to 19 kHz but to 57 kHz (3rd harmonic). Check your wir-

ing and C17, C18 in this case. Return R31 to its minimum setting after this step.

5. Connect the scope or audio indicator to the junction of C26 and R50. Set R35 temporarily all the way to one side (CCW or CW, it doesn't matter). Audio input to L and R channel should be zero. Set R9 to minimum. Adjust L2 for maximum output or indication. Now *very slowly* adjust R35 for a null. This should occur somewhere near center. The null should be *deep* and *definite*. If no null, check IC3 and associated components.
6. Reset R9 to center. Apply audio as in Step 1 to L, then R inputs. You should get an indication on the scope or level indicator of signal at the junction of R50 and C26. If the L and R inputs are connected together and audio is simultaneously applied, no output should be obtained. Touch up R35 again if necessary. Return R9 to zero.
7. Now connect an oscilloscope (If no scope, skip this step and go to Step 7A) to the output of IC4B. This is pin 1. With audio applied to *both* L and R, adjust R7 for a sinewave equal to the input in amplitude. Next, remove the jumper from L to R input so as that only the L input is receiving audio. Adjust R9 for the waveform shown in Fig. 3-45D. Now remove audio from L and connect to R. You should get a waveform as you did when audio is connected to the L input. If not, adjust R7, R9, and R35, and *slightly* adjust R3 if needed. It will probably take several repeats, since the adjustments, interact. You should get the waveforms in Fig. 3-45D and they should be very close to those shown. Set R64 to $^{1}/_{3}$ maximum.

7A. If no oscilloscope is used, set R7 about one third of maximum. Set R9 about $^{1}/_{4}$ maximum. Set R31 about halfway, and set R64 to about $^{1}/_{3}$ maximum. Now go to Step 8.

8. Tune a stereo FM receiver known to be in good working order to an unused channel near 88 MHz. Set it for mono reception. Connect a stereo program source to the L and R inputs of the MPX transmitter. Turn C32 until the FM receiver picks up the program fed to the L and R inputs. A 12″ wire can be connected to the junction of C40 and C41 to serve as an antenna. If the program is not heard, check the setting of R64 and Q3 and Q4 circuitry, also that L4 and L5 are correctly wound.
9. Adjust R64 until the volume of the FM receiver is about the same as that if a commercial FM station is tuned in. Now, switch the FM receiver to stereo. Adjust R31, the pilot level pot, until the stereo indicator lights, then advance it about 50% further. If no stereo light, check to see if the Q1 stage is producing 76 kHz and then try readjusting R31. You should now hear some stereo from the receiver. (Make sure your stereo light on the receiver is working OK.)

10. Adjust R9 for best stereo separation. Now adjust L3 for best separation. If you used a scope, little adjustment will be required. If you did not use a scope, adjust R9, L3, R7, and R3 a *little* at a time for best stereo separation. This takes some patience, but it is not too difficult. A pair of stereo headphones and a tone on one channel at a time is a great help.
11. Adjust L5 for maximum signal strength on your FM tuners' indicator. This step is not critical and if you do not have an indicator, forget it unless the range of this transmitter is inadequate.

This completes construction and alignment. This transmitter will be a good education in FM stereo techniques and should prove handy in many applications. A suitable power supply and mounting is shown in Fig. 3-46 using Radio Shack parts. See Fig. 3-47.

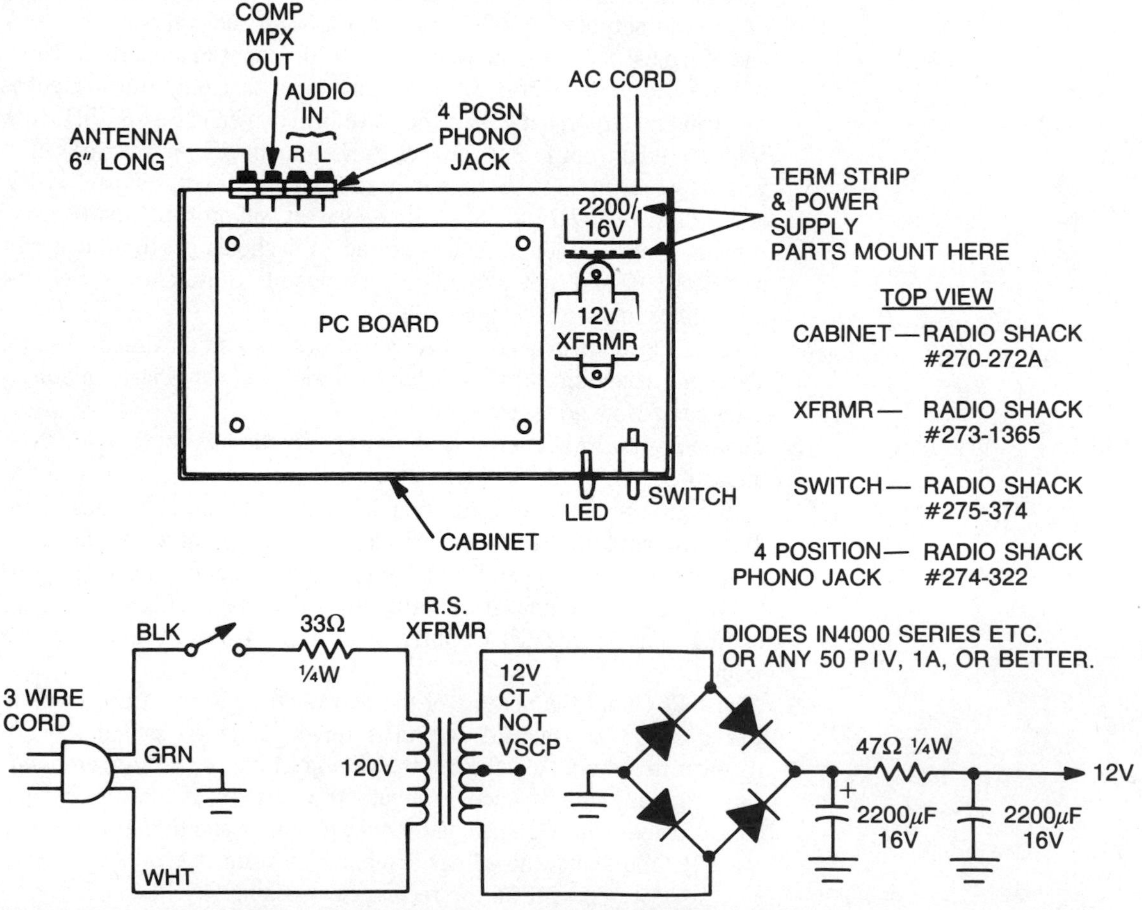

Fig. 3-46. Suggested cabinet mounting and ac power supply of FM MPX transmitter.

RESISTORS — All 1/4W ±10% Tol.

R1, R5, R14, R15, R20, R59	4.7k
R2, R4, R8, R11, R18	33k
R3, R7, R9, R10, R31, R35, R64	25k variable pot
R6, R12, R13	47k
R16, R17	100 ohm
R19	68k
R21, R25, R40	470 ohm
R22, R32, R33, R49, R50, R51, R52, R53	100k
R23	2.2M
R24, R54	10k
R26, R27, R41, R42	2.2k
R28, R29, R37, R47	220 ohm
R30, R39, R57, R63	1k
R34, R36	22k
R38	680 ohm
R43, R48	3.3k
R44, R65	10 ohm
R45	1.5k
R46	220k
R55	1 Meg.
R56, R66	330 ohm
R58, R62	15k
R60	100 ohm

INDUCTORS

L1, L2, L3	P/N 212103 (No. Country Radio) 2–12 mH variable
L4, L5	5 1/2T #22EN on Cambion Blue 8–32 x 1/4″ slug

ICs AND SEMIs

IC1, IC4	LM1458N
IC2	CD4027
IC3	LM1496N
D1, D3	1N757
D2	MV2107
Q1, Q2	2N3565
Q3, Q4	2N3563

CAPACITORS

C1, C2, C3, C4, C5, C6, C12 C14, C20, C21, C22, C23, C24, C25, C36, C42	47μF 16V radial lead Electrolytic
C7, C8	470pF SM or NPO ceramic
C9, C10	.0068 Mylar cap
C11, C26, C27, C29	1μF 20V radial lead elec.
C13, C39	470pF ceramic ±20%
C15	.047 Mylar ±10%
C16	.0033 Mylar ±10%
C17	.01 Mylar ±10%
C18, C19	.1 Mylar ±10%
C30	5pF SM or NPO
C28	470μF 16V radial lead elec.
C31, C35, C40	22pF SM or NPO
C32	2–10pF trimmer
C33, C34	82pF SM or NPO
C37, C38	.01 GMV disc
C41	68pF SM or NPO

Fig. 3-47. Parts list, Project 6.

A kit of parts consisting of the printed circuit board and all parts that mount on the board is available from:

North Country Radio
P.O. Box 53, Wykagyl Station
New Rochelle, NY 10804

Price: $57.50 plus $2.50 postage and handling.
New York residents must include sales tax.

PROJECT 7: FREQUENCY SYNTHESIZER

There are instances where a number of discrete frequencies must be generated and accurately referenced to a standard. An example is a frequency control application for a transmitter or receiver in which a large number of discrete channels must be covered.

The synthesizer to be described covers a 2 MHz range of frequencies in 10 kHz steps. The actual PLL (phase-locked loop) covers 2 – 4 MHz, but this is mixed with a crystal controlled heterodyne oscillator that can be placed at any frequency from about 5 MHz to over 200 MHz. Our particular application used two crystal oscillators for the generation of rf output frequencies from 39.7 to 41.7 MHz and 39.6 to 40.6 MHz. The crystals used therefore were 37.7 and 36.7 MHz respectively. The synthesizer output frequency was used as a local oscillator for an amateur radio repeater/10 meter FM application. When used with a 10.7 MHz i-f frequency, the receiver and transmitter used with this synthesizer would cover 28.0 to 30.0 MHz, the ten meter amateur band. A μPD858C IC (Uniden) is used in the synthesizer loop.

By changing the crystals suitably and modifying the coils, other ranges, of course, can be covered. Suitable coil data for 40 MHz is given here and should be valid for neighboring ranges, say 35 to 42 MHz. This would be suitable for coverage of 25 to 30 MHz (with a 10.7 MHz i-f), which also includes CB and the 12 meter amateur band. We will also give sample data for other frequency ranges.

A look at the block diagram of Fig. 3-48 explains the synthesizer system layout. A voltage controlled oscillator covering about 38 – 42 MHz feeds a buffer amplifier and rf output jack. Some of this signal is fed to a mixer using a 3N140 or 40673 field effect transistor. The mixer is also fed with a signal at 37.7 or 37.6 MHz (used in this application as an offset oscillator to offset the synthesizer 100 kHz, for use with a repeater application). The output of the mixer is in the 2 – 4 MHz range. The mixer output is filtered to remove spurious high frequency mixer signals, and only the 2 – 4 MHz component is used. This component is fed to an amplifier stage where it is amplified and shaped, and level shifted to the +5V level required to feed the phase-locked loop IC. A pair of 2N3563 transistors is used to do this job. The signal is fed to μPD858C the PLL IC.

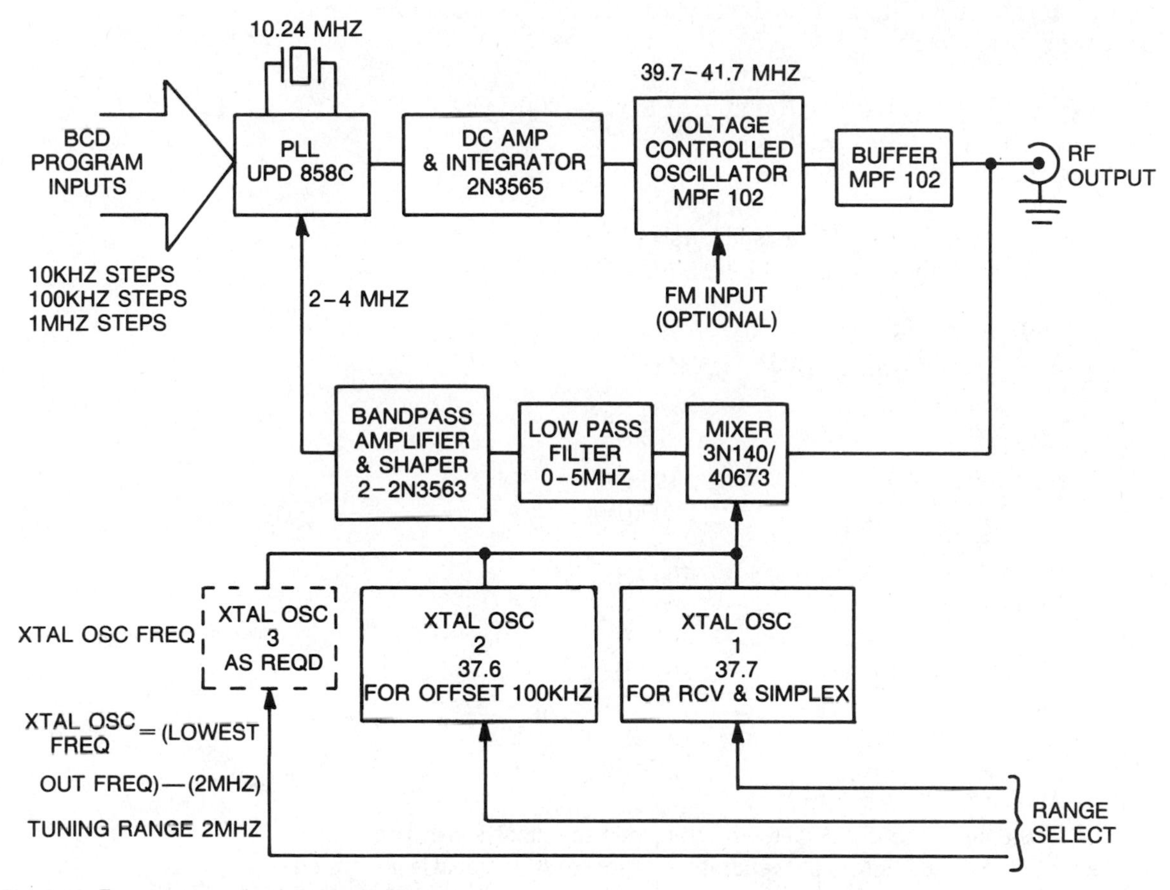

Fig. 3-48. Frequency synthesizer block diagram.

The μPD858C IC contains a variable modulus frequency divider, phase detector, dc amplifier, and reference oscillator operating at 10.24 MHz. The 10.24 MHz signal is generated with an internal crystal oscillator, using an external 10.24 MHz crystal. This is internally divided by 1024 to produce a 10 kHz signal reference to the phase detector.

The 2–4 MHz signal from the mixer and bandpass amplifier is fed to an internal variable divider *inside* the PLL IC (μPD858). The division ratio is determined by the data present on the ten program inputs. The input data is in BCD form. There are two leads for the MHz digit, and four for each of the 100 kHz and 10 kHz digit. Programming is straight (binary) BCD type, with "a" a high (+5 volts) and "0" a low (0 volts). The inputs are either CMOS or TTL compatible. For example, if an output frequency of 40.3 MHz is required (in our application corresponds to 29.6 MHz transmit/receive), the input program is found using this formula:

$$\text{DIVIDE RATIO} = 100 \times (\text{OUTPUT FREQ}_{\text{MHz}} - \text{XTALOSC FREQ}_{\text{MHz}})$$
$$= 100 \times (40.3 - 37.6) = 270$$

In BCD 270 = 0010 0111 0000

Therefore the input data would be as follows:
(Pins correspond to those on the PLL IC (μPD858)

PIN # of μPD858	*DATA (to Divide by 270)*	*REMARKS*	
13	Low	LSB 10 kHz)	
14	Low	20 kHz)	Tens
15	Low	40 kHz)	Digit
16	Low	80 kHz)	
17	High	100 kHz)	
18	High	200 kHz)	Hundreds
19	High	400 kHz)	Digit
20	Low	800 kHz)	
21	Low	MSB 1 MHz)	MHz
22	High	2 MHz)	Digit

Inside the chip, the phase detector compares the input signal (frequency divided by 270) with the internally generated reference. A dc signal is generated from the result. This dc signal is amplified both inside the PLL chip and the 2N3565 dc amp and integrator. It is applied to the VCO. If the frequency of the VCO is too low, a positive going signal is applied to the VCO. If too high a negative going signal is applied to the VCO. This has the effect of moving the VCO frequency in such a direction as to force it to be exactly that required to produce a 2,700 kHz signal at the input of the μPD858C. When divided by the externally programmed. (In our particular example) 270, a 10 kHz signal will be produced and this will exactly match and lock to the internal 10 kHz reference. Thus, the synthesizer output must be 40.3 MHz for this to occur. The output of the synthesizer is therefore phase-locked to the internal (10 kHz) reference.

The synthesizer is built on a 3″ × 6″ PC board. The VCO can be separately shielded from the main synthesizer board if desired by cutting the PC board at 1½″ from the VCO end and using strips of brass, galvanized steel, copper, or even G-10 PC board to form a wraparound case (see Fig. 3-49). If only one range is needed only one crystal oscillator circuit need be built. There is space for two on the PC board (see layout Fig. 3-50).

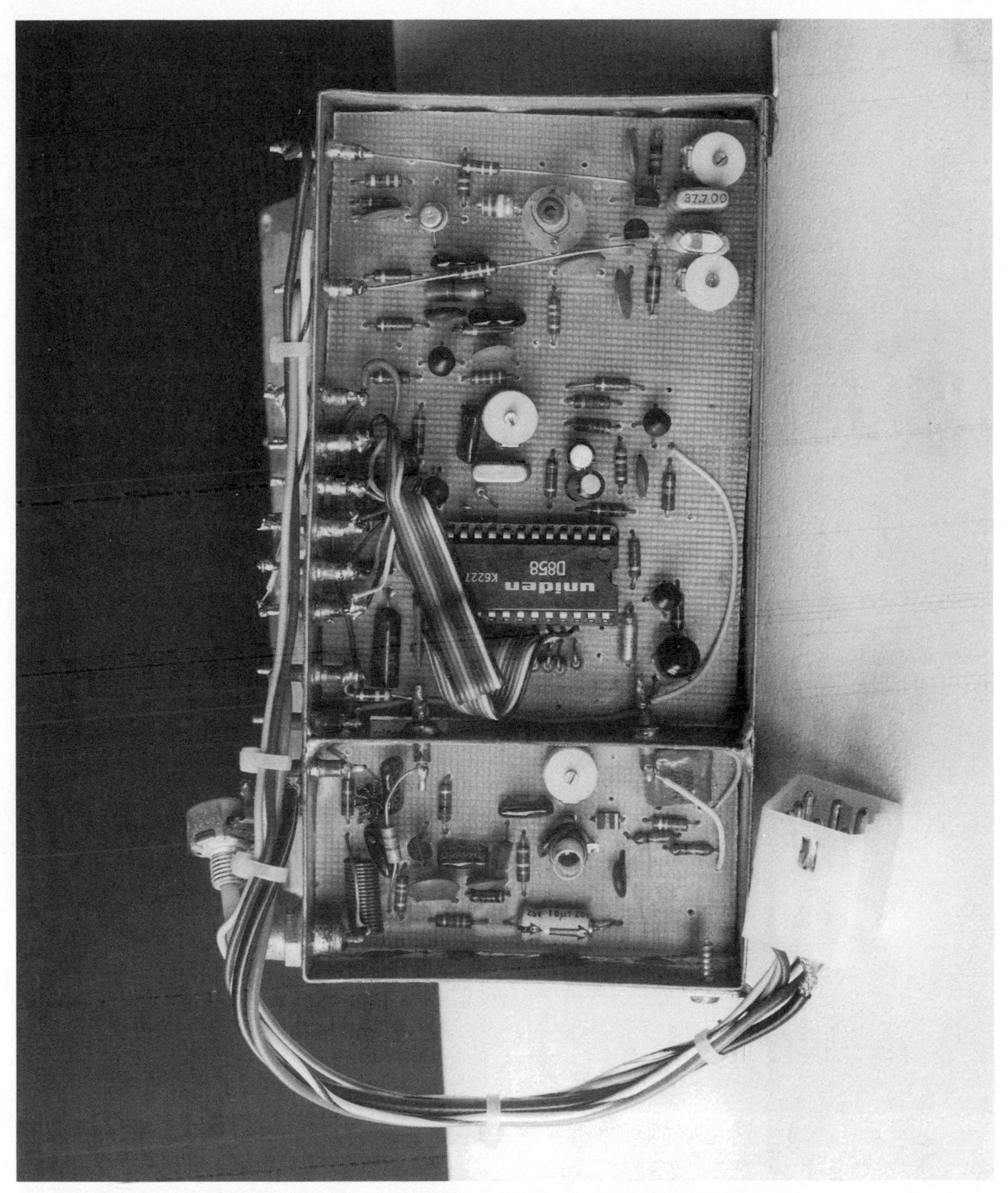

Fig. 3-49. Frequency synthesizer assembly as used in a repeater for 10 meter amateur use.

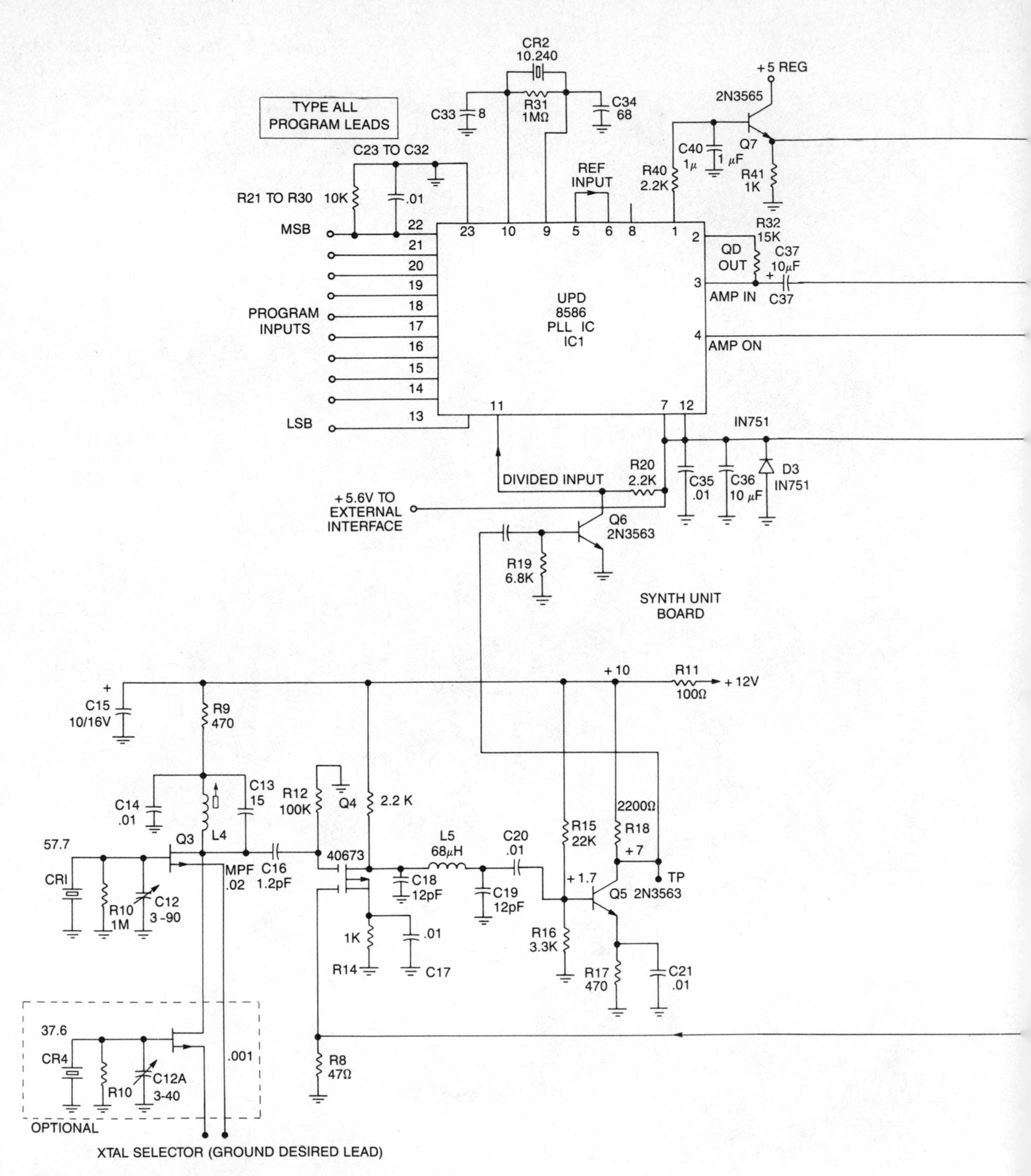

Fig. 3-50. Project 7 schematic.

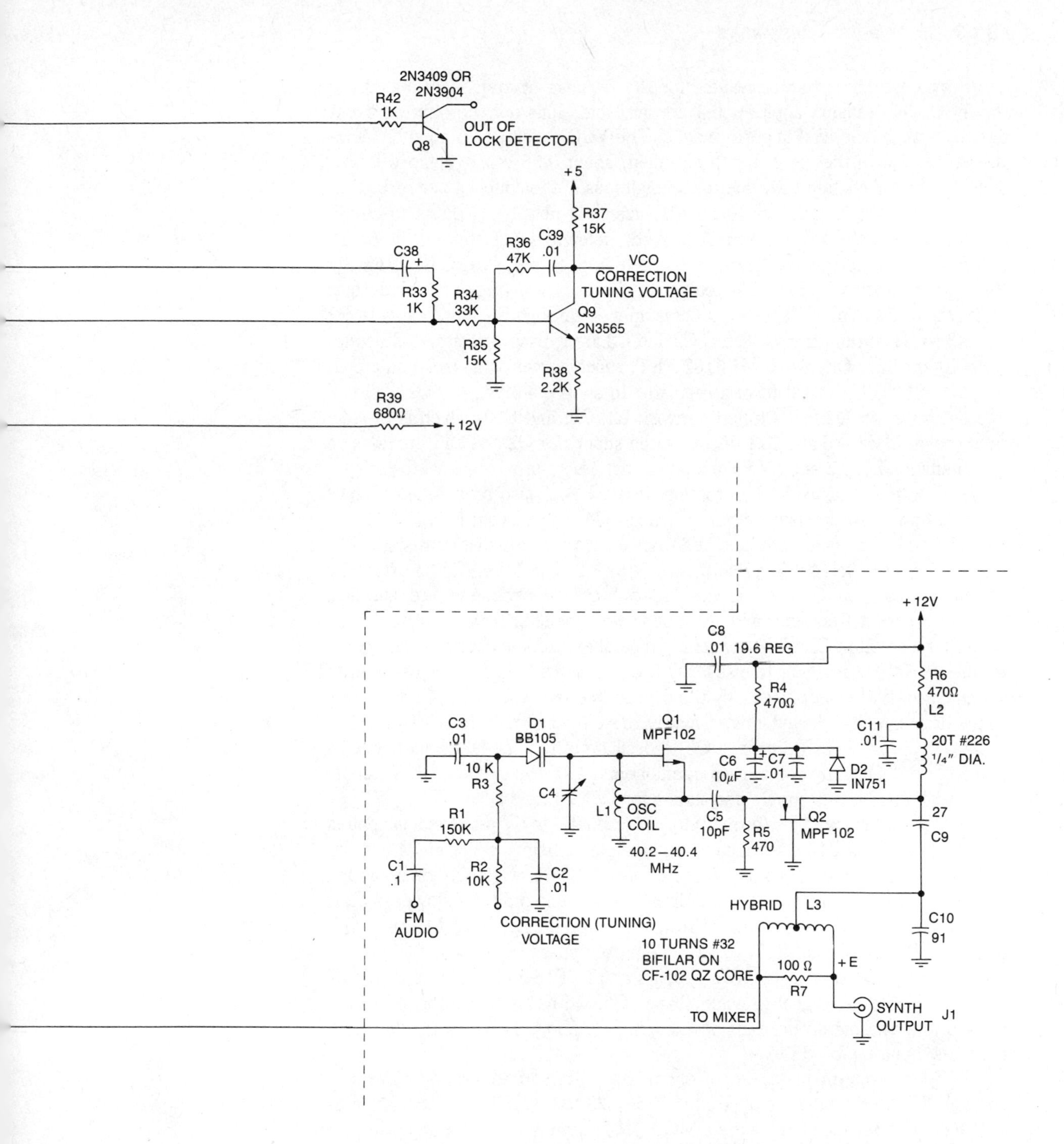
2N3409 OR
2N3904
R42
1K
Q8
OUT OF
LOCK DETECTOR
+5
R37
15K
C38
R36
47K
C39
.01
VCO
CORRECTION
TUNING VOLTAGE
R33
1K
R34
33K
Q9
2N3565
R35
15K
R38
2.2K
R39
680Ω
+12V
+12V
C8
.01
19.6 REG
R4
470Ω
R6
470Ω
L2
C3
.01
D1
BB105
Q1
MPF102
C11
.01
20T #226
1/4" DIA.
10 K
R3
C4
C6
10μF
C7
.01
D2
IN751
L1
OSC
COIL
C5
10pF
Q2
MPF 102
27
C9
R1
150K
R5
470
40.2–40.4
MHz
C1
.1
R2
10K
C2
.01
FM
AUDIO
CORRECTION (TUNING)
VOLTAGE
HYBRID
L3
C10
91
10 TURNS #32
BIFILAR ON
CF-102 QZ CORE
100 Ω
R7
+E
TO MIXER
SYNTH
OUTPUT
J1

Provisions for introducing audio for the purpose of frequency modulating the VCO (for transmit applications) are available. This may be omitted if not needed with no sacrifice of performance. The VCO sensitivity is about 1 MHz/volt so that, with the components specified, about 0.25 volt of audio will produce ±5 kHz deviation FM, due to the high loss in the coupling network.

A detailed description of the synthesizer will now be given as to circuit operation. First the VCO will be described. Refer to schematic Fig. 3-50. Q1 (MPF102) is a Hartley oscillator, covering about 38 to 42 MHz. L1, trimmer C4, and the varactor diode D1 determine the oscillator frequency. A dc bias (correction or tuning voltage) from Q9 is applied through isolation resistors R2 and R3 to D1 to tune the oscillator. C2 and C3 are bypass capacitors. C5 couples rf to amplifier Q2, also a MPF102 FET, which serves as an isolation amplifier. R4, C6, C7, and D2 form a regulator to supply +5 volts to Q1. C8 is a bypass capacitor. Q2 feeds tuned network L2, C9, and C10, which is broadly tuned to 40 MHz. R6 and C11 decouple the supply for Q2 and C11 provides ac grounding for L2. L3 and R7 form a power splitter to provide signal to J1 (synthesizer output) and to the heterodyne mixer. L3-R7 also provide isolation of the load on the synthesizer output and mixer Q4. Parts layout is Fig. 3-51.

The 40 MHz signal from L3 is fed to mixer Q4 and termination resistor R8. Q4, a 40673 (or 3N140) is a dual gate MOSFET. The MOSFET is driven by crystal oscillator(s) Q3 (or Q3A). The two crystal oscillators are identical except for crystal frequency and only one is described here. Q3 is a tuned drain type FET oscillator. CR1 is equivalent to a parallel resonant circuit at the crystal frequency. R10 provides gate bias for Q3. C12 is a fine frequency adjustment to set the crystal oscillator at exactly the crystal frequency. L4 and C13 are tuned to the drain circuit locks inductive (slightly above the crystal parallel resonance). Feedback internally in Q3 causes Q3 to oscillate at the crystal frequency. C14 and R9 are decoupling and bypass components. C15 bypasses the 12V supply lead and R11 provides supply decoupling for Q3, Q4, and Q5.

R12 provides gate bias (ground) for Q5. About 4 to 6V p-p signal is applied to gate 2 of Q4 via C16 a small coupling capacitor. Mixing occurs in Q4 and the resultant signal appears across R13. R14 and C17 are source resistor and bypass to bias Q4. The signal at the drain of Q4 is taken through lowpass filter C18, L5, C19. C20 couples signal from the filter (limited to 5 MHz) to Q5, biased by R15, R16, and R17. C21 is an emitter bypass for amplifier Q5. A few volts of signal between 2–4 MHz appears at TP (R18 and collector Q5). C22 and R19 couple this signal to pulse shaper Q6. Q6 has its collector fed to the input of the variable divider part of IC1 (pin 11). About 5V p-p is needed. R20 provides a pull-up for Q6 to +5V.

Divider program pins, ten of them, are returned to ground reference through R21 to R30 and are ac grounded by C23 to C32. These networks provide a definite termination for the CMOS (HI-Z) input pins of IC1 and also help to eliminate transients. C33, C34, R31, and CR2 form a crystal oscillator at 10.24 MHz together with internal oscillator circuitry of IC1 (pins 9 and 10). The

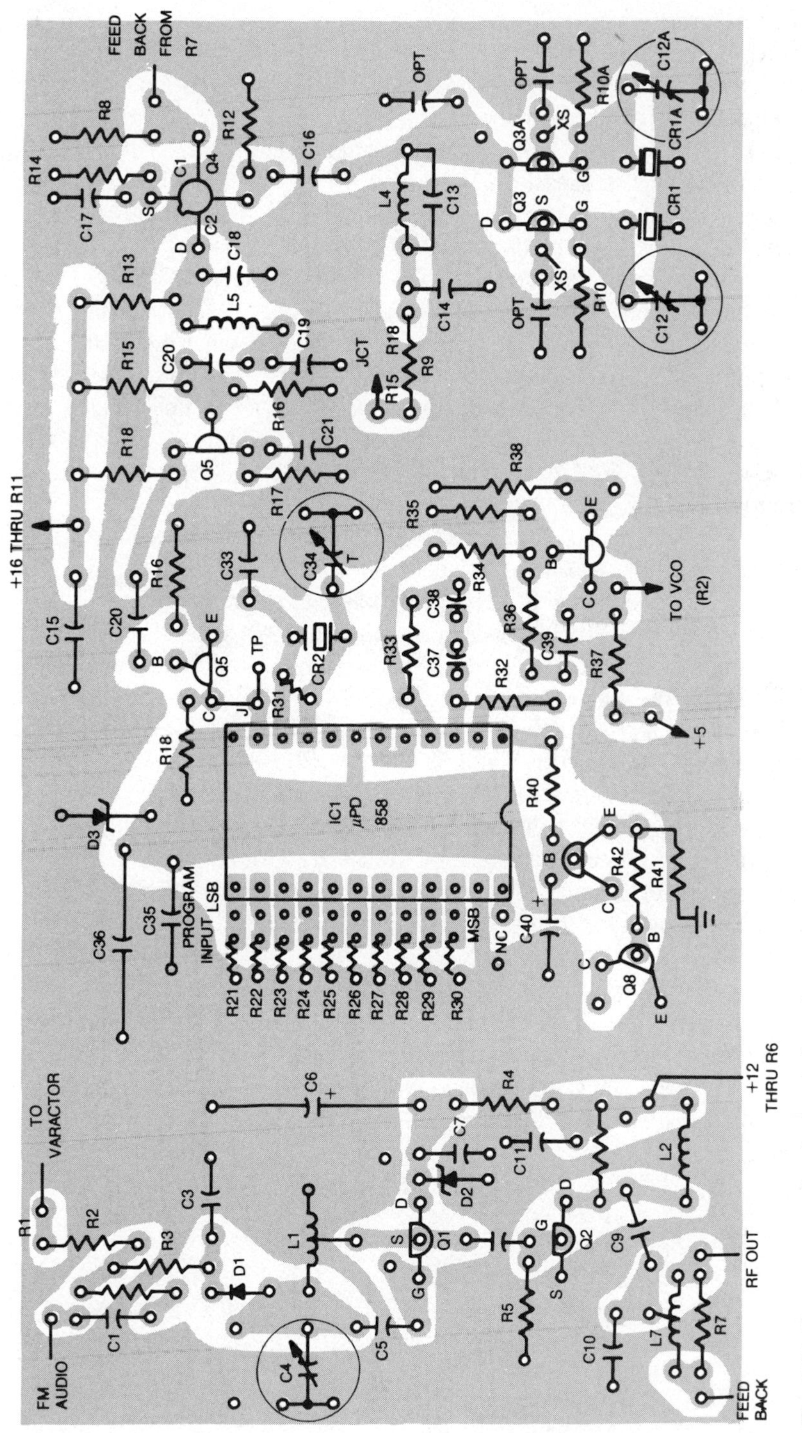

Fig. 3-51. Frequency synthesizer parts layout.

10 kHz reference signal appears at the jumper between pins 5 and 6. Phase detector output appears at pin 2 of IC1. It is coupled to an internal integrator (pins 3 and 4). R33, C37, C38, and R32 are used for this internal integrator as compensation network components.

IC1 is powered by a +5.6V supply formed by R39, D3, a zener diode, and C36 and C35 bypass capacitors. This point (pins 7 and 12) is brought out for external interface purposes, such as a supply for a BCD thumb wheel switch connected to pins 13–22, for frequency selection purposes.

Dc voltage from the internal amplifier is fed to dc amplifier Q9. R34, R35 are level shifting resistors while R38 is a bias resistor for R38 and R37 providing the collector load. C39 and R36 provide ac feedback to tailor the ac loop characteristics for satisfactory lock-up time, stability, and phase noise level.

The collector of Q9 feeds R2 the required dc correction voltage to tune VCO Q1 to the correct frequency so loop lock conditions are met. Q7, Q8, R40, C40, R41 and R42 form an out of lock detector (Q8 collector goes low) for detection of a fault or problem in the loop.

The construction of this synthesizer is relatively simple. Power requirements are +10 to +12 volts at about 50 milliamperes or less. The Uniden μPD858C can be obtained from parts suppliers to the radio-TV repair trade. It was used in many CB radios made during the late 1970's and early 1980's and it is a common item. A CB radio repair shop may stock it. It is also supplied by Phillips ECG. We suggest it is mounted in a 24 pin DIP socket and *not installed until the rest of the synthesizer is checked out*, as it, being CMOS, is somewhat fragile. The PC layout is on Fig. 3-52.

First, mount all resistors, then D1, D2, and D3. Next, all capacitors, Q1 through Q9, then coils L1 through L5. L1 through L4 are made by the constructor. L5 is a commercial rf choke. Coil details are given in Fig. 3-53, as well as typical values of associated components for other frequencies. These are intended only as an estimate—some ''cut and try'' may be required, depending on the application.

Next, install crystals CR1 and CR2 (and CR1A if used). Carefully check all wiring, soldering, and construction for shorts, opens, bad joints, or wrong or incorrectly inserted components. *Do not* yet insert IC1.

After all has been checked, apply power to the +12V rail and check for the following voltages. Use a 20k/volt VOM or DVM:

Pins 7 & 12 IC1 (IC1 *not* installed):	+5 to +6V
Pin 11 IC1	+3 to +6
All other pins IC1	Zero
Collector Q1	+5 to +6V
Drain Q2	+9 to 11.5V
Drain Q3, Q3A (if used)	+4 to 10V
Drain Q4	+6 to +11V
Collector Q5	+7V
Collector Q9	+5 to +6V

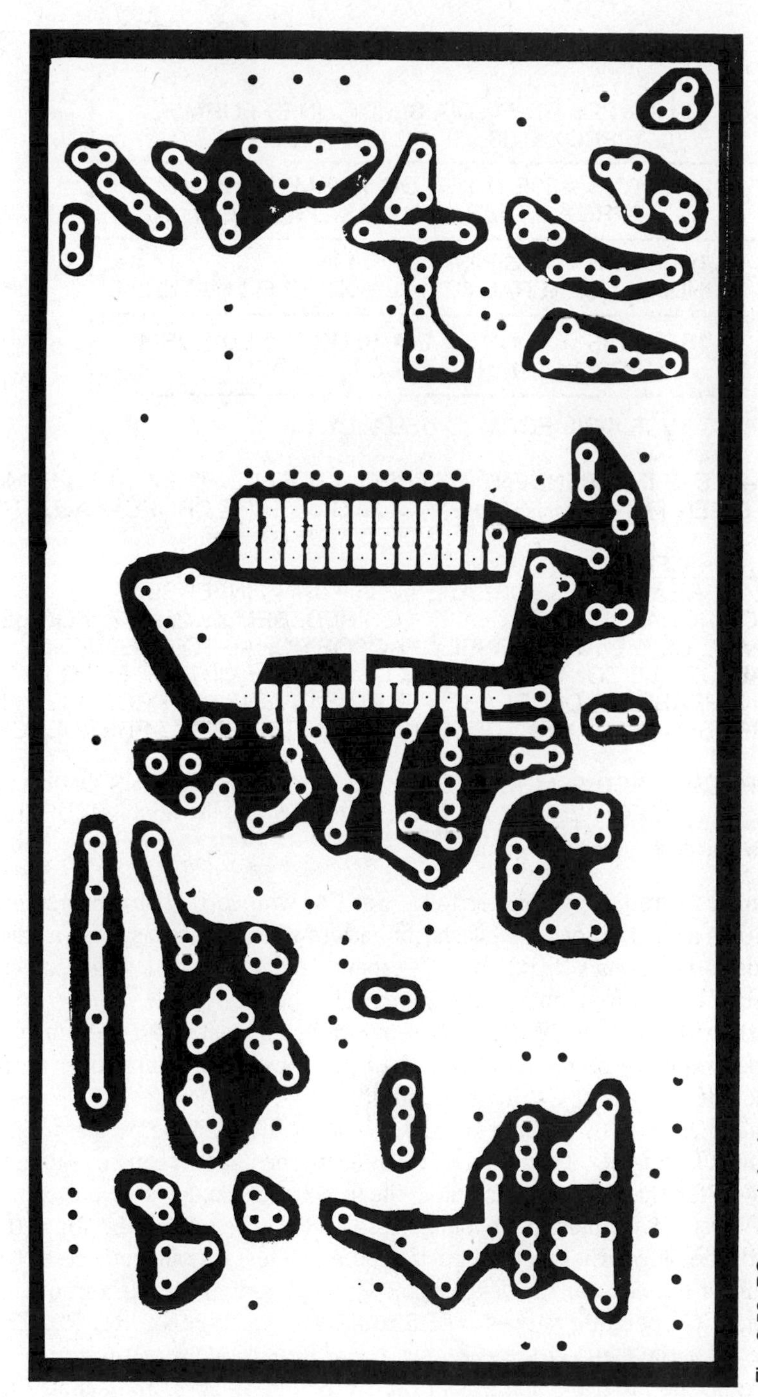

Fig. 3-52. PC artwork for frequency synthesizer.

COIL DATA

L1	7½ T #22E ON ¼″ DIA SLUG-TUNED FORM* (L APPROX 0.33μH), TO FIT PC BD
L2	20 TURNS #20E ON ¼″ DIA FORM* NO SLUG L APPROX 0.70μH (AIRWOUND COIL OK)
L3	10 TURNS #32 BIFILAR PAIR ON INDIANA GENERAL CT 101 TOROID Q2 MATERIAL
L4	15 TURNS #32 ON ¼″ DIA SLUG-TUNED FORM* L APPROX 1.0μH

*SURPLUS TV IF COIL FORM IS USUALLY OK

NOTE—L3 IS INDEPENDENT OF FREQ—USEFUL FROM 10 TO 50MHZ
DATA GIVEN FOR 40MHZ OUTPUT. LS IS 68μH RF CHOKE—ALL FREQS.

FOR OTHER FREQUENCIES:
L1, L2 INCREASE OR DECREASE # OF TURNS INVERSELY PROPORTIONAL TO FREQUENCY DESIRED. BELOW 25MHZ, FOR L2 INCREASE C9 & C10 INVERSELY PROPORTIONAL TO FREQUENCY AND ADJUST L2 TO RESONATE AT CENTER OF OUTPUT FREQ RANGE
L4—INCREASE OR DECREASE # TURNS INVERSELY PROPORTIONAL TO FREQUENCY OF XTAL USED, NOT SYNTH. OUTPUT FREQUENCY

EXAMPLES: a) FOR 30MHZ USE 40/30 OR 33% MORE TURNS ON L1,L2
b) FOR 25MHZ XTAL INSTEAD OF 31.5MHZ XTAL USE 20% MORE TURNS ON L4

Fig. 3-53. Coil data for frequency synthesizer.

Connect a frequency counter to J1. Set C4, with correction voltage set at +1.5V (use a single D or C cell flashlight battery to establish this, with a pair of clip leads to junctions C2, R2, R1 and ground. Adjust C4 for lowest expected synthesizer output frequency. (Approximate.)

Next, with junction C2, R21, R1 temporarily connected to +5V check to see if the frequency on the counter is the highest expected synthesizer output frequency. If not, adjust C4. This checks the VCO, Q1 and Q2.

Enable Q3 or Q3A (ground source) and adjust L4 until the oscillator operates. Adjust C12 to set the crystal on frequency. The counter can be connected across R12 for this test. If no results, check Q3, Q3A and associated components. Adjust L4 for maximum voltage across R14, then adjust L4 for a 10% drop with the slug turned in the direction so as to decrease inductance of L4.

Connect the counter to TP (collector Q5) and verify 2–4 MHz frequency, with the VCO operating between +1.5 to +5V at junction C2, R1, R2. Then also check for the same at the collector of Q6. Remove power.

Wire up the programming leads to IC1 (see Fig. 3-54 for the suitable test setup). Check wiring and then install IC1.

Apply power. Check for 10 kHz of signal at pins 5 and 6 IC1. Connect the counter to J1. Set the program (data) to IC1 at the middle of the desired range. Connect the VOM to collector Q9.

At this point you should get an output frequency very close to that programmed into IC1. Vary C4 *slightly*. The reading at the collector of Q9 should vary but the output frequency should not. If it does, the loop is not locked. Go back and check over the wiring.

Finally, adjust C4 so that the entire frequency range of the synthesizer is covered with the *midpoint* producing +3.5 volts or so at the collector of Q9. This voltage will vary over the tuning range of the synthesizer. If it goes over +4.8 volts or under +1.5 volts readjust C4. If you cannot get the desired coverage add a few turns to L2, readjust C4 and try again.

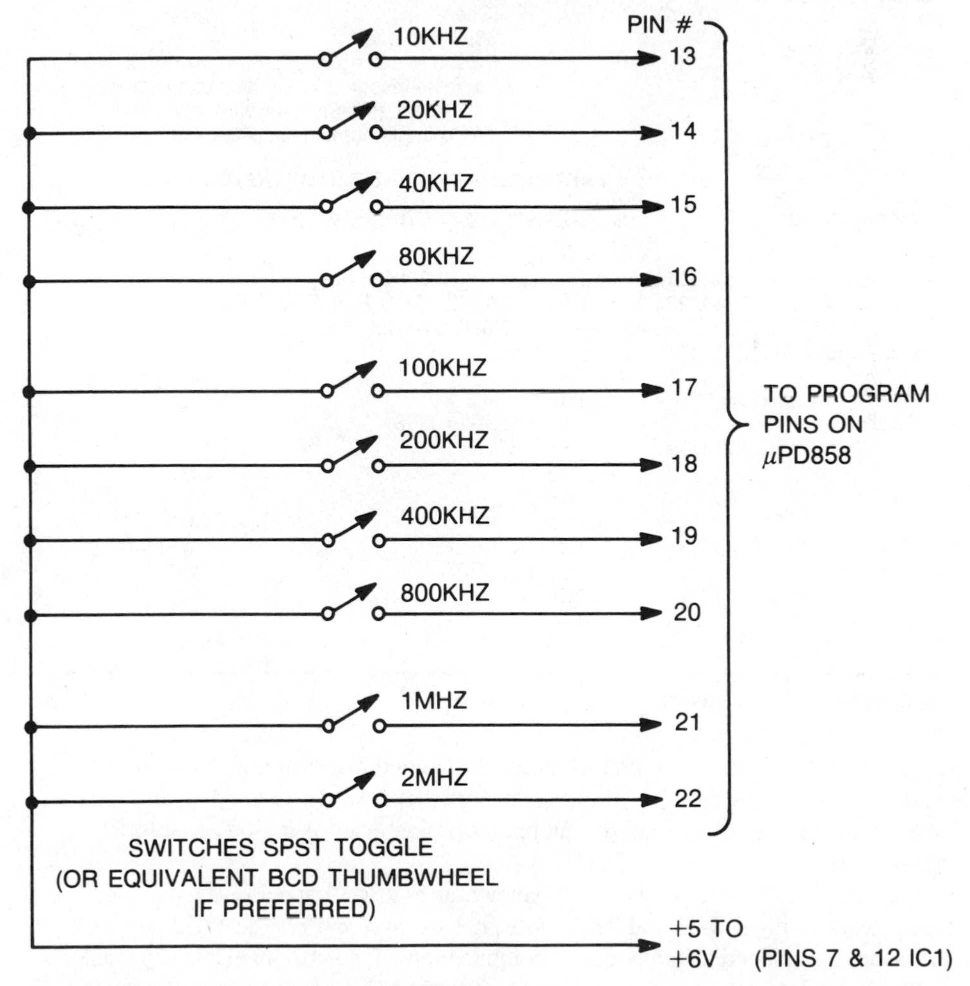

Fig. 3-54. Programming of frequency synthesizer.

SEMICONDUCTORS

SYMBOL	PART NUMBER	MANUFACTURERS
IC1	μPD858C	Uniden, NEC Philips ECG
Q1, Q2, Q3, Q3A	MPF102	Motorola
Q4	40673 or 3N140	RCA/GE, Motorola
Q5, Q6	2N3563	National, Fairchild, etc.
Q7, Q9	2N3565	National, Fairchild, etc.
Q8	2N3569 or 2N3904	National, Fairchild, etc.
D1	BB105	Motorola
D2, D3	IN751	Any

COILS AND MATERIALS

SYMBOL	NOTES
L1, L2, L3, L4	(See text)
L5	68μH rf choke
CR1, CR1a	Crystal to suit application 3rd overtone, parallel resonant, 32pF load capacitance, .005% tol. (International Xtal, etc.)
J1	Any suitable output jack (RCA, BNC, etc.)

PARTS LIST—SYNTHESIZER USING μPD 858C

CAPACITORS	RESISTORS — 1/4W +5%		
C1	R1	150k	.1 Mylar 50V ±10%
C2, C3, C7, C8, C11, C14,	R2, R3, R21, R22, R23, R24	10K	.01μF 50V disc
C17, C20, C21, C22, C23,	R25, R26, R27, R28, R29, R30		
C24, C25, C26, C27, C28,	R4, R5, R6, R9, R9A, R17	470 ohm	
C29, C30, C31, C32, C35,	R7, R11	100 ohm	
C39	R8	47 ohm	
C4, C12, C12A (opt.)	R12	100k	3–40pF trimmer
C5	R13, R18, R20, R38	2.2k	10pF NPO/SM ± 1pF
C6, C15, C36, C37, C38	R14, R33	1k	10μF/16V elec
C9	R15	22k	27pF NPO/SM ±5%
C10	R16	3.3k	91pF NPO/SM ±5%
C13	R19	6.8k	15pF NPO/SM ±5%
C16	R31	1 Megohm	1.2pF ceramic ±.5pF
C18, C19	R32, R35, R37	15k	12pF NPO/SM ±5%
C33	R34	33k	8pF NPO/SM ±5%
C34	R36	47k	60pF NPO/SM ±5%
C40	R39	580 ohm	1μF 50V elec

Fig. 3-55. Parts List for Project 7.

The synthesizer should be mounted in a shielded box or metal enclosure to reduce possible "birdies" in the transmitter or receiver it is used with. It is also a good idea to bypass all dc and programming leads going out or into the synthesizer.

A photo of the completed unit is shown in Fig. 3-49 and PC artwork and parts layout in Figs. 3-50 and 3-51. Note that we have cut off the VCO part of the layout and placed it in a separate compartment, for extra shielding. This is desirable but not necessary. See Figs. 3-52 through 3-55.

Part 4
Video Projects

There are many instances in video work in which it is desirable or necessary to modify a video signal in one way or another. Such modification may include color correction with regard to tint or level. Possibly the color is okay but contrast is low due to a weak luminance signal component. Another modification may include deliberate distortion of the signal for the purpose of creating special effects. These modifications may include color and/or luminance reversal (positive-negative), posterization, in which the video signal is altered such that there are only a few discrete values of luminance (usually 2, 3, or 4), or solarization, in which the gray scale (luminance) is "folded" on itself. Solarization gives rise to an eerie, surrealistic picture which contains both positive and negative tones. Those readers who are also photography buffs might have seen "solarized" prints and can visualize this effect. Of course, several of these effects may be performed simultaneously.

PROJECT 8: VIDEO EFFECTS GENERATOR—VIDEO PALETTE

A device described here will perform these video modifications on any standard NTSC video signal. It has many applications and is also fun to play with strictly as a toy. Some of these applications are:

1. *Video Recording*
   a. Tint correction
   b. Chroma boost/cut
   c. Luminance/sync boost and cut
   d. Additions of special effects to recordings
   e. Posterization and solarization for effects and titling

2. *Photographic Uses*
   a. Viewing color negatives as positives (a separate camera is required)
   b. Advance predictions of finished appearance of photographic special effects
   c. Negative inspection and analysis
3. *Video Production*
   a. Simulation
   b. Special effects
   c. Artistic effects

These examples are not the only things that can be done. By proper manipulation of the front panel controls many other effects can be achieved. This will be evident after an hour or so experimentation, using a video source (VCR, TV tuner) and a video monitor to see the results. A kit of parts and/or PC boards is available. See details at the end of this article.

A block diagram of the Video Palette is shown in Fig. 4-2. Standard baseband (0-4) MHz) NTSC having a nominal amplitude of 1 volt peak-to-peak, negative sync (0.5 to 1.5V is permissible with gain adjustment) at a 75 ohm impedance level is applied to the system via the VIDEO IN jack. A bypass switch is provided to bypass the system, and a loop through connector is also provided for loop through setups. Two switches on the rear panel enable ac or dc coupling and 75 ohm or HI-z input impedance (useful in loop-through applications or where another 75 ohm device is already terminating the input line). An input video amplifier boosts this signal up to 3 volts peak-to-peak and at the same time, inverts it so that the sync tips are positive. The video signal is now still unaltered but is larger and inverted. It is now fed to a sync separator and a single pole double throw CMOS video switch.

The SPDT video switch is used to split the video signal into two components.

1. Sync, blanking and burst pulses only;
2. Video information and chroma information without sync, burst, or blanking.

The reason for the splitting up of the signal is to allow separate processing and treatment of four signal components—sync, burst, luminance (black and white component) and chroma (color-difference) component. Figure 4-3 shows the ''anatomy'' of a standard NTSC color signal.

It is necessary to split up (decompose) the signal for other reasons. The sync signal is used for timing and cannot be modified unless intentional ''scrambling'' of the picture is required. If we, for example, want an inverted (negative) picture, we simply cannot just reverse the phase of the whole signal, since the video monitor needs sync pulses of a given polarity. Therefore, a simple

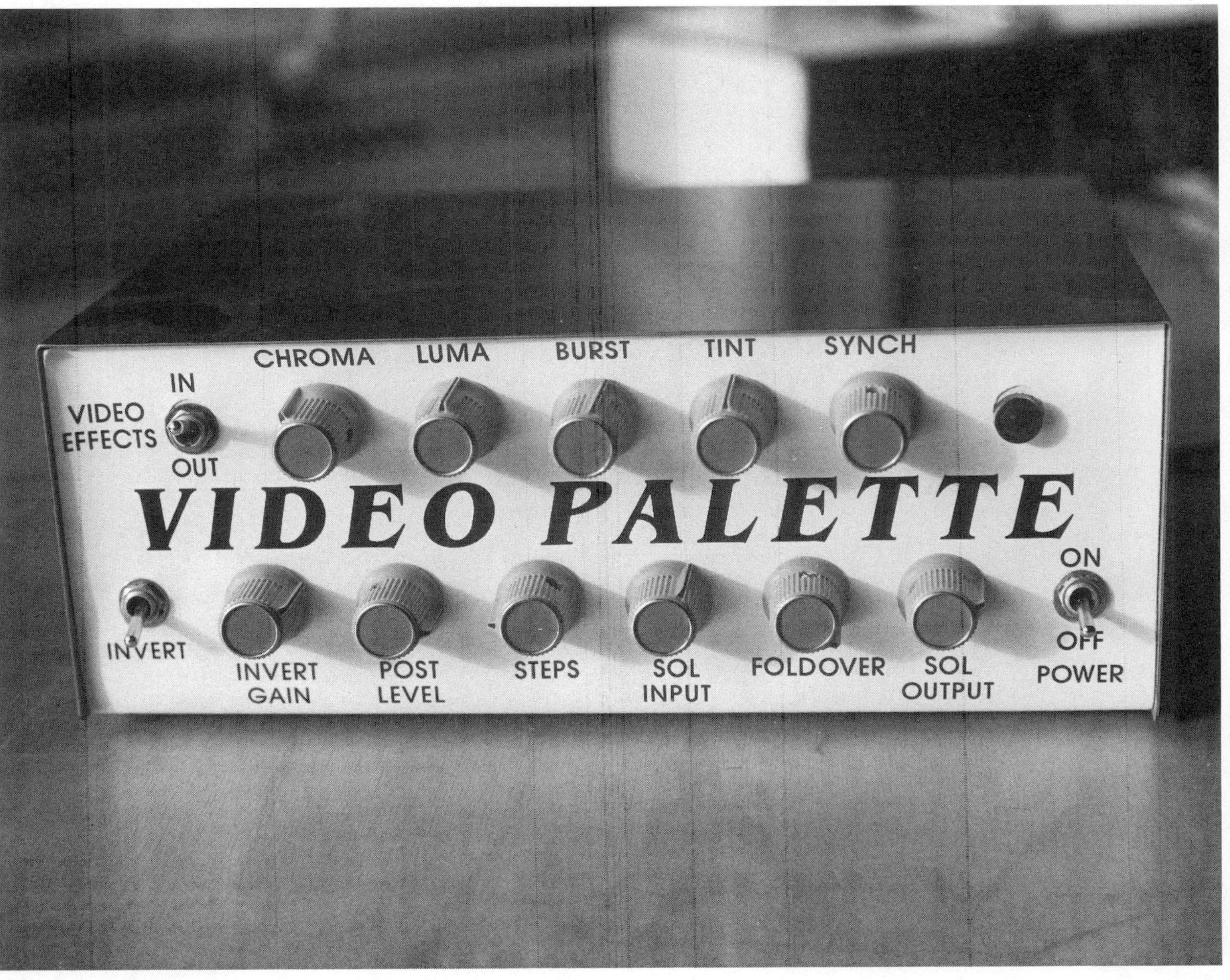

Fig. 4-1. Video palette—a system for analog video effects.

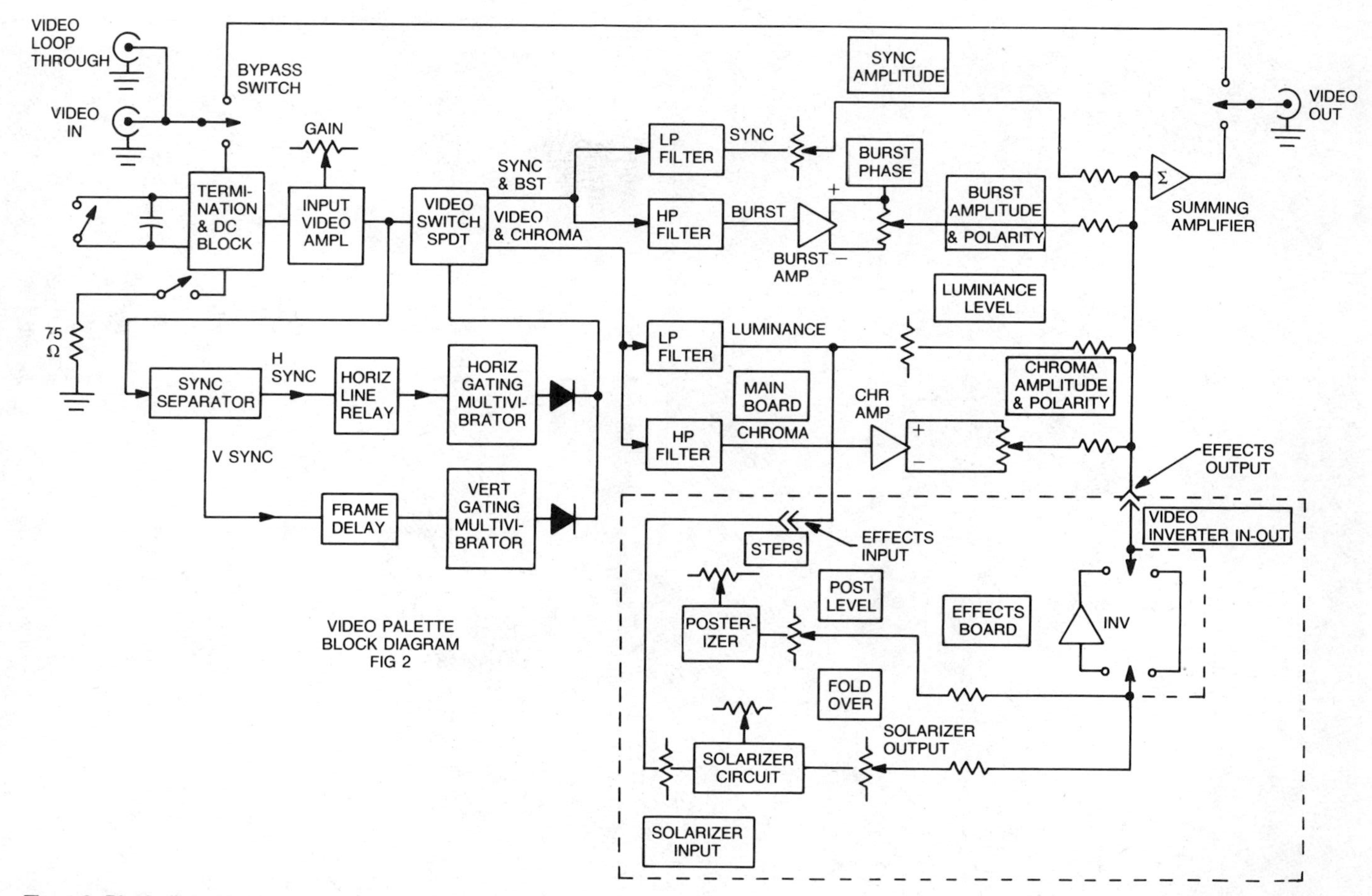

Fig. 4-2. Block diagram of video palette.

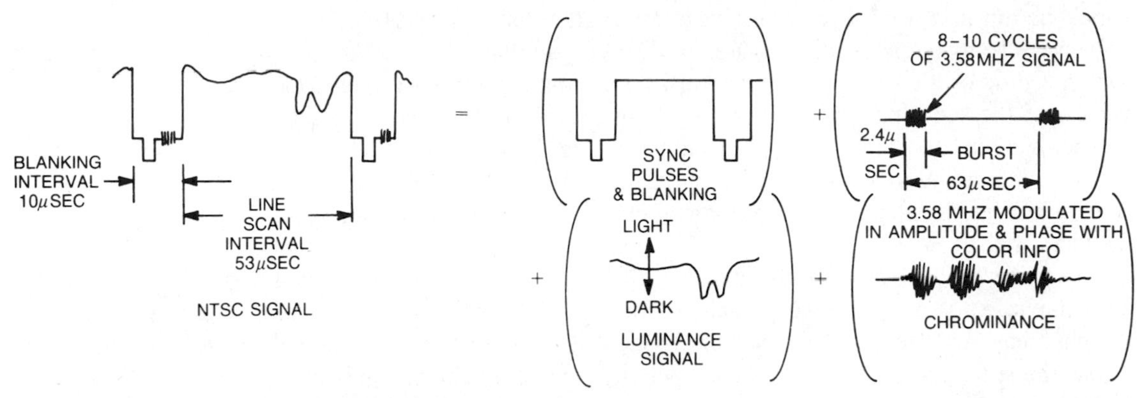

Fig. 4-3. Resolutions of NTSC signal into four components.

inverter will not generally do a good job. If we want to introduce special effects, we may want, for example, to just operate on the chroma signal and leave everything else the same. Therefore, this splitting up of the signal is obviously necessary. Otherwise, improper operation of the sync and color burst circuitry in the monitor, VCR, or TV receiver will result.

Sync and blanking are separated from the burst pulse with filters. The sync and blanking pulses are fed into a low pass filter having a cutoff frequency of 2.5 MHz with a max rejection frequency of 3.58 MHz.

Luminance and chrominance are separated in the same way. Strictly speaking, a comb filter would be the ideal way to separate luminance and chrominance, but it would complicate construction and increase cost. For this application, simple M derived filters do the job adequately.

A switching signal to control the video splitting is derived from the sync information on the original signal. A sync separator strips the horizontal sync and vertical sync information from the original signal. Note that the original signal must be a legitimate NTSC signal, with sync tips most negative. Therefore, a ''scrambled'' signal will not operate this unit. It is possible to ''descramble'' a pilotless, suppressed sync scrambled signal with this device if an extra sync regenerator circuit is used. For now, a standard NTSC signal is assumed.

The horizontal sync drives a monostable multivibrator of nominal 53 microseconds pulse width. This multivibrator drives a second multivibrator that generates a gating pulse of about 10 microseconds width. This pulse is timed to be coincident with the blanking pulse on the original NTSC signal. This is done by adjusting the 53 microseconds (nominal) multivibrator pulse width. This 53 microsecond pulse causes the video switch to route the video signal into the sync and burst side of the switch. When the pulse is not present (during horizontal line scan) the video is routed to the video and chroma side of the switch.

Vertical sync is separated in a similar manner. A monostable having a delay of one frame (nominally 16 microseconds) is used to trigger a second monostable multivibrator of nominally 600 microseconds pulse width. This pulse should be timed to occur during vertical retrace. Therefore, the entire composite vertical blanking pulse will be gated to the sync and burst side of the video switch. Setup of the switching circuit consists of adjusting the delays and pulse widths of the four multivibrators so as to switch the sync pulses, blanking pulses, and burst pulses to the synch and burst side of the video switch.

The sync is fed to a LP filter so as to remove the burst. It then goes to an amplitude adjustment pot, then into a summing amplifier, which reassembles the video signal.

The burst goes to a high pass filter to get rid of the sync. Then the signal is fed to a phase correction network, and amplified in a differential amplifier. This amplifier has two outputs 180° out of phase. A potentiometer is connected across these outputs to act as a gain and polarity control. At the center setting of the pot, zero output (no burst) occurs. At one side or the other, full positive or negative burst is available. The burst can be varied from plus three times normal to minus three times normal. By using a negative burst phase, the monitor or TV screen will show colors reversed (complementary) to the originals.

In this manner, the burst phase and amplitude can be altered or corrected as required, independently of everything else. The burst is then fed to the summing amplifier for "reassembly" into a complete video signal.

Luminance is treated in a similar manner as the sync. It goes through a LP filter similar to the sync filter and is then fed to a level control to the summing amplifier along with the sync and burst information.

Chrominance is extracted with a high pass filter. It is fed through a differential amplifier with positive and negative outputs. (Similar to the burst amplifier.) A control enables amplified chroma, either positive or negative, to be fed to the summing amplifier at any desired level. This feature enables weak chroma to be amplified, or inverted. Since some 2.5 MHz or higher luminance components will be present along with the chroma signal, the chroma control functions as a "sharpness" control in black and white applications. The capability of using either positive or negative burst with positive or negative chroma will at first seem redundant. Negative chroma with a negative burst will yield a positive picture. However, the phase of the high frequency luminance components will not be the same. This effect can be used as a "sharpness" control for color applications. The chrominance component is fed to the summing amplifier.

The summing amplifier has a nominal gain of unity. Since we amplified the video signal by three in the input amplifier, and (allowing for circuit loss) about 2 volts of composite video components is present at the input of the summing amplifier, a 2 volt peak-to-peak video signal, reassembled, is present at the output of the summing amplifier. Therefore, the gain controls for chroma, burst, sync, and luminance will typically be about halfway open for unity gain. You may be wondering what we have really accomplished so far. What we have now, is the ability to take apart the video signal, control each of four components on an individual basis, and recombine them into a new video signal. This new video signal may be one that gives an improved picture, degraded picture, or altered picture depending on control settings. By this method, we can custom correct individual faults in the video signal.

At this point, we have a "graphic equalizer" for video. However, it is very easy to add some special video effects by operating on the luminance portion of the signal. This will now be discussed.

A few analog special effects that are incorporated in this video palette are known as inversion, posterization, and solarization.

Inversion (Fig. 4-4) is simply what it means. The video signal levels are inverted about a given reference axis. For example, if zero volts represents white and 1 volt represents black, at the two extremes, passing the signal through an inverter such that the output is 1 volt when the input is zero volts and the output is zero volts when the input is 1 volt will produce an inverted video signal. Note that a true inverter would just change the sign. For example, let us take the case of a signal at +.5 volt, which would represent a middle gray. Ideally, if we invert a gray tone, it is gray. (Gray is its own complementary color.) However, a true inverter would give −.05 volts output. This would be whiter and white. Therefore, we must add a dc offset of +1 volt to the output so a zero volt input produces 1 volt, and therefore restore the gray (average level) to original. This can be done by adding a dc offset level to the input. In our illustration note that one-half of a volt is needed since the op amp configuration has a gain of two at the noninverting input.

Inverting the luminance signal produces a "negative" picture. If the picture is black and white, the picture will appear like a black and white photo negative. If color, the colors will still be correct but the tones reversed. Actually, since the "color" is really the sum of luminance, and chrominance, the hues are unchanged but the saturations may be different.

In order to get a negative similar to a color photo negative (unmasked type) you would also invert either the chroma or the burst. A photo negative also generally has an orange cast, but this is deliberately introduced into the film to compensate for the inherent dye (color) deficiencies in the negative and has no counterpart in video.

If you looked at a color photo negative with a video camera/monitor setup and ran the resulting video through this unit, you would see a positive color image. This assumes the correct light source is used for viewing the negative. In fact, commercial color photo printing uses this method to predetermine correct color printing exposure and filtration (light source color balance) so the final print is correct the first time. This photo printing technique is known as "video analysis." For more details, consult a textbook on photography.

"Posterization" is a term used to describe the process of converting a photo or video image from a scene containing a wide range of tones to a scene containing only a discrete few, sometimes only two (black and white). No intermediate "shades" between tones are present. Commonly, four tones are used

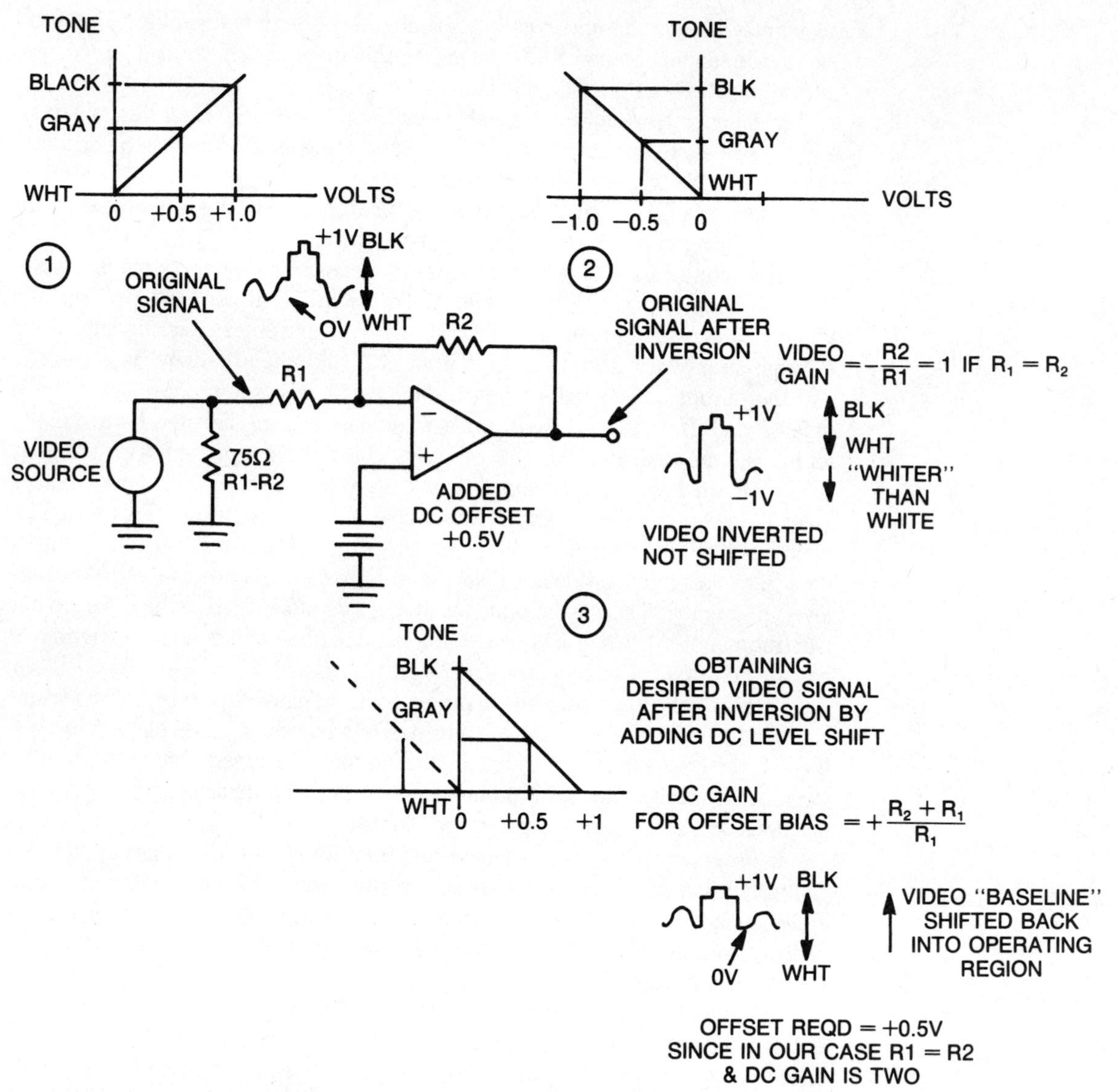

Fig. 4-4. Video inversion techniques.

to practice—white, light gray, dark gray and black. Colors are sometimes left in their natural state, or may be saturated (chroma is limited to one value—maximum). In practice, a nice visual effect is obtained by posterizing only the luminance.

An illustration of this effect is shown in Fig. 4-5A. A posterized video scene has a distinct "computerized" or "cartoon" effect since there are only a few

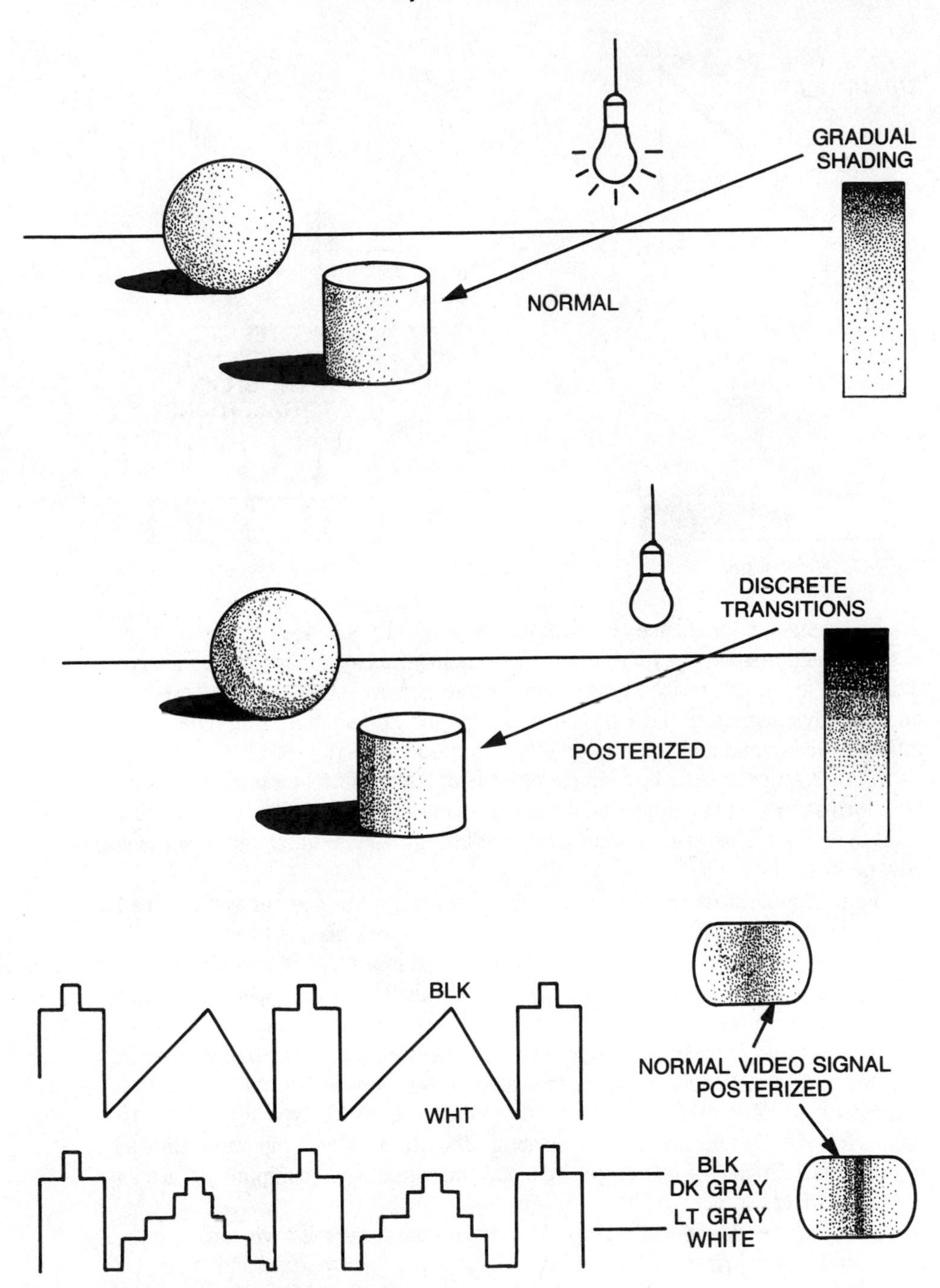

Fig. 4-5A. "Posterization" of video signals.

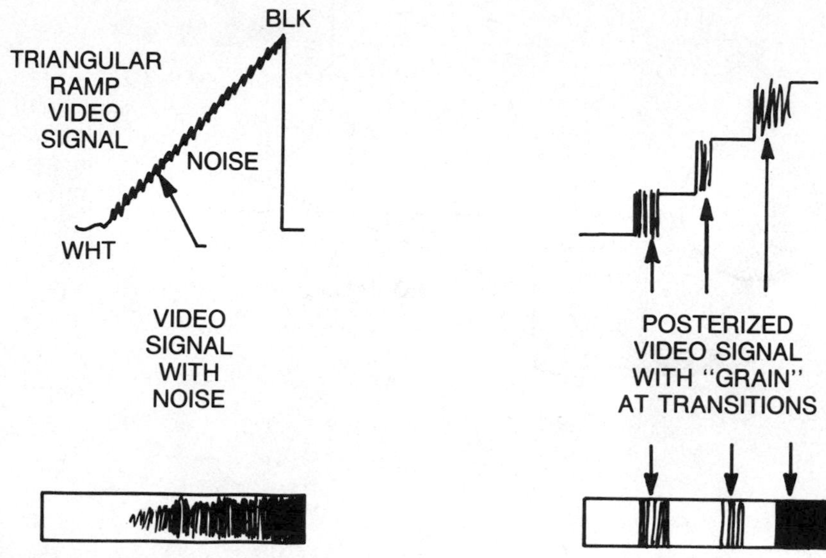

Fig. 4-5B. Video Signal Types.

discrete values of luminance. Posterization is useful to enhance contrast. Since, at the transition points, a very small difference in luminance may produce a large difference in the posterized output small noise signals, snow, and glitches are enormously magnified. This produces a "grainy" effect but sometimes it is rather pleasing and adds an artistic effect to the picture (Fig. 4-5B).

Posterization is done by a simple process of converting the signal to a "digital" format in a very simple A-D converter and then immediately converting back to analog. This can be done as shown in Fig. 4-6. A simple approach will now be described.

Four comparators are set up and biased with a reference voltage obtained from a divider string (four resistors R1). A4 has +1 volt bias, A3 has $+^3/_4$ volt, A2 has $+^1/_2$ volt, and A1 has $+^1/_4$ volt bias on the inverting input. All four non-inverting inputs are connected together to a nominal zero volt (white) to 1 volt (black) video source.

The outputs of each comparator can be either zero or some positive voltage V_O. Assume V_O = 5 volts. Each comparator output goes into a resistor R2 to a common load resistor R_L, which is much smaller than R2 typically, R1 = 10 ohms, R2 = 4.7 ohms and R_L may be about 220 ohms. The comparator amplifiers must be capable of fast response since the video input components are as high as 3 MHz.

Assume the video level is zero. All four comparators will have zero output. As the video level rises to $^1/_4$ volt A1 will suddenly change state and about $^1/_4$ volt will appear across R_L (5 volts from A1 divided by R2 and R_L). As the video level exceeds $^1/_2$ volt A2 will change state and now A2 will contribute a current

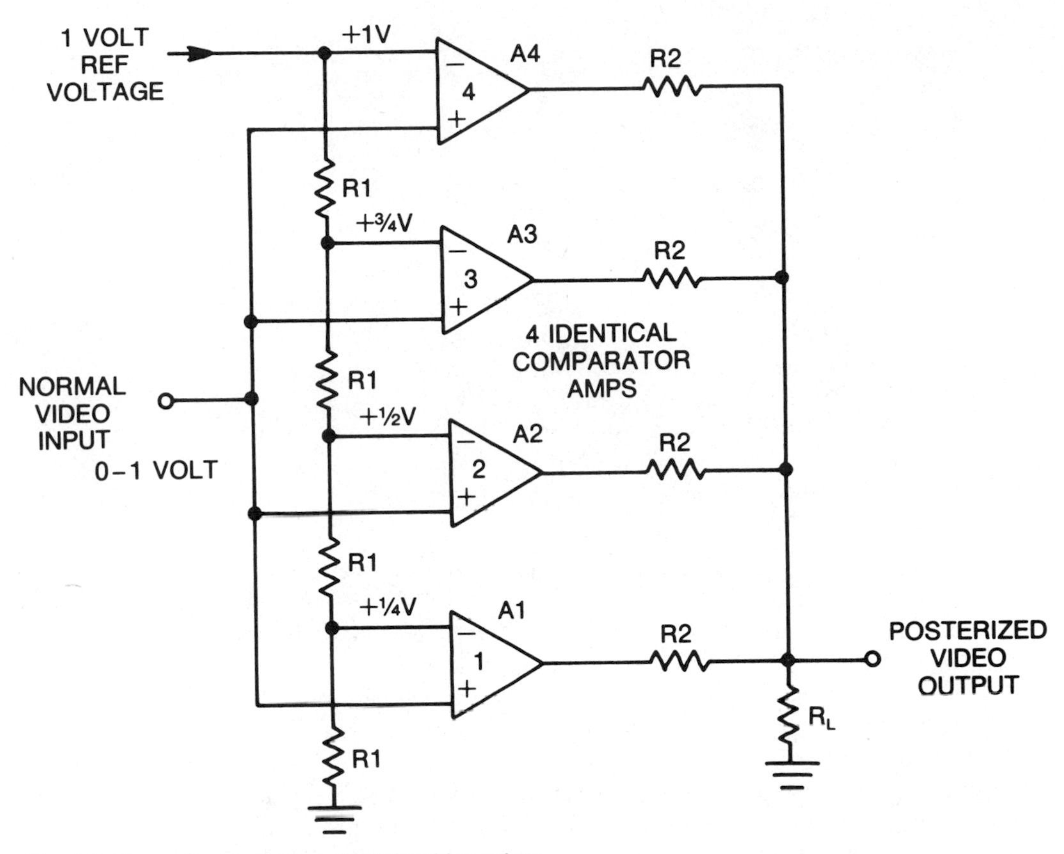

Fig. 4-6. Elementary posterizer circuit for video palette.

through its R2 so now R_L has 1/2 volt across it. With a high gain comparator, a few millivolts change in the video level produces 1/4 volt (250 millivolts) change abruptly in the output (across R_L). At 3/4 volt A3 conducts and at 1 volt A4 conducts. Therefore, a ramp input voltage produces a staircase output with four discrete levels +1/4, +1/2, +3/4, and +1 volt (actually five since zero volts is a level as well). In this way, we can produce only several discrete levels from a continuously varying level.

By varying the reference voltage we can vary the spacing between these steps and for a given input video level, the number of steps can thereby be changed. R1 (reference voltage divider) should be kept low in impedance to provide a "stiff" reference source.

In contrast to posterization, solarization is still a "linear" distortion technique. However, what is done is to fold the gray scale back on itself. White becomes white, light gray becomes dark gray, gray turns to black. Now, as the tones tend further toward black, the video output goes back towards white.

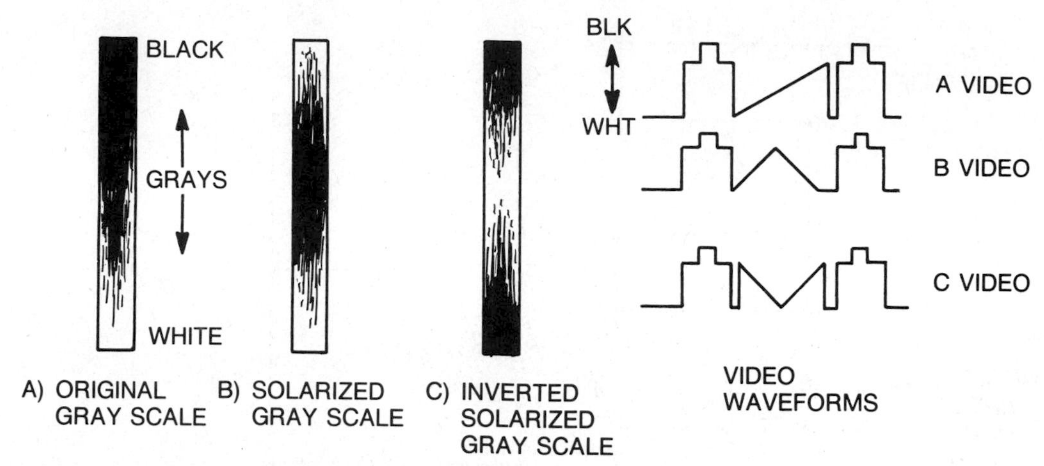

Fig. 4-7. Graininess in posterization.

Therefore, the highlights of the picture are positive and the lowlights (shadows) turn negative. Inversion can be used in conjunction, so that the shadows are positive and highlights are negative. See Fig. 4-7 for an illustration of this and also for an example of what a solarized picture looks like.

Those readers that are also photo buffs have undoubtedly seen or even made solarized photographs. Solarization is done by exposing a partially developed image to a light source. The dark developed areas shield the undeveloped emulsion (light sensitive coating) of the film (or paper) more than the light areas. On completion of development these highlights turn negative, since they were exposed more to light than the previously developed dark areas. It is tricky and difficult to control in photography, but with video it is easily controlled and manipulated to suit the taste of the videographer.

Solarization is accomplished by using an amplifier that has a transfer characteristic shown in Fig. 4-8. As can be seen, two amplifiers, one with a gain of two and one with a "delayed" gain of minus four, with their outputs combined, result in the necessary transfer characteristic. Although gains of two and four are shown, we can use any other 2:1 combination, resulting in different gain figures. In our case, we used gains of 0.5 and 1 giving an overall loss. This was done so that a larger input signal could be used. This made the transfer characteristics of CR1 more "ideal," since its knee (at 0.6V) is rather rounded, and small signals would not abruptly switch the diode.

Examining Fig. 4-9, the solarizer amplifier uses a simple op amp configured as follows: Assume CR1 is reverse biased. The gain if R1 = R2 equals 0.5, since R3 provides unity gain and since R1 = R2 only half of the input signal appears at the inverting input.

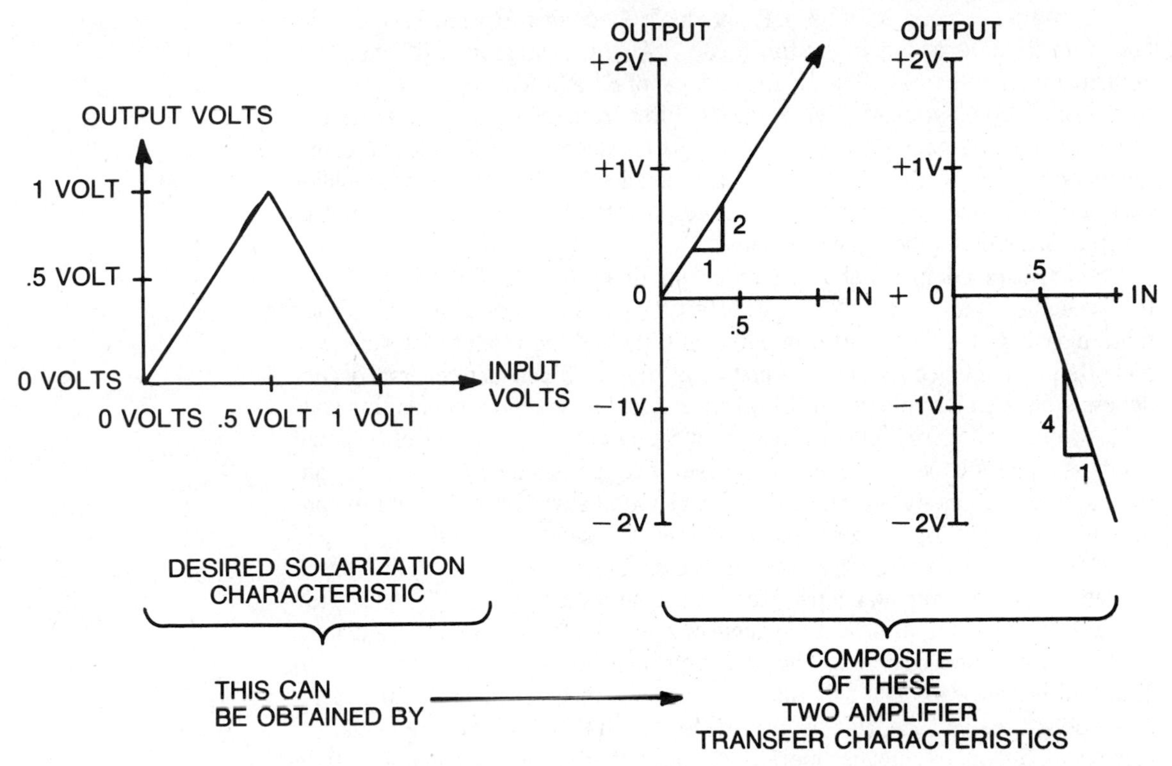

Fig. 4-8. Solarization and amplifier characteristics.

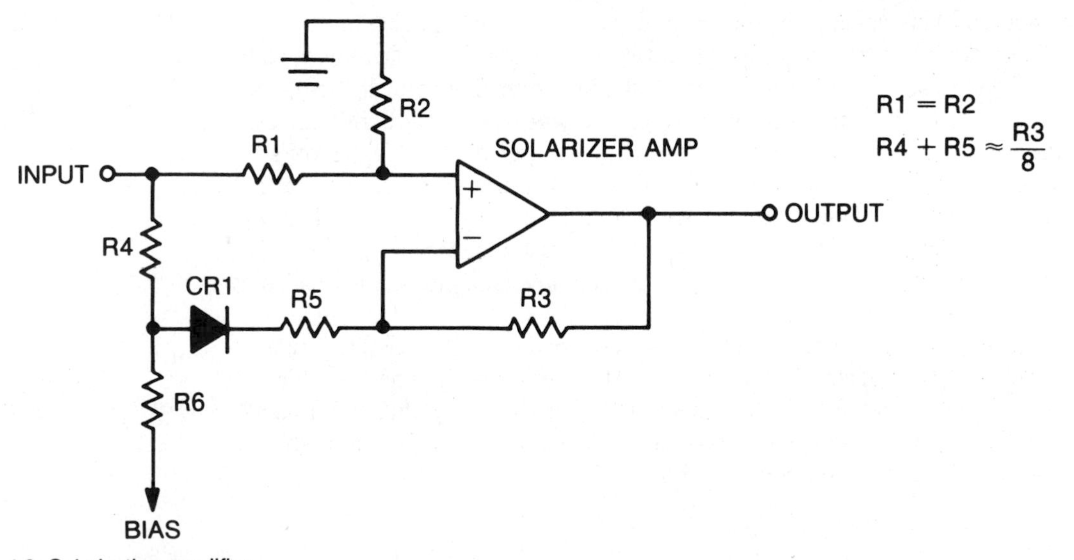

Fig. 4-9. Solarization amplifier.

As the input goes positive, CR1 gradually becomes forward biased. The bias from R6 determines where this occurs. Since the junction of R5 and R3 is at virtual ground (same voltage as the junction of R1 and R2) which is also equal to the output voltage until CR1 conducts. First, remember that for small signals, the amplifier gain is positive. When CR1 conducts, R3, R5, and R4 form a new feedback network. The feedback here is such that the resultant amplifier gain becomes much higher and of opposite polarity. This causes the output to reverse direction as the input gets larger.

Strictly speaking, R3 should equate four times R4 and R5. We got "better" results when R3 equalled eight times R4 and R5. Better means visually more pleasing. This is a matter of personal taste and the reader may very well find otherwise if he constructs this unit. The point at which gain reversal occurs depends on the bias voltage on R6 which (Foldover Control) supplies this voltage. "Optimum" visual effect requires adjusting both the drive level (solarize input) and the foldover control. The output level determines the picture contrast and also is separately adjustable. It was found advisable to use three controls in this circuit for the best results and most flexible operation.

Referring back to Fig. 4-2, the three circuits on the effects board can be operated together (simultaneously) for various interesting effects. The inverted output or normal output processed luminance is fed into the summing amplifier. When using the special effects, the main luminance level control is usually set to zero (but not always). The only thing the operator must watch is that the video levels do not exceed the sync tip or blank levels. In this case, picture instability may occur. The sum amplifier merely combines everything and can only handle two to three volts peak before clipping occurs. So bear this in mind. If a scope is available (not necessary) you can watch the waveforms while viewing the video on a monitor. This helps give you a clearer picture of what is going on.

The video palette is constructed on two PC boards. The main board contains the video signal splitting and recombination (summing amplifier) circuits. The special effects board contains the solarizer, posterizer, and inverter circuitry, as well as the power supply circuits. The power required is 6.3 volts to 12 volts 60 Hz ac. This can be obtained from a plug-in wall transformer or an inexpensive 300 mA filament transformer. Dc will not work, as the rectifier circuits need both halves of the ac sine wave to deliver the necessary positive and negative voltages. More than 12 volts is not recommended unless the 7805 and 7905 regulators are heat sinked.

The power supply consists of two half wave rectifiers, 2200μF filter capacitors, and two IC regulators (7805 and 7905) delivering +5Vdc and −5Vdc. It can be powered with 6.3 to 12.6 volts *ac*, which can be easily obtained from an inexpensive filament transformer, wall plug-in transformer, or even a hardware store doorbell transformer (8 to 10 volts).

Circuit Description

Refer to the schematics of Fig. 4-10 and Fig. 4-11 of the Main Board. Video input at J1 (1 volt peak-to-peak negative sync) is coupled through C1 to the video amp consisting of VR1, R2, IC, R3 and C2. A switch across C1 can short this capacitor if dc coupling is necessary. R1 can be switched across the input to provide a 75 ohm termination. VR1 is a gain control. At least 0.5V peak-to-peak video is necessary for proper operation. IC1 is a video op amp, an LM318. R3 provides feedback and C2 provides frequency compensation for IC1. R18 and R19 together with C3 and C4 provide ±5 volts to IC1 and serve as decoupling networks, reducing video crosstalk through the power supply lines. About 2 to 3 volts of inverted (pos sync) composite video appears at pin 6 of IC1.

Composite video (inverted) is fed from IC1 to both IC2, a CA4053 analog switch, and to the sync separator system through R4 and C5 to Q1. Q1 is normally nonconducting. Bias is generated across R5 that keeps Q1 cut off except during positive sync types. Negative sync pulses appear between the collector of Q1, R6 and ground. R6 provides a collector pull up for Q1. R7 and R8 couple the sync pulses to Q2. R9 is the collector load for Q2. R10, C6, R11, and C7 form a vertical integrating network to extract vertical timing pulses from the composite sync at the collector of Q2. C8 couples these pulses to Q3, which squares them up and shapes them. These negative going vertical sync pulses are used to trigger multivibration IC4.

Pulses at the collector of Q1 trigger dual multivibrator IC3, a CD4528. This is connected as two cascoded monostable multivibrators. VR2, R14, and C9 determine the pulse width of the first section—about 53 microseconds. VR3, R15 and C10 determine the pulse width of the next section about 10 microseconds. C14 is a +5V bypass capacitor. A positive going pulse appears at pin 10 of IC3. By proper adjustment of VR2 and VR3 this pulse can be made coincident to the horizontal blanking pulse, and of the same width. Similarly, vertical sync pulses at the collector of Q3 trigger IC4. Both IC4 sections function identically to IC3. VR4, R16, and C12 determine the pulse width of the first section (16 milliseconds nominal). VR5, R17, and C13 determine the pulse width of the second section. By proper adjustment of VR4 and VR5, the pulse appearing at pin 10 of IC4 can be made coincident with the vertical sync interval of the video input signal. A negative pulse at pin 9 of IC4 cuts off IC3 (horizontal gating) during vertical retrace intervals. The horizontal and vertical gating pulses are summed across R20, CR1 and CR2 isolate IC3 and IC4 outputs. The pulse across R20 is nominally +5 volts and is low during line scan and high during sync intervals. It is fed to pin 9, the control pin, of video switch IC2.

During line scan intervals, video (luminance and chroma) from pin 4 of IC2 appears at pin 5, since pin 9 is low (zero volts). L1, C16, and C17 form a low pass filter and C15, R22, and L2 a high pass filter. R23 and R24 terminate the high pass filter. VR6, the luminance gain control terminates the low pass filter.

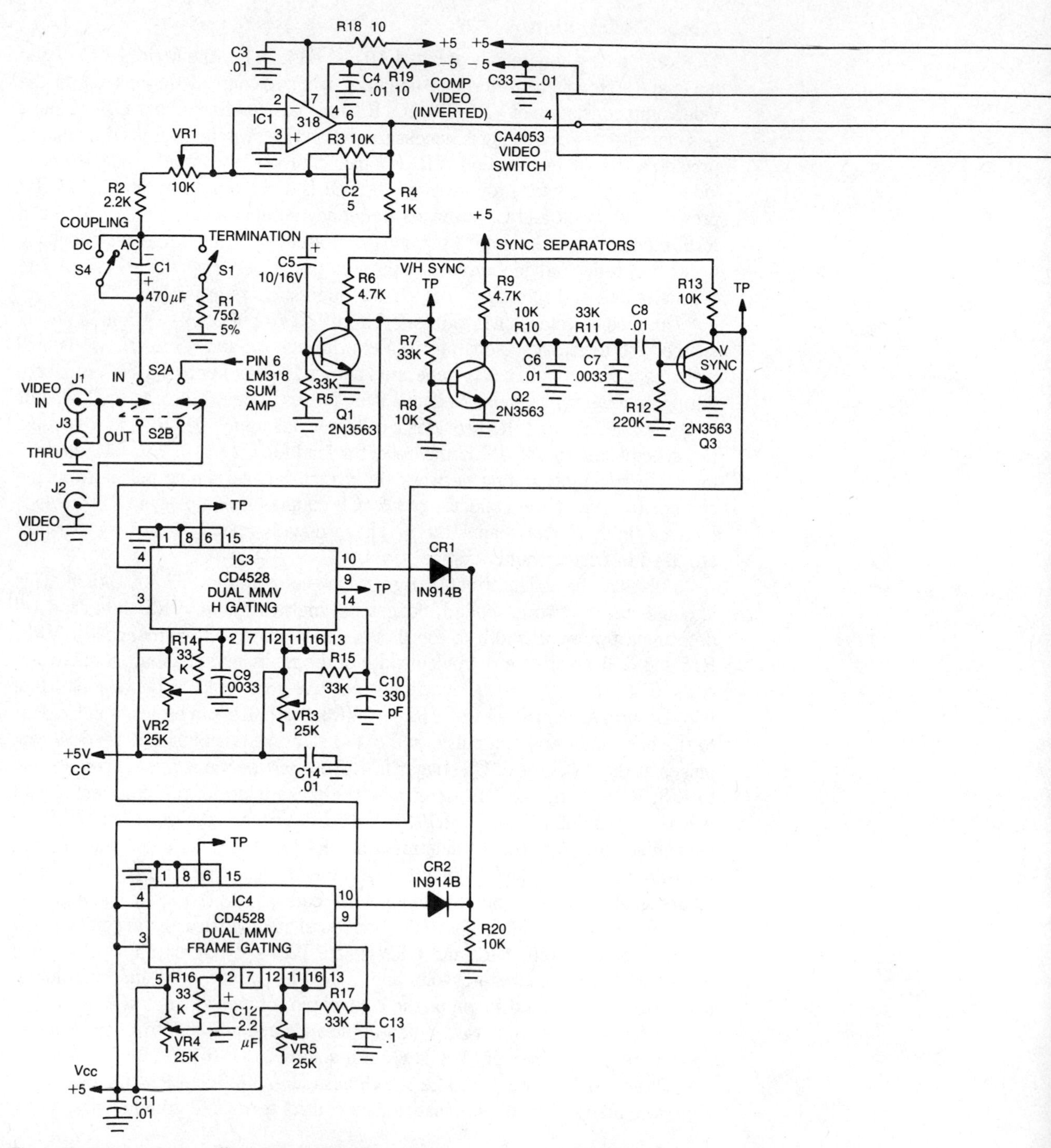

Fig. 4-10. Video palette—effects board schematic.

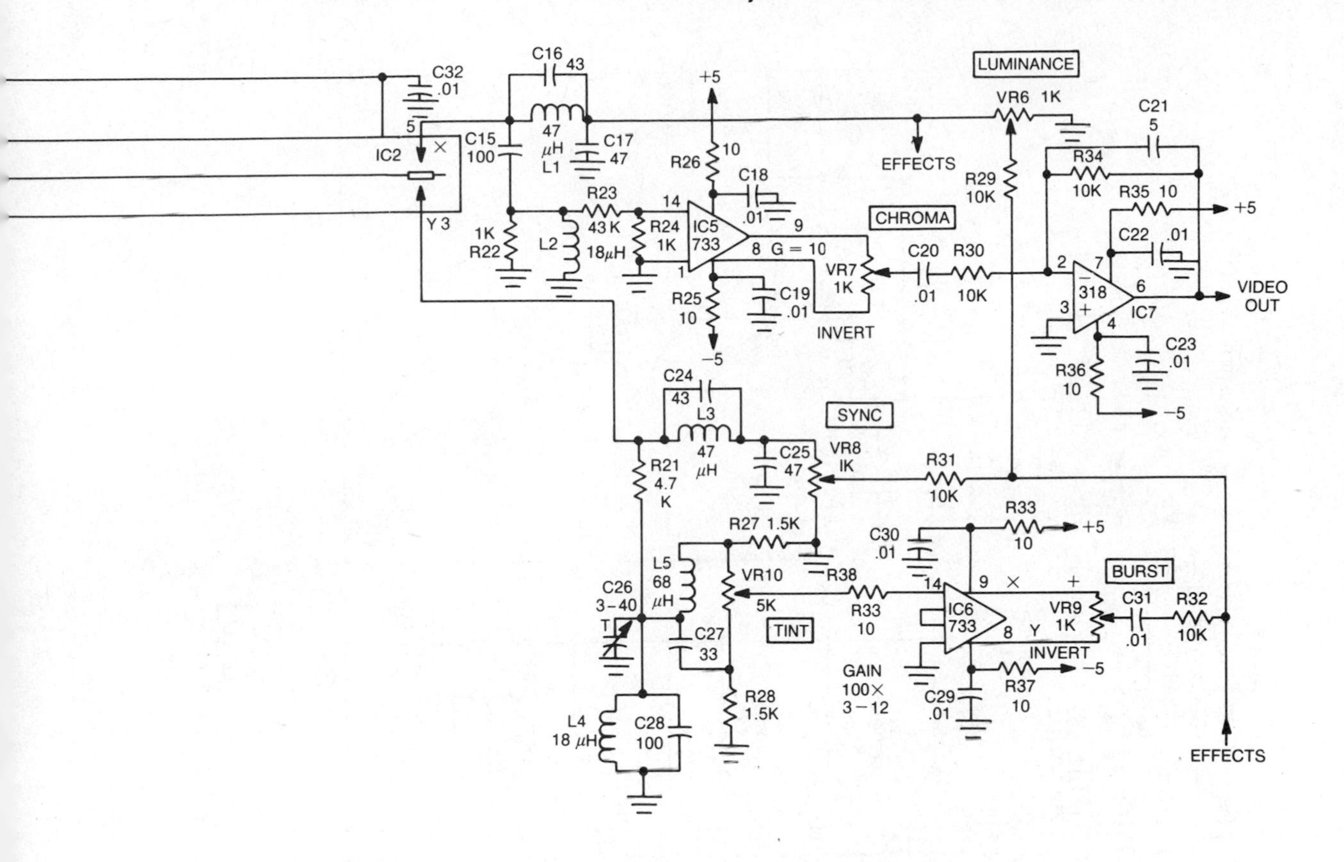
C16 43
C32 .01
+5
LUMINANCE
VR6 1K
C21 5
IC2
C15 100
47 μH L1
C17 47
R26 10
EFFECTS
R29 10K
R34 10K
R35 10
+5
Y 3
1K R22
L2
R23 43 K
18μH
R24 1K
IC5 733
C18 .01
9
CHROMA
C22 .01
8 G = 10
VR7 1K
C20 .01
R30 10K
318
IC7
VIDEO OUT
R25 10
C19 .01
INVERT
C23 .01
R36 10
−5
−5
C24 43
L3 47 μH
SYNC
C25 47
VR8 IK
R21 4.7 K
R31 10K
R33 10
+5
C30 .01
R27 1.5K
L5 68 μH
VR10 5K
R38
R33 10
IC6 733
BURST
VR9 1K
C31 .01
R32 10K
C26 3−40
TINT
C27 33
INVERT
−5
GAIN 100× 3−12
C29 .01
R37 10
R28 1.5K
L4 18 μH
C28 100
EFFECTS

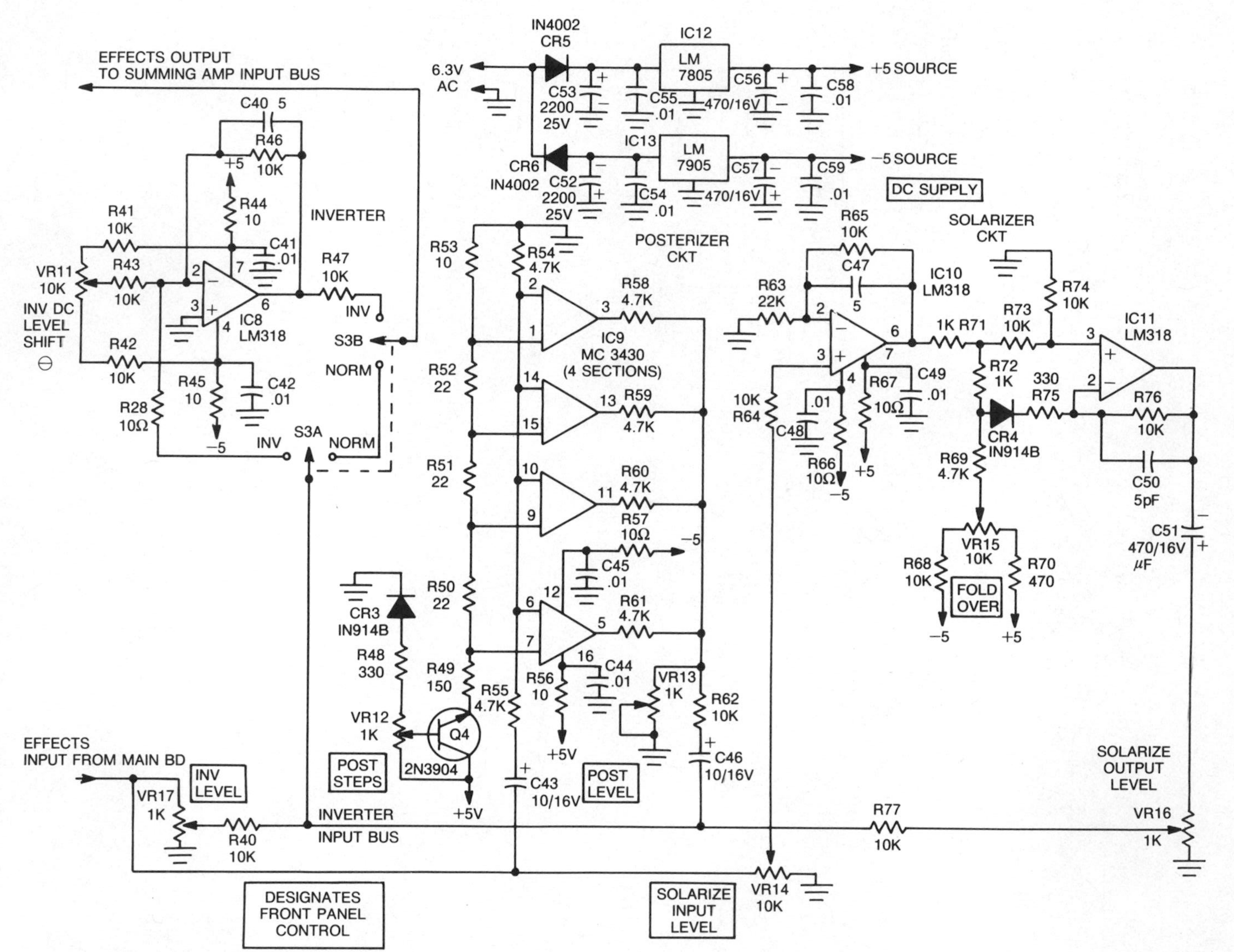

Fig. 4-11. Video Pallette Effects board schematic.

Video from VR6 wiper goes to IC7, the summing amplifier, through R29. IC5 is a chroma bandpass amplifier. It has a nominal gain of 10. R26, C18, R25, and C19 supply ±5 to IC5 and decouple the supply lines. Chroma appears at pins 8 and 9 if IC as equal level, 180° out of phase signals. VR7 is the chroma gain or level control. Either positive or negative chroma signal can be supplied to IC7 through R31 depending on the setting of VR7.

During sync intervals, pin 9 of IC2 is high sync, burst, and blanking appear at pin 3. C24, L3, and C25 form a LP filter feeding sync and blanking to VR8, the sync level control. The wiper of VR8 feeds C20, R31, and IC7, the summing amplifier. R21, C26, L4, and C28 are used as a burst take off filter. C26 is adjusted so the tint control circuit (L5, C27, R27, R28, and VR10) produces correct tints when BR10 is set at center. The burst from VR10 wiper goes to IC6, a burst amplifier with a gain of 100 (there is considerable loss in the tint control circuit). VR9 controls the burst level. The wiper of VR9 goes to C31, R32, into IC7. Processed video from the effects board is fed to IC7 input pin 2 IC7, R34, C21, R35, R36, C22, and C23 make up a unity gain summing amplifier. It is similar in its operation to IC1 except there is no gain control. IC7 reinverts the video so it appears at J2 as negative sync, 1 volts peak-to-peak NTSC (nominal) up to 2 volts into 75 ohm of video is available, depending on individual control settings.

The effects board receives its input from the top of VR6. Luminance is supplied to VR10 and VR14. Video is applied to the posterizer circuit through CR43 and R55. R54 provides a ground return for the comparators in IC9. Q4 supplies an adjustable reference bias determined by the setting of VR12 to a divider network consisting of R49, R50, R51, R52, and R53. R48 and CR3 provide temperature compensation of the reference voltage. Comparator outputs R58 through R61 are summed across level control VR13 and appear on the inverter input bus to S3A (norm/invert) IC10 and IC11 form the solarizer circuit previously discussed. IC10 is an amplifier with a gain of four. The input signal is taken from the wiper of VR14 through R64. The amplified video (up to 4 volts peak-to-peak) appears at pin 6 of IC10. This feeds IC11. VR15 is the ''foldover'' control. R68 and R71 limit the range of the control VR15 for easier operation. Solarized video is fed through C51 to VR16, the solarizer output level control. The wiper of VR16 feeds the inverter input through R77. Unprocessed video luminance is fed to this bus from VR10 through R40. S3A is used to switch in the inverter circuit consisting of IC8 and its peripheral components. You may have noticed by now that the circuits using the LM318 are all very similar, hence we are not discussing them in detail except where significant differences are encountered. VR11, R41, R42 and R53 feed an adjustable dc offset to the inverter IC8 so as to maintain correct dc baseline levels when inversion is used. R47 feeds inverted output to IC7, the summing amplifier, through switch S3B (Inverter in-out). As in the other amplifier circuits using the LM318, a 10k ohm feedback resistor and 5 pF shunt capacitor (R46 and R40) are used to set the gain and provide frequency compensation.

CR5, CR6, and C52 through C55 form two half-wave rectifiers supplying ±8 to ±16Vdc to IC12 and IC13. These two voltage levels those that result when a 6.3 or 12.6Vac transformer is used to supply the ac input. It is preferable to use 6.3Vac as this minimizes dissipation in regulators IC12 and IC13. IC12 provides +5 volts out and IC13 provides −5 volts. C56 through C59 are filter and bypass capacitors for the supply lines. About 300 mA ac current is required. Suitable transformers are available from sources such as Radio Shack or Mouser Electronics. A small 6.3V plug-in wall transformer can be used. Make sure that the output is ac, *not* dc.

Construction and Alignment

This device may be constructed from a kit of parts, which provide all the materials necessary for the two PC boards. A suitable cabinet is the Radio Shack P/N 270-274. Knobs, switches, jacks, plugs, etc., are at the discretion of the builder and are not supplied with the parts kit. Figure 4-12 shows the PC artwork for the main board. Figure 4-13 shows the artwork for the effects and power supply board. The boards may be etched on single sided .031 or .062 material phenolic or fiberglass epoxy G10 (preferred). They are single sided, which is much easier to use for home construction. A few jumpers are required, but the ease of fabricating a single sided board more than makes up for this. Figures 4-14 and 4-15 show parts placement. The effect board is not necessary for the video palette if only video correction capability is needed. In this case, the +5 and −5 supplies can be hard-wired to an external piece of G-10 board, or the power supply parts can be chassis mounted. However, the special effects capability is well worth the extra PC board and components. Note that the special effects board will *not* work by itself, since it requires the main board to separate out the luminance video components.

Construction should be started by first stuffing the PC boards. Resistors first, then inductors, then capacitors. Finally, install controls, transistors, and IC devices. Lead lengths of connecting wires are not critical but should be as direct as possible. Leads carrying video signals at a 75 ohm level (input and output) should be coax if possible. The leads carrying video signals at higher impedance such as those to and from the effect board, should be dressed away from grounded metal or power supply leads, to reduce both stray capacitance and induced 60 or 120 Hz hum pickup. Control shafts for all the front panel controls should be strain relieved. This can be done by passing them through holes in the front of the cabinet that are about .005″ larger than the shaft diameter (nominally 1/4″). If desired, bushings can be used around the shafts. Plastic control shafts are adequate.

There are 11 front panel controls, a pilot light and three switches so do not crowd things too much, or operation will be somewhat difficult, unless you have very small hands. RCA phono plugs UHF, BNC or F type video connectors are suggested for external interface. Switches can be the small 1/4″ hole mount

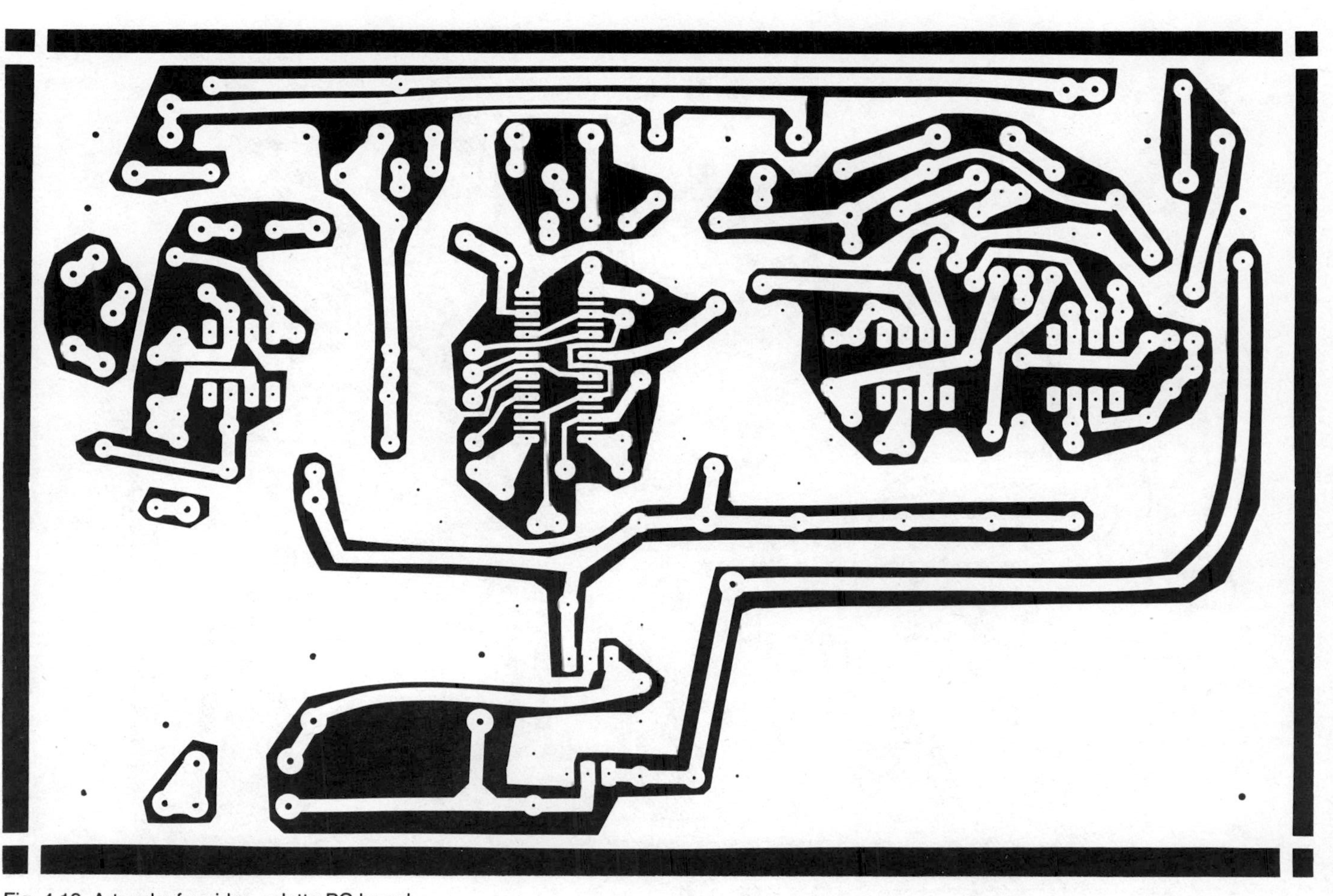

Fig. 4-12. Artworks for video palette PC board.

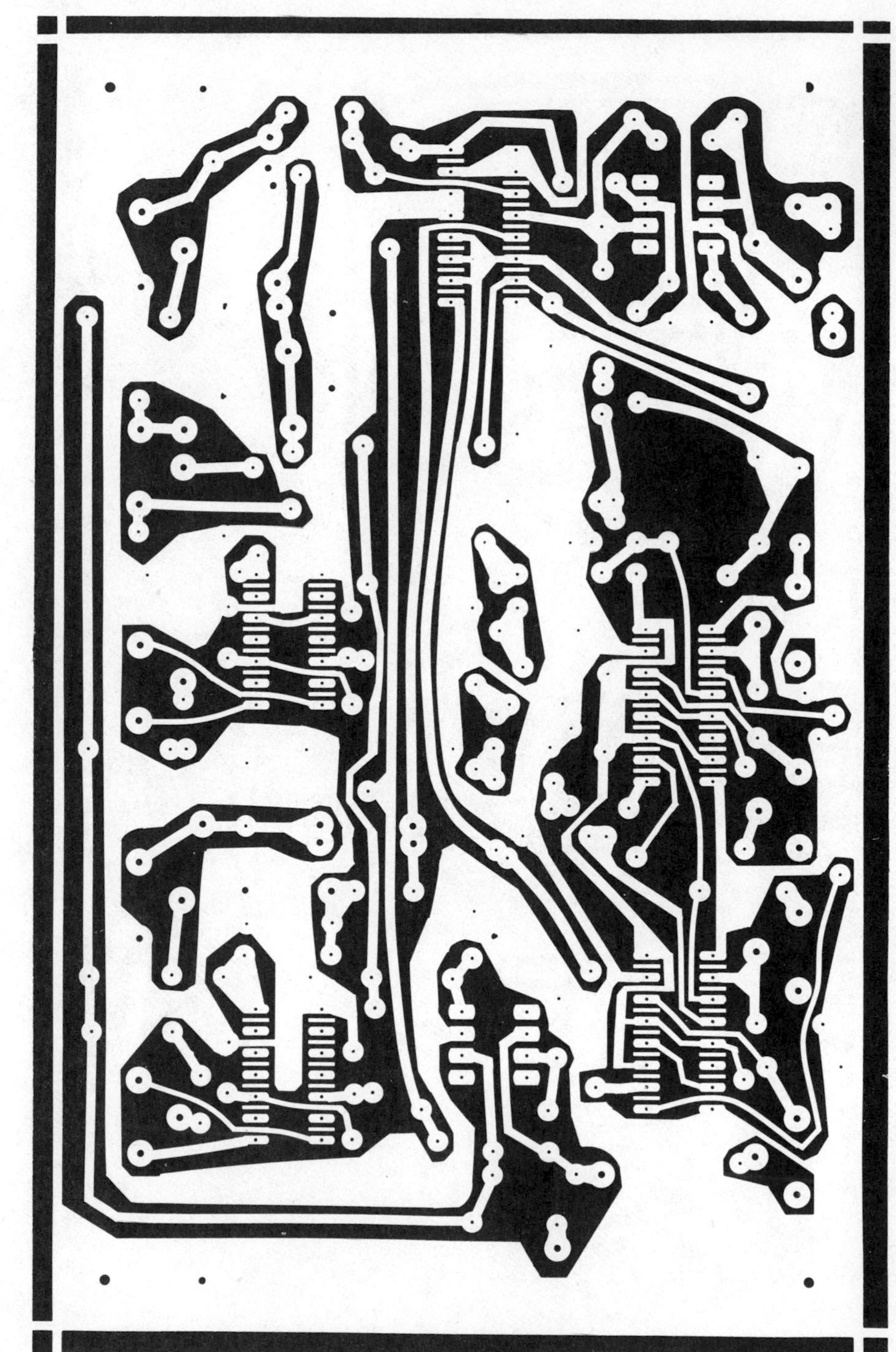

Fig. 4-13. Video palette mainboard PC artwork.

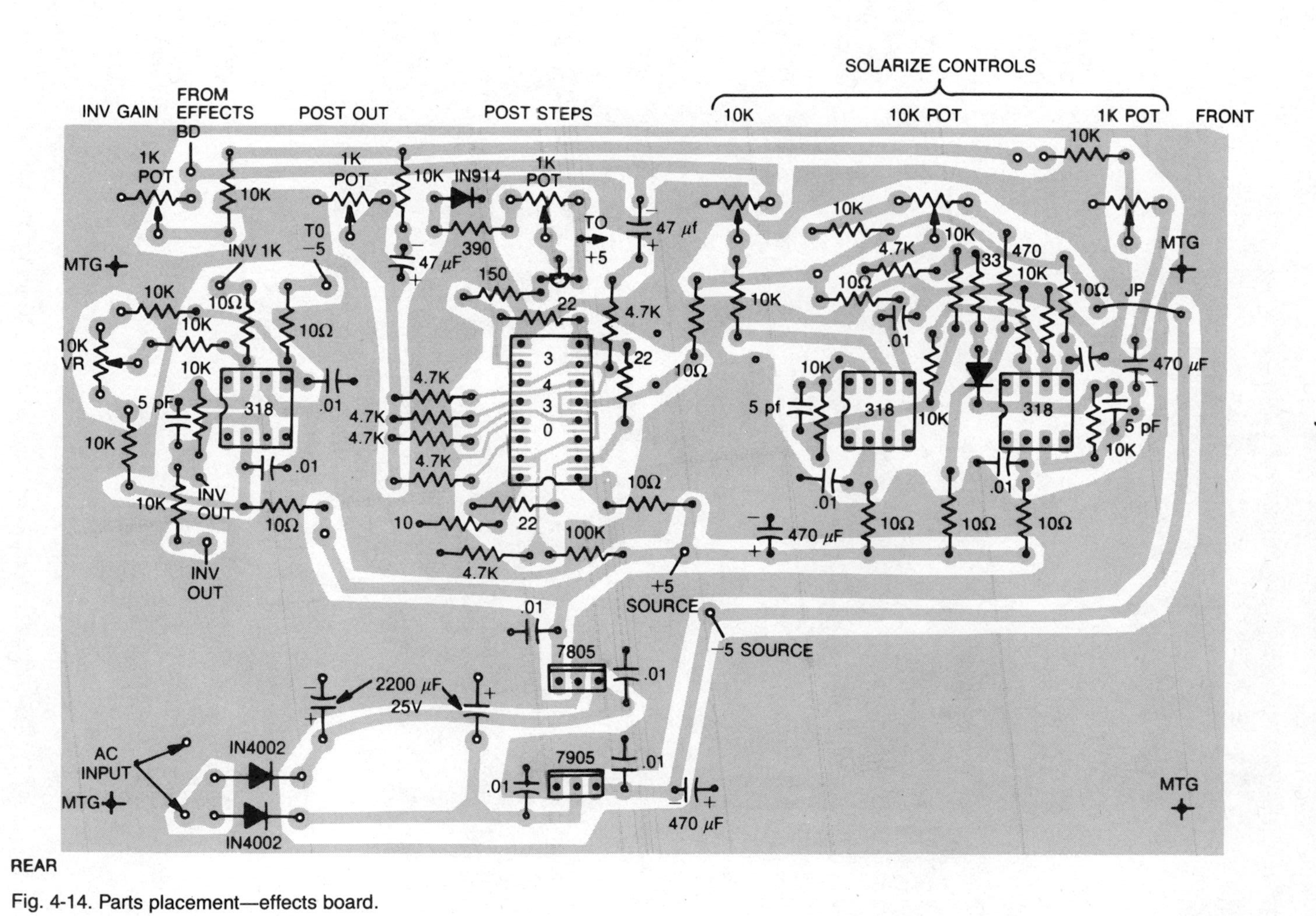

Fig. 4-14. Parts placement—effects board.

Fig. 4-15. Video palatte boards mounted in cabinet.

imported toggle switches. Your Radio Shack or Mouser Electronics catalog contains many types of suitable parts. If desired, the two PC boards can be "piggy-backed" or mounted end-to-end, depending on your choice of cabinet or mounting arrangements. Signal connections to and from the video palette should be 75 ohm coax cables with appropriate connectors.

Alignment is simple. If possible, use an oscilloscope (>5 MHz BW). If a scope is not available simply set VR1, 2, 3, 4, 5 and VR11 so that they are in the center of the range (midway). By understanding what is going on, you can use the TV monitor to adjust these controls. A scope does make the initial job easier, but do not let the lack of a scope discourage you. Final "tweaking" will be found easiest to do by watching the picture anyway.

Connect the video palette as in Fig. 4-16. At this point, you should have previously checked your wiring and PC boards for correct component insertion and pin orientations, unwanted solder bridges, and completeness. If any wiring or assembly errors exist, correct them before proceeding further.

Next, have a 20k ohm/volt or higher VOM set to read about 10 volts full scale. Connect the ground lead to the PC board ground foil. Apply power (120 Vac) to the power transformer primary. *Very quickly* check the voltages across C56 and C57—they should be 5 volts each (C56 has negative side grounded, and C57 has positive side grounded). Then *very quickly* make the following checks.

IC1	Pin 6	0V (±0.5 OK)
IC7	Pin 7	+5V
IC8	Pin 4	−5V
IC10		
IC11		

If okay, then do the next set of checks on IC5 and IC6. Be very brief but accurate.

IC5	Pin 5	−5V
IC6	Pin 6	+5V
	Pin 8, 9	0V (±1V OK)

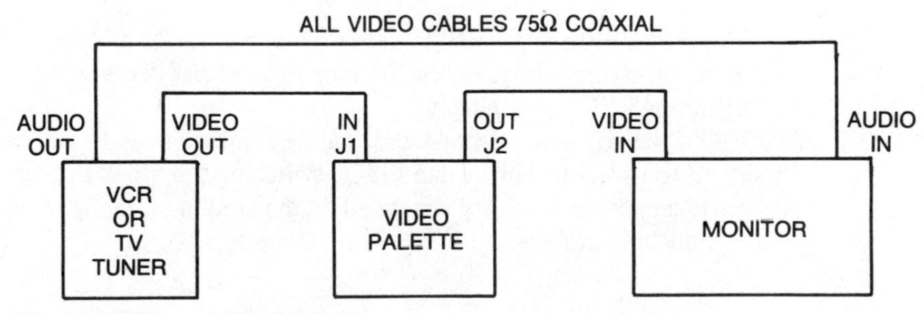

Fig. 4-16. Video palette—hook up.

IC3	Pin 10	0V
IC4	Pin 9	+5V
IC1	Pin 16	+5V
	Pin 8	0V
IC2	Pin 7	−5V
	Pin 16	+5V
IC9	Pin 16	+5V
	Pin 12	−5V
Q1	collector	+5V
	base	0V
Q2	collector	0V
	base	+.6V
Q3	collector	+5V
	base	0V
Q4	collector	+5V
	base	2 to 5 volts (depends on setting of VR12)
	emitter	0.6V less than base

NOTE: THESE VOLTAGES ARE WITH NO SIGNAL INPUT TO J1 or J3

Nothing should get hot—if it does, there is a problem. Locate and correct it before proceeding further. Refer to Fig. 4-17. Since nothing is now smoking, apply 1V peak-to-peak negative sync NTSC video to J1. Close S1 (75 ohm termination). Open S4 (ac coupling). Place S2 in the "IN" position. Adjust VR1 for 3V peak-to-peak at pin 6 IC1. Check collector Q1 for negative going pulses (Fig. 4-17). Collector Q3 should also exhibit negative going pulses. Next adjust VR2 for about a 53 microsecond pulse at pin 7 IC3. Then set VR3 for 10 microseconds pulse at pin 10 IC3. Adjust VR4 for a 16 microsecond pulse at pin 7 IC4. Then adjust VR5 for a 600 microsecond pulse at pin 10 IC4. If no pulse, readjust VR4 for a slightly narrower pulse width. If the 16 millisecond multivibrator is set for too long a pulse, no 600 microsecond pulse will be generated. So make sure this is the case before looking for "trouble" that is not there. Make sure that IC3 and IC4 circuits work properly as *everything* else depends on their operation. Refer to Fig. 4-18 for the following, as required, if no scope is available and a TV monitor is used.

1. Set VR6, VR8, VR9, and VR10 to center. You should get a black and white or weak colored image on the TV monitor. Set all effects controls (VR12, through VR17) at minimum.
2. Vary VR2—you will see a "transition" on the right (Fig. 4-18) and or left side of the screen. This is due to IC2 switching the video through the sync channels. If instability is noticed on the screen of the monitor, correct control operation.

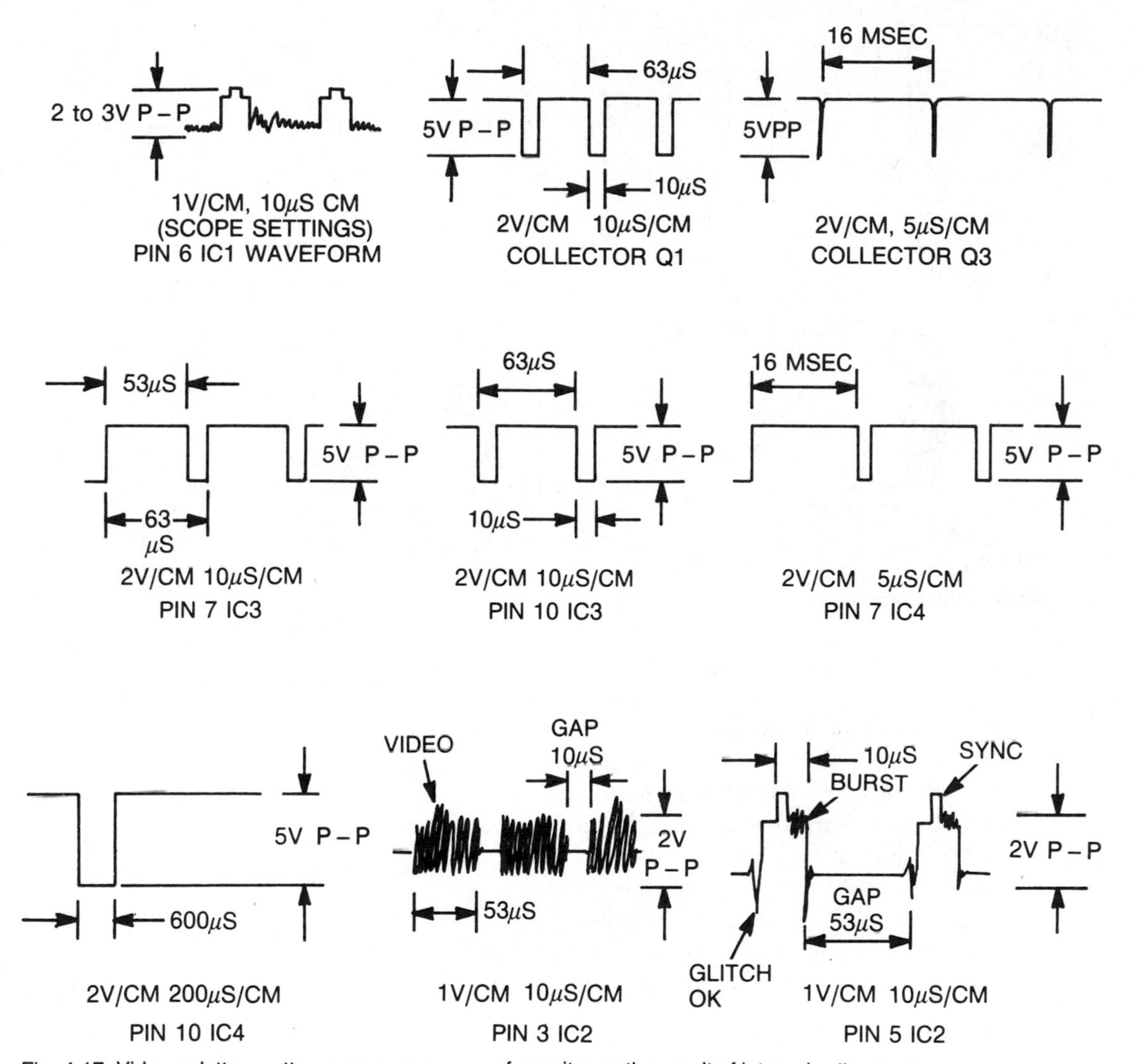

Fig. 4-17. Video palette—patterns seen on screen of monitor as the result of internal adjustments.

adjust VR8. Adjust VR2 and VR3 to move these transitions just off the right and left edges of the screen. The picture may roll vertically—this is okay for now.

3. Vary VR4 and VR5 for a stable vertical locked picture. VR4 and VR5 if set right, should result in a stable picture with no "transitions" on top or bottom.
4. Set all effects controls to zero. Set S2 to "OUT." Adjust the TV monitor for a normal picture. Now place S2 in the IN position. Check for

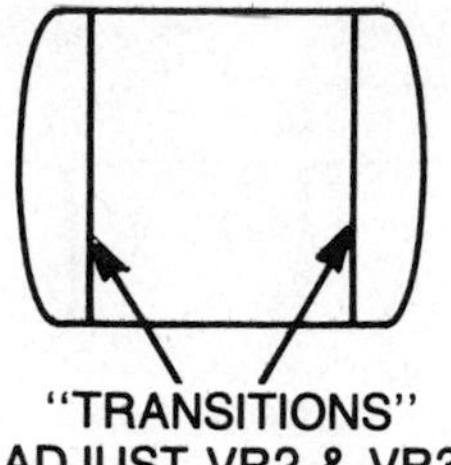

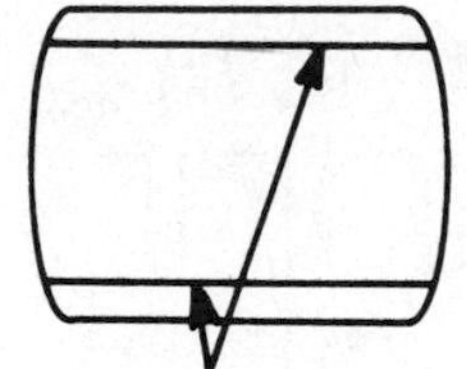

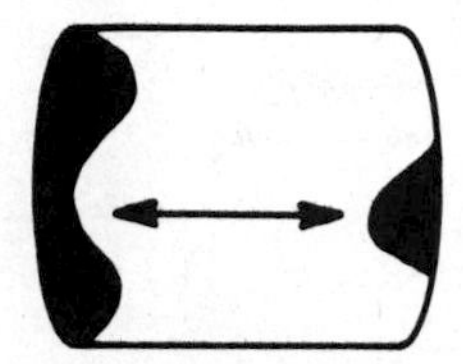

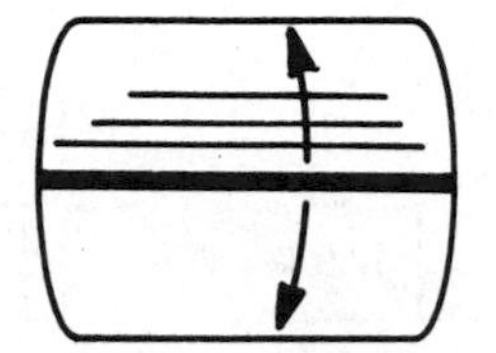

Fig. 4-18. TV monitor patterns.

VR6 should vary picture contrast (luminance).
VR8 should vary picture brightness. When VR8 is toward minimum, the picture should lose lock (sync).
VR9 may vary the color saturation and reverse colors (burst).
VR7 should operate in a similar manner to VR9 (chroma).
VR10 should vary the tint. Adjust C26 to produce normal tint with VR10 at center.

If you are using a scope, you can place the scope on IC7 pin 6 and observe the action of each control on the video signal. Next, if you are building the special effects board as well, this board has to be checked out.

5. Set up VR6 through 10 for a "normal" TV picture as in Step 4. Then turn VR6 to minimum. All effects controls should be at zero so far. You should have a plain raster with only splotches of color, or if black and white, just a raster with only a very weak, faded picture.
6. Set the inverter switch to "Normal" (S3). Rotate VR17 clockwise. The picture should return.
7. Place S3 in the "Invert" position. A negative picture should be seen.
8. Adjust VR11 for a satisfactory "negative" picture. You may have to touch-up VR8 (SYNC) at first. When VR11 is properly adjusted, VR8 can be left alone. Now place S3 in "NORMAL" position.

9. Rotate VR17 fully counterclockwise. Set VR12 and VR13 halfway. Observe the effect. You should see a posterized image. It will be obvious. Now vary VR12 and VR13 to see their effect. Finally, return VR12 and VR13 to zero (fully CCW).
10. Set VR14 and VR16 about halfway open. Slowly rotate VR15—you will see the solarization effect. Adjust VR14 and 16 as necessary for best results. VR8 may have to be adjusted (sync) at some settings. (By definition, ''best results'' is what you like best of all.)
11. Set S3 in both ''NORM'' and ''INVERT'' and observe the solarization effect as in Step 10.

This completes check out. The rest is up to you. A few hours of just plain experimentation is the best way to learn what this video palette can do. Undoubtedly, you will have a lot of fun with this device and find many applications for it. If you have curious children, they will really like to play with this video palette. It can be very interesting to view video tapes you are familiar with (or even sick of watching), with the video palette between your VCR and monitor.

The authors are certain that you will not be sorry you built this unit. Happy viewing. See Fig. 4-19.

VIDEO PALETTE EFFECTS BOARD

RESISTORS

R40, R41, R42, R43, R46, R47, R62, R64, R65, R68, R73, R74, R76, R77	10k
R44, R45, R53, R56, R57, R66, R67	10 ohm
R48, R75	330 ohm
R49	150 ohm
R50, R51, R52	22 ohm
R54, R55, R58, R59, R60, R61, R69	4.7k
R63	2.2k
R70	470 ohm
R71, R72	1k

POTS

VR11	10k TW
VR12, VR13, VR16, VR17	1k shaft
VR14, VR15	10k shaft

SEMICONDUCTORS

CR3, CR4	1N914B
CR5, CR6	1N4002

CAPACITORS

C40, C47, C50	5pF
C41, C42, C44, C45, C48, C49 C54, C55, C58, C59	.01 disc
C43, C46	10μF/16V
C51, C56, C57	470μF/16V
C52, C53	2200μF/25V

ICs

IC8, IC10, IC11	LM318N
IC9	MC3430
IC12	LM7805
IC13	LM7905

MISCELLANEOUS

PC board — Effects

Fig. 4-19. Parts list, Project 8.

VIDEO PALETTE MAIN BOARD

RESISTORS

R1	75 ohm or 82 ohm
R2	2.2k
R3, R8, R10, R13, R20, R29, R30, R31, R32	10k
R4, R22, R24	1k
R5, R7, R11, R14, R15, R16, R17	33k
R6, R9, R21	4.7k
R12	220k
R18, R19, R25, R26, R35, R36, R33, R37, R38	10 ohm
R23	3.3k
R27, R28	1.5k

CAPACITORS

C1	470/16V elec.
C2, C21	5pF SM
C3, C4, C6, C8, C11, C14, C18, C19, C20, C22, C23, C29, C30, C31	.01 disc
C5	10μF/16V elec.
C7, C9	.0033 Mylar
C10	330pF SM or NPO
C13	.1μF Mylar
C12	2.2μF Tantalum
C15, C28	100pF SM
C16, C24	43pF SM
C17, C25	47pF SM
C26	3 – 40 trimmer
C27	33pF SM

POTS

VR1	10k TW
VR2, VR3, VR4, VR5	25k TW
VR6, VR7, VR8, VR9	1k shaft
VR10	5k shaft

SEMICONDUCTORS

CR1, CR2	1N9143
Q1, Q2, Q3	2N3563

ICs

IC1, IC7	LM318N
IC2	CD4053
IC3, ICr	CD4528
IC5, IC6	LM733N

COILS AND CHOKES

L1, L3	47μH
L5	68μH
L4, L2	18μH

MISCELLANEOUS

PC board — Main

The following are available from:

North Country Radio
P.O. Box 53, Wykagyl Station
New Rochelle, NY 10804

1. Main Board (PC only) $12.50
   Main and Effects Boards (PC only) $25.00
   A kit of parts consisting of the printed-circuit board(s)
   and all parts that mount on the board(s).

2. Main Board and all parts that mount on board $49.95
   Main and Effects Boards and all parts that mount on boards $84.95

   Add $2.50 per order for P & H.

   New York residents must include sales tax.

   NOTE — IC sockets used on prototype boards not supplied,
   left to builder discretion

   NOTE — Effects Board sold only in conjunction with Main Board.

Fig. 4-19. Project & parts list.

PROJECT 9: VIDEO CONTROL SYSTEM

In video work, the need for switching between two or more video channels often arises. In particular, a video program may be in the progress of assembly where material has to be sequentially put together. The problem now arises as how to make the transitions between scenes or between sources as smoothly as possible, without visually or esthetically disturbing transitions. The system to be described will assist in this process.

In order to switch between video channels with a minimum of disturbance, several technical requirements must be met.

1. Sources must be identical in polarity and as to type i.e., both NTSC, negative sync, for example. In practice this is usually met, as it is necessary for interfaces between various other equipment.
2. Sources must have the same voltage impedance levels. This can be met with matching gain adjustments.
3. The time phases of the sources must be constant and have a fixed relationship. The sync pulses must coincide both in time of occurrence and frequency, both vertical and horizontal.
4. Color burst phase must match in order to reduce color shifts between scenes. This can be accomplished by suitable matching devices, such as delay lines, matching of cable lengths, etc.
5. Terminations and impedance matching must be considered in order to reduce reflections and ''ghosting.''

Most of the time there is no problem in meeting items a, b, d, and e. These are under direct control of the system operator. However item c, the requirement that video sources have sync pulses in phase sometimes presents a problem. For example, the use of two separate VCRs, or a VCR and a camera, VCR and over the air program present a problem in the fact that generally there is no fixed phase relationship between sync phases. The term ''genlock'' is used to describe the act of using a master syncronization source to control the sync phase of other sources. Some video equipment has genlock inputs which allow this, but most of the time, for home video work, the availability of two genlocked sources cannot be relied on.

When a video monitor, TV receiver, or VCR is suddenly switched from one source to another the syncronizing circuits of the video device under consideration experience a discontinuity of input. The discontinuity may be in frequency, phase, or both, (they are related) depending on the moment of switching. If, on chance, the vertical and horizontal sync pulses of both sources are coincident in time (in phase) at the moment of switching, there will be no noticeable disturbance. If, however, they are not (the usual case), a momentary loss of syncronization may occur. Depending on the characteristics of the sync system in the video device that is in question, a momentary flicker, jump, tear, or roll will occur in the picture. This is objectionable and generally esthetically undesirable. It gives an "amateur" look to a program, and should be eliminated. If this is not possible, it can be made part of the transition, and done during a certain sequence where it is least noticeable.

A common way to deal with this problem is to fade the picture to a level, usually black. The video fades out gradually to a black level. During the black interval, switching takes place. Since the screen is black, transient effects are less noticeable. After a predetermined time, the new video is switched in and then the fade from black to the program is performed. In some cases, no fade is used, but this can be at the discretion of the producer.

There are other methods that can be employed. A black level can be "keyed" into the picture. This shows up as, for example, a black level wiping over the picture, much like a curtain, either horizontally or vertically, or both (diagonally). Also, the black level can be broken up like a series of strips that gradually enlarge, covering the picture. This gives the effect of a venetian blind, either horizontally or vertically. By doing this both vertically and horizontally at the same time, dots which are black appear in the picture and expand in size to first overlap and then completely obscure the picture. Figure 4-20 shows these patterns. Many other patterns of course are possible.

The act of "keying" is actually video switching using waveforms synchronous in some way with the sync pulses or other elements of the picture, such as

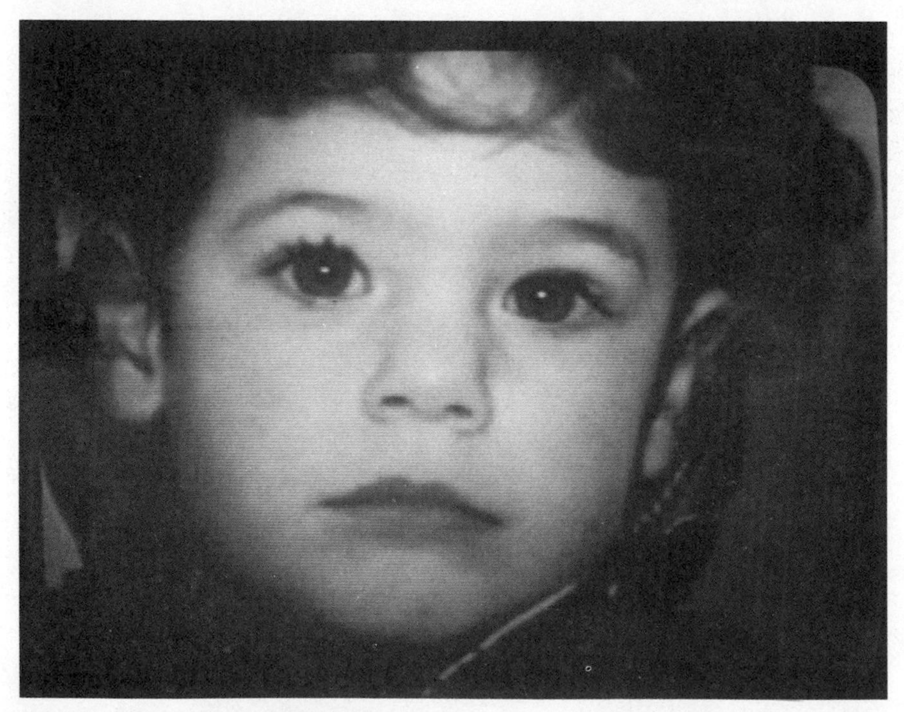

Fig. 4-20. Picture—before keyed fades.

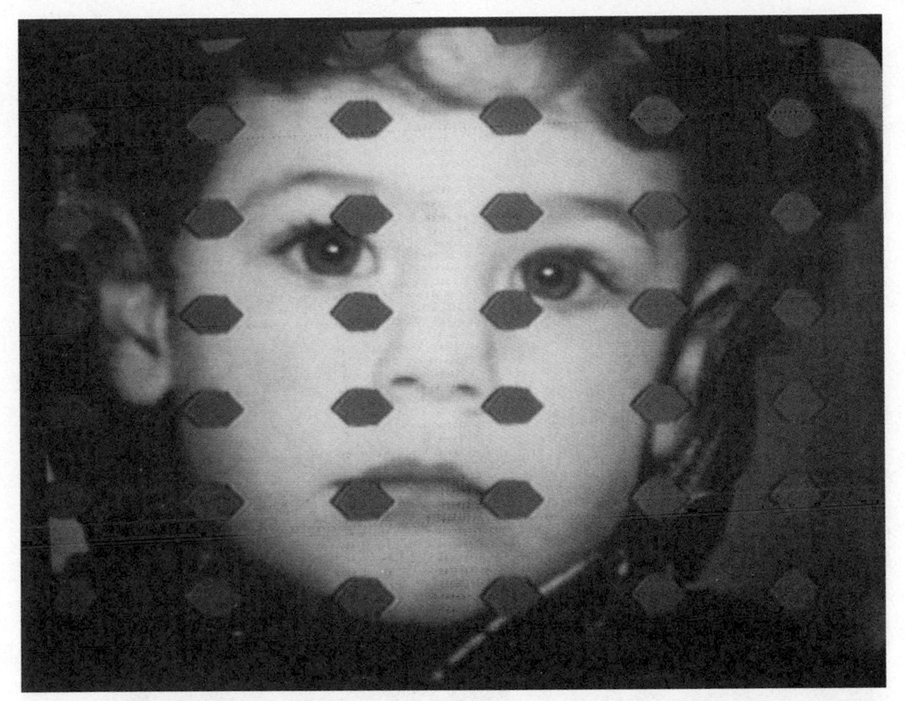

Fig. 4-21. Picture—start of keyed fade to black.

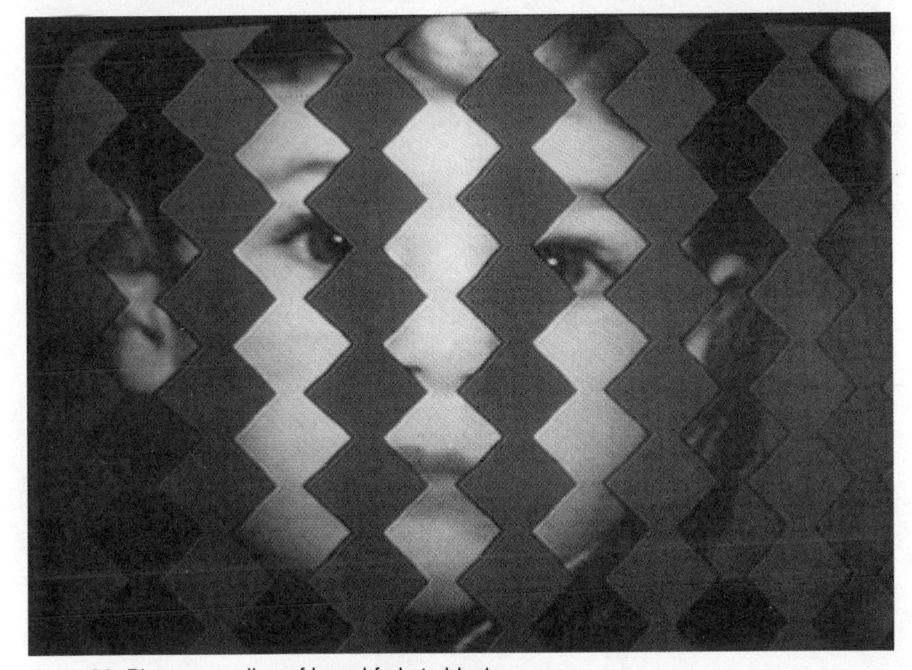

Fig. 4-22. Picture—ending of keyed fade to black.

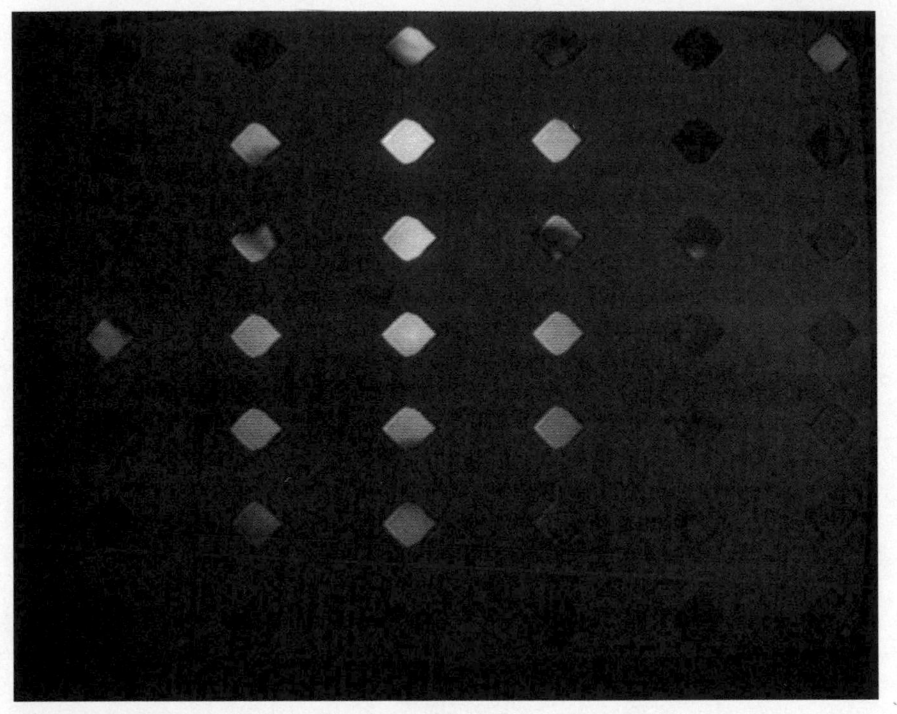

Fig. 4-23. Picture—middle of keyed fade.

the luminance level (luminance keying) or chroma (chrominance keying). By producing waveforms that are syncronized in some way to the picture, a great variety of switching and special effects can be produced.

Note, that these effects are performed solely on the video. The sync pulses must remain unaltered during the switching process.

For wipes, keying, or other switching between two sources without an intermediate fade, the two sources must be genlocked or synchronous. There is no easy way around this, save for a large video buffer memory, or some form of syncronizing system employing storage. However, this is and should not be a serious limitation, as a practical consideration, since many techniques employing a fade to black intermediate method are very pleasing as to result and effect, and in some cases actually preferable, since it gives a more defined differentiation between scenes.

In order to perform operations on a video signal certain precautions must be observed. Most of all, the restrictions of maintaining system levels, such as sync, blanking, black and white levels, and burst level and phase must be dealt with. Figure 4-23 shows a typical NTSC signal. Within limits of keeping maximum black and white levels we can do anything we please within system limitations during the interval called ''line scan.'' This interval contains the elements

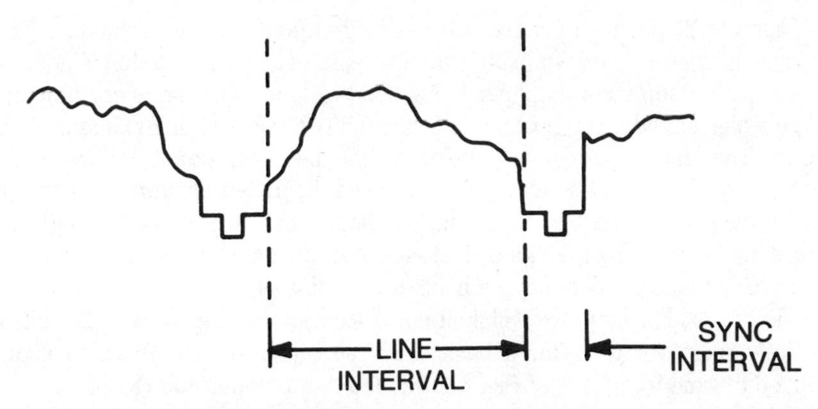

Fig. 4-24. Normal NTSC video waveform.

that form the picture. However, during sync intervals, nothing much can be disturbed without causing problems. This generally can be accomplished by separating the video and sync components, so that each can be operated on separately, after which the components can be recombined. Figure 4-24 shows this technique. A video signal is fed to the input (X) of a SPDT switch. The SPDT switch is electronic and is capable of switching speeds of several tens of nanoseconds or better. A switching signal, syncronized to the input video, controls the SPDT switch. During line scan intervals between sync pulses, the signal is routed to the upper switch pole (A). During sync intervals, the signal is routed to the lower switch pole (B). Therefore "syncless" video appears at (A), and raw composite sync, with no video information, appears at (B).

The SPDT switch can be a transistor, vacuum tube, or IC. The preferred method uses an analog switch IC. The CD4053 device, a CMOS IC, made by several manufacturers, is ideal for this application. It has adequate switching speed and low internal resistance, and good isolation. It handles 1 volt p-p 5 MHz BW video with little crosstalk. The CD4053 contains three independent SPDT switches in one 16 pin DIP IC package and is easily available and low in cost.

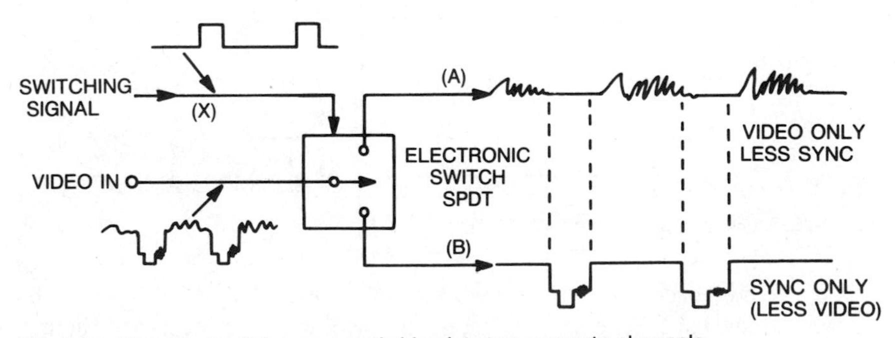

Fig. 4-25. Method of splitting sync and video into two separate channels.

Figure 4-25 shows a general processing system using this technique. Note that sync is unprocessed. In many cases, the use of an output switch is not necessary, with a summing amplifier being used instead. The summary amplifier can be a high speed op amp run at low gain (unity to several times) and must be able to drive the output (video) interface. For the discussion to follow, unless noted, we will assume the sync is unprocessed, being left unaltered in any way.

Suppose we wanted to "fade out" a picture of a scene. We can fade the picture out by reducing the video level such that the image loses contrast and/or brightness, gradually becoming a blank raster of white, gray, black, or even a specified color. For now, we will limit our discussion to fades to white or black.

Referring to Fig. 4-26, a basic fader circuit is shown. A video signal, stripped of sync (output A of Fig. 4-25) is fed to the upper end (X) of the fader control R_F. The wiper arm is connected to a buffer amplifier (ideally, gain = 1 (unity), and Z in = infinity and Z out = zero). This practically may be an op amp hooked up as a voltage follower. A variable dc level of zero to 1 volt is connected to the other side of the pot. If the video input source has a low impedance (lesser than 20 ohm or so) and the resistance of R_F is high (greater than 1000 ohms, when the wiper arm Z is at the top of the pot (X), the variable dc level has negligible effect on the signal and the output of the buffer amplifier ideally is the signal at X. If the wiper arm is at the bottom (Y) of the pot, the output of the buffer is equal to the dc level at Y. At in between settings, there is a mix of the dc level and the video signal. For example, at the center setting of the control, we would have a dc component of half that of the supply voltage and half of the video signal. By turning the control from X to Y, we have gradually replaced the (pulsating dc) video signal with a constant +0.5Vdc level, which represents a blank raster of a gray tone. If this level had been 1 volt, we would end up with a white raster. If 0 volts, we would end up with a black raster. Therefore, by varying both R_F and the dc level, we can fade, at will, any picture out to any shade between white and black.

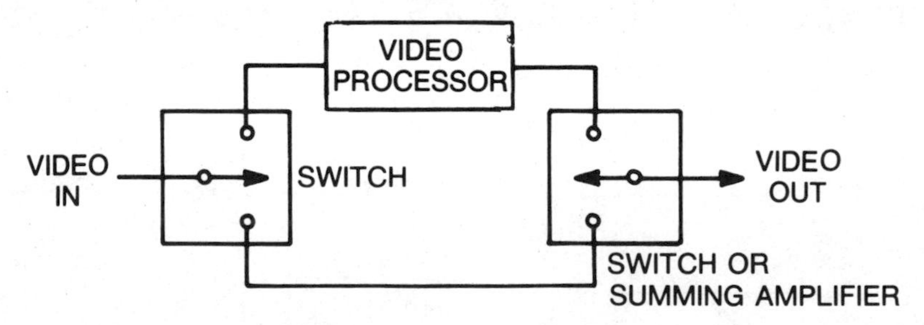

Fig. 4-26. Basic fade to level circuit.

If a colored background is desired, we can add a 3.58 MHz sinewave to the variable dc level at Y. The saturation (mixture with white or black) and the hue (red, orange, yellow, etc.) can be determined by the level and relative phase (with respect to the burst) of this signal. For simplicity, we will not go into this any further, as the extra complexity is not really justified unless a fade to a color is absolutely necessary. It is mentioned in passing so the reader will know how it can be done if desired.

Suppose we want, instead, to "wipe" a colored area, like a curtain or a shade, over the picture, either from the left or the right, instead of a fade. With video switching of a little more complexity this can be done. Refer to Fig. 4-27. Here the situation is different, but similar principles are involved. Assume a black curtain (horizontal) or shade (vertical) is wanted. What we do is send the video signal through a SPDT switch and a buffer amplifier. Also connected to the SPDT switch is a variable dc level, which determines the "color" of the curtain or shade (black, gray, white, or other if a 3.56 MHz signal is available).

Now, say we want the wipe to be a curtain that travels from left to right gradually across the screen, and it is to be gray in color. Video is sent to the video switch terminal X. At first, no switching (hereafter to be referred to as "keying signal") signal is present to the switch, and video passes from X to Z. Now suppose that, for a very short time after each sync pulse (horizontal in this case), the video switch is keyed so the nominal +0.5V level (required for gray) is switched into the buffer amplifier instead of the video at X. What will happen in a narrow gray border will appear at the left side of the picture. Suppose that gradually the time that the key signal is "high" is increased to half the line scan period (about 53 microseconds) or in other words, about 26 microseconds. At this point, the entire left half of the picture will be a blank gray raster, the right half normal video. Eventually the key signal can be high during the entire line scan. In this case, the whole raster will be a flat, blank gray. By slowly varying the time the key signal is high, the screen can be "wiped" over with the gray "curtain."

A "shade" effect can be obtained by synchronizing the key signal with the vertical sync signal instead of the horizontal signal.

By combining the two effects, a diagonal wipe can be achieved. By inverting the phase of the keying signal, we can reverse the direction of the wipe. In this case, we can replace the variable dc reference source with video input as shown in Fig. 4-27.

Suppose, rather than wanting to fade or wipe to a blank level, we wanted to replace the level with another video source or channel. For instance, a gradual dissolution of one scene into another, or a "split screen" (half one channel, the other half another channel). There is no basic difference since all the dc reference source serves is as a substitute video source corresponding to a single signal level.

By using a complex keying signal that is synchronized with one or preferably both video signals, many switching patterns can be obtained.

Suppose we switched several times during a line scan between a video channel and a dc level. A series of vertical "bars" would appear to overlap the picture. If we switched several times during a field, horizontal bars would cross

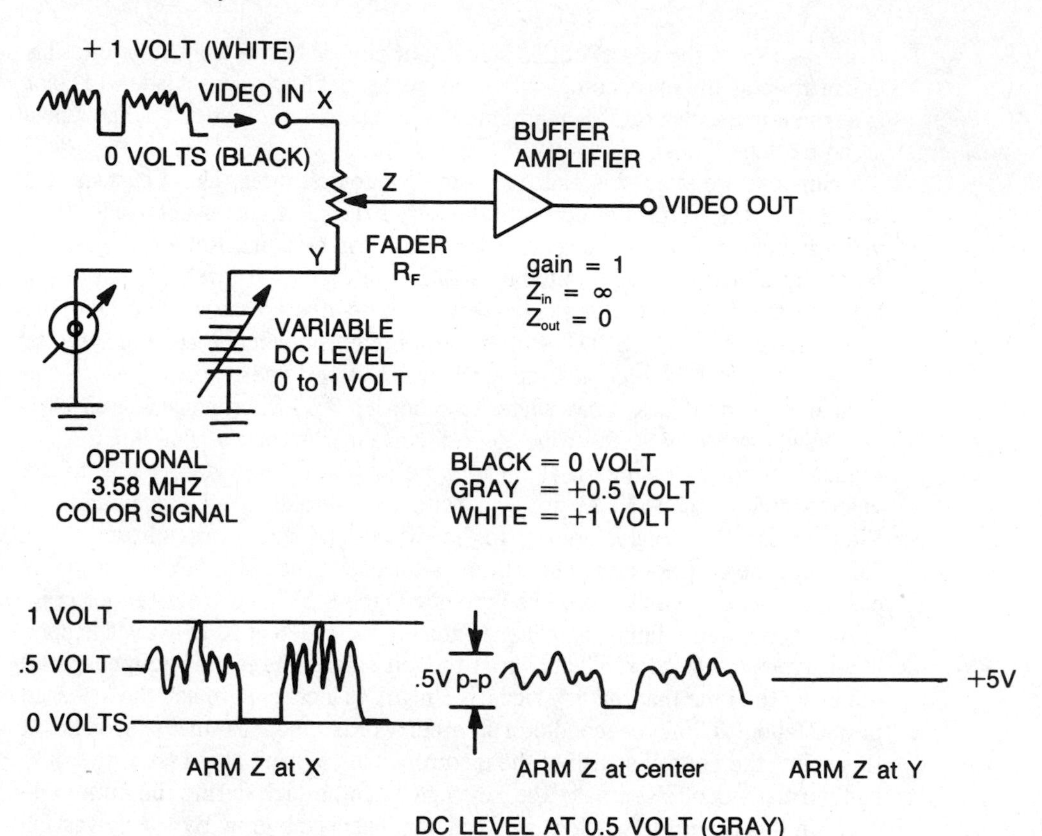

Fig. 4-27. Insertion of video processor. Sync pulses are not processed.

the picture, much like that of a venetian blind. A keying (switching) waveform that is a square wave several times higher in frequency than either the vertical or horizontal scan frequencies would be the correct waveform in this case. This waveform might be obtained from a PLL (phase-locked loop) circuit locked to a harmonic of either the horizontal or vertical rate, 15.7 kHz or 60 Hz respectively.

It may be evident why both video sources are usually required to be synchronous (gen-locked) in any switching operations. This ensures a smooth transition, free of rolling, tearing, or other problems. For split screen wipes, this is a must. For other effects, such as fades or wipes to black or white or other levels, or fades, wipes or keep to video effects such as produced by our Video Palette where one video channel is derived from the other, synchronism is assured since the same video sync source is used. As an example, the Video Palette can produce inverted or "posterized" pictures. By using the output of the Video Palette as a second channel, and for instance using the Video Palette to reverse the video to produce a "negative picture" we can wipe, key, or fade into and out of this. Figure 4-30 is a photo of this effect.

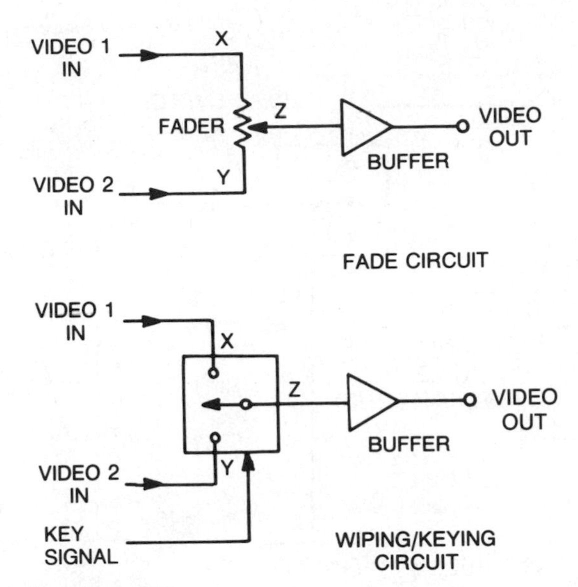

Fig. 4-28. Basic wipe circuit.

Also, many video cameras have a "genlock" input to syncronize them to an external source or other camera, VCR, etc. This is another way to guarantee two synchronous sources.

By the way, to clear up the differences between a wipe and a key, a "wipe" is generally an operation where one switching operation per line (or field) takes place. A keyed operation is where a complex waveform with multiple switchings per line, frame, or both takes place. Actually, wiping is simply a specific form of keying.

Another form of keying, requiring two synchronous sources, uses a second video source to key the main video source. A particular characteristic of (brightness component), or chrominance (color) component or even a certain pattern on the screen is used to key the first video source. Figure 4-31 is an example of how a man might insert his image in a scene. Simple superimposition would not really be satisfactory, since the man would appear somewhat "transparent" and the resultant picture would be a double image. In the case of Fig. 4-31, assume black level = 0 volts and white level = 1 volt. The video from the man scene can be anywhere from zero to one volt. By televising the man against a black background and if no really black clothes are worn by the man virtually all of the video component represented by the man will be above zero. If we arbitrarily define anything above 0.1 volt as the man and we feed the "man" video to a comparator set to trip at 0.1 volt, a key signal will result during the portions of those scan lines corresponding to those pixels making up the "man" image.

A "scene" image is also available. Video from the "scene" is fed to the video switch input X. "Man" video is fed to "Y." The output of the video switch is "Z." In the absence of a keying signal, the scene video appears at Z.

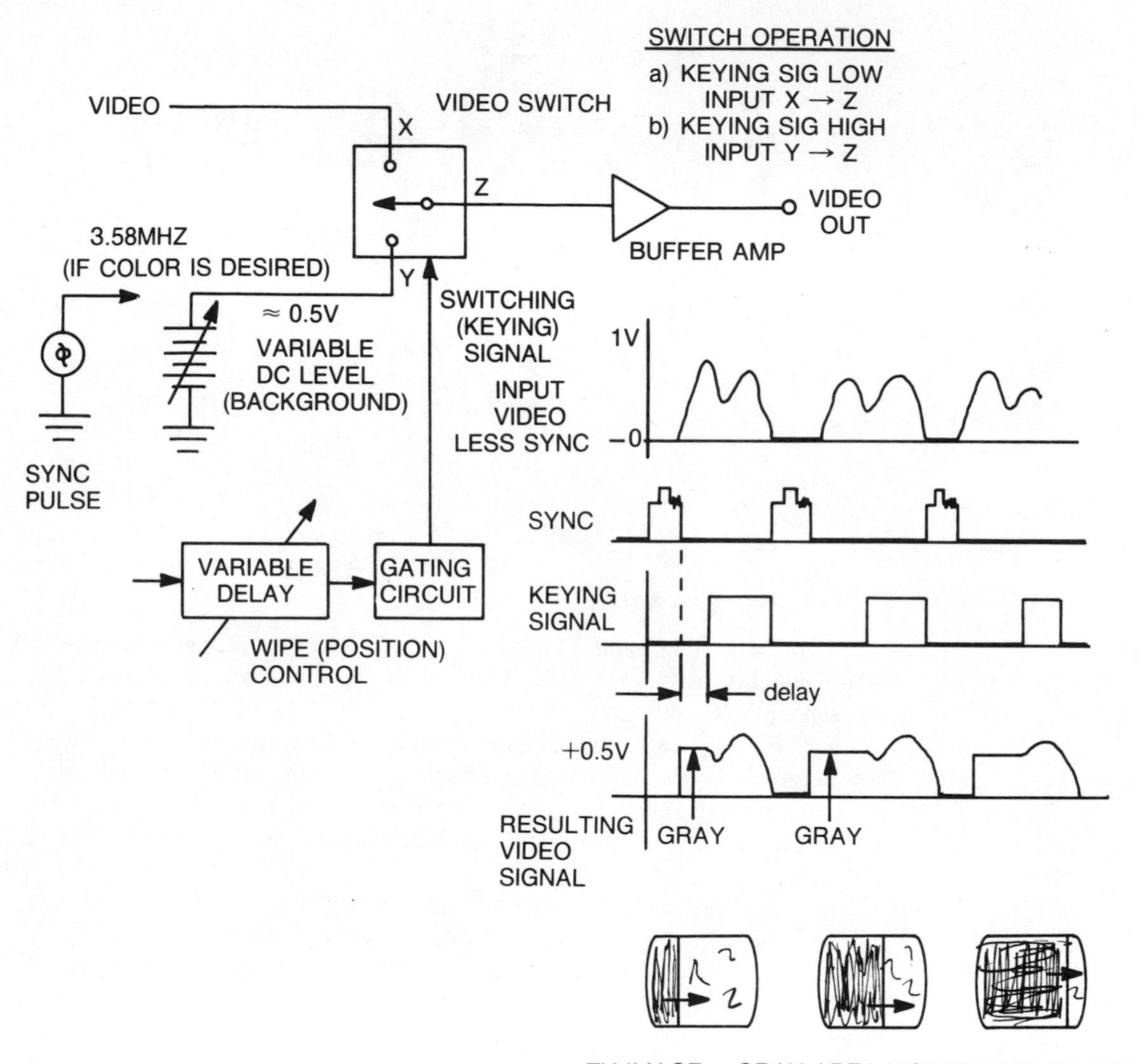

Fig. 4-29. Fading/wiping/keying between two video sources.

When the "man" video is above 0.1 volt in level, the video comparator outputs a key (switching) signal to the video switch. This causes the man video to appear at Z. This has the effect of making the man appear in the scene. By judicious camera placement, the man can actually "walk around" in the scene. By this method, a completely imaginary image can be produced with no danger to the man. This principle is used in TV production for both special effects and normal everyday things, such as news and weather programs, etc. Both sources must be synchronized for this "trick" to work.

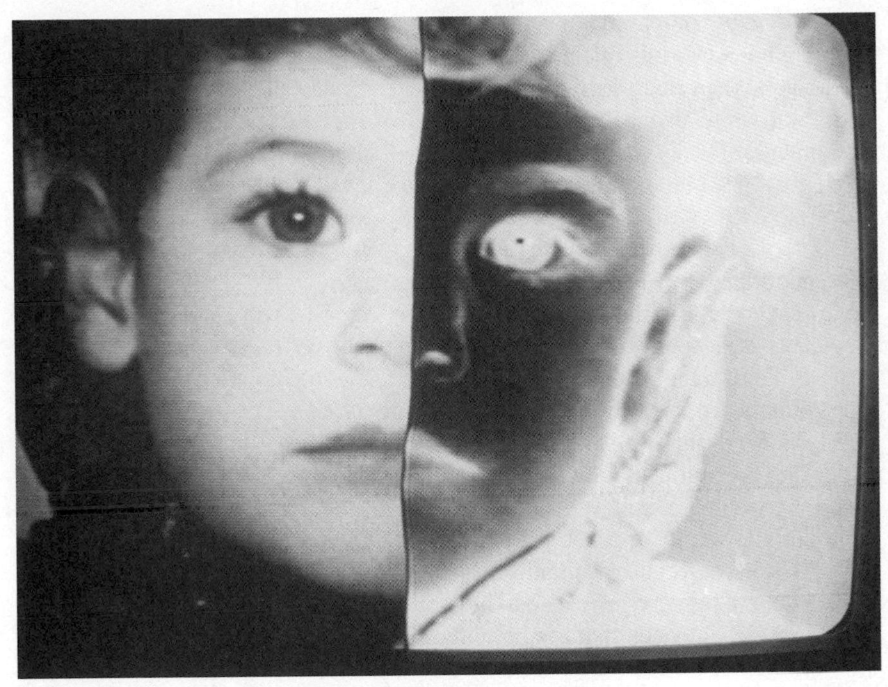

Fig. 4-30. Horizontal wipe to effects—1/2 picture as negative.

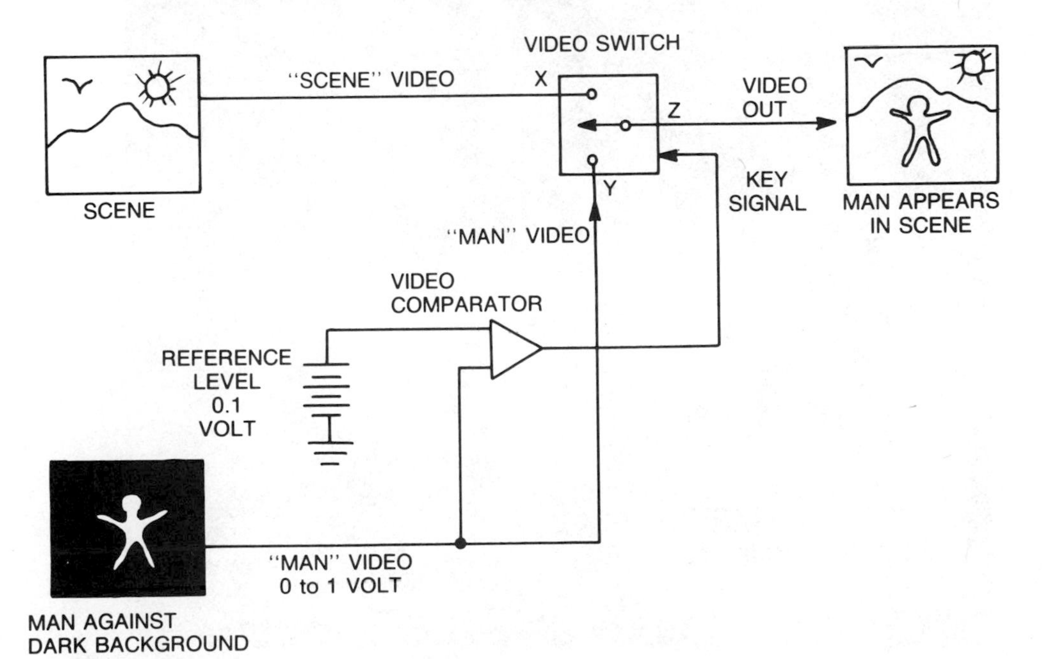

Fig. 4-31. Video keying to combine two scenes.

The video comparator must have fast switching times for best results. As mentioned before, this effect can also be done by using chrominance as a reference quantity. In this way, portions of a color signal can be superimposed on a black and white image, or selected areas of the picture can be blacked out (scrambled, etc.) as is often seen in televised interviews where a person's face or identity must be concealed. The possibilities are only limited by ingenuity and the creativity of the producer.

Figure 4-32 shows examples of various effects that can be produced by the aforementioned techniques. The video enthusiast with some experience in electronic construction can build a suitable system that will produce many of the previously discussed video effects. The complexity will depend on the features desired. For simple fades, a few IC devices and components will be enough. If more effects are desired, and capability to generate keyed effects, more complexity is necessary. A video effects unit can easily cost hundreds of dollars if purchased ready-made. Some video effects units use sophisticated microprocessor controlled circuitry. While nice to have, this requires software development and the would be constructor either has to have this capability or else has to buy black-box preprogrammed (usually proprietary) ROM devices. Actually, quite a large number of effects can be produced with twenty or so inexpensive, off the shelf IC devices that are not difficult to find, together with rather simple

Fig. 4-32. Various keyed effects.

circuitry. The part of the circuitry that handles the video frequency components (up to five megahertz) is actually rather simple, consisting of analog switches and buffer amplifiers operating at low gain. The circuitry necessary to generate the keying waveforms employs common off the shelf digital CMOS chips and ordinary TV receiver IC chips, and no high frequency circuitry is involved. While good layout is important in order to keep things ''clean'' and to minimize transients and switching noise in the picture, it is not overly critical. Ordinary good workmanship is all that is required and no special magic tricks or precautions are necessary.

This chapter will describe a system and present construction details on a video control device that can produce fades and wipes and keyed effects. It has two independent channels and can fade one channel to the other, or wipe and key between channels, as well as fade either channel to white or black, or any gray level in between. It also can, in conjunction with an external effects device like the Video Palette perform various other effects. It can be built for less than $150, depending on your stock of parts and also what you do yourself. In order to simplify construction, a complete kit of parts and/or PC boards are available from:

North Country Radio
P.O. Box 53
Wykagyl Station
New Rochelle, NY 10804

The kits contain the PC boards and all parts that mount on them. Power supply components and switches, jacks, and certain pots are not supplied, since they are left to the discretion of the constructor and depend on the application intended, but these external components are readily available at Radio Shack, etc.

A block diagram of the system to be described is shown in Fig. 4-33. It basically consists of two parts. One part is a video switching system to switch in various video effects, fade levels, and to select CH1 or CH2, and a special waveform generator to generate keying waveforms to drive the analog switches at precisely timed intervals.

Referring to Fig. 4-33, there are two video channels. Each channel (we will describe channel 1 since channel 2 is identical) then is fed to a splitter circuit. The splitter separates the video components and sync components. It is a high speed switch that routes the sync portion (sync plus color burst) to one pole, and video (between sync intervals) to the other pole. In this way operations can be performed on the video separately from the sync, which is not processed in any way.

The video from CH1 next passes through an analog switch. The analog switch either passes it or selects CH1 video that has been altered by an external special effects unit. When the video from the special effects unit (such as

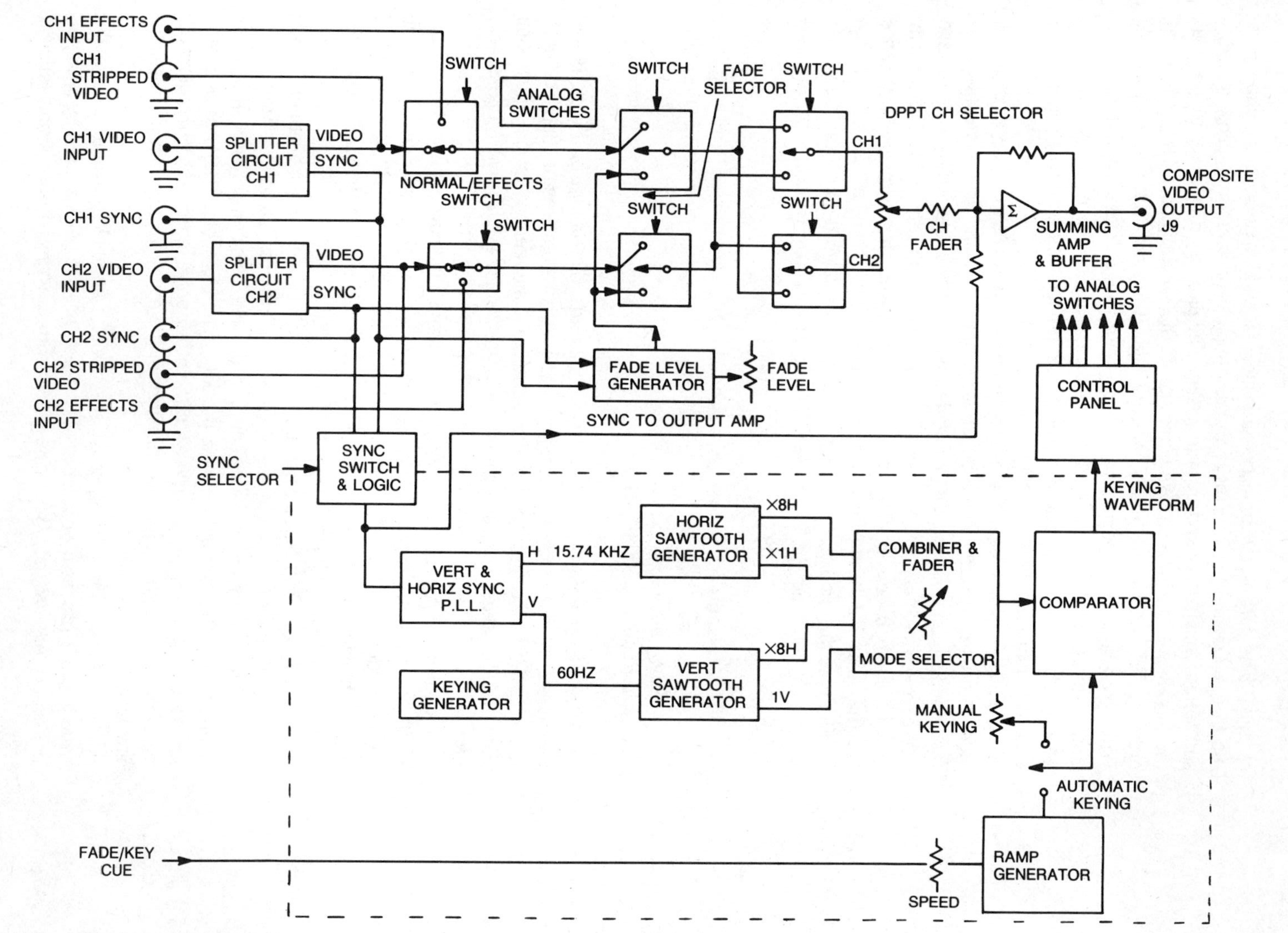

Fig. 4-33. Video switching system block diagram.

the video palette) is inherently synchronous with the CH1 video, direct switching is possible.

Next, the video is fed to another analog switch. The output of this switch (Fade Selector) is either unaltered video or a dc level from the Fade Level generator, which is variable between black (about zero volts and white (1 to 1.5 volts). This is determined by the setting of the fade level control. During a line scan interval, several switching actions may take place, which causes various pattern configurations to be generated on the monitor screen. The background level is set by the fade level control. The analog switch gets its switch signals from the control panel and keying generator.

Next, the video goes to a switch network which routes it to either side of the fader control, or selects CH1 or CH2. Both analog switches in this network are driven by the keying waveform from the keying generator and control panel, and switching may take place several times during a line scan, depending on the effects desired.

The output from the fader control is fed to a summing amplifier and mixed with appropriate sync. The output is composite video, and this is the system output.

The keying generator consists of two phase-locked loops, pulse shapers, and comparators. Sync from CH1 or CH2 (selectable) is fed to a phase-locked loop, and a constant output of 15.74 kHz and 60 Hz is generated, phase-locked to the CH1 or CH2 video input waveform. These outputs are fed to horizontal and vertical phase locked loop sawtooth generators. These generators each produce two waveforms—a sawtooth at eight times the input frequency and a sawtooth at the input frequency. These sawtooth waveforms are mixed in a combiner depending on effects desired and then fed to a comparator, whose ''trip'' level is adjustable. The sawtooth is compared to this trip level from the keying control. When the sawtooth exceeds the trip level, the comparator switches. Since the sawtooth level varies synchronously with the horizontal, or vertical, or both sweeps, varying the trip level causes the comparator to switch at varying points in either the horizontal or vertical scan. The comparator output is the keying waveform. Furthermore, we can control the position of the switching at any desired point in either the horizontal or vertical scan cycle.

The comparator trip level can be varied by a front panel control or by a variable speed slowly changing dc level generated by the ramp generator. Manual control makes it possible to ''freeze'' the video switching at a particular point, while automatic gives a smooth transition. This is front panel selectable.

The switching waveform is fed to the control panel switches, which determines the analog switches acted upon. If no keying is desired, a steady dc level to the analog switches is also available. By using the four sawtooth waveforms or combinations of these waveforms, various switching patterns can be generated.

The circuit features access capability to the switch signals, and also sync outputs via emitter followers. This permits using an external computer or microprocessor system to generate other switching patterns than we have here, if desired. This is left as a project for the experimenter or computer hobbyist.

Due to the repetitive nature of these circuits, very detailed descriptions of every circuit will not be given. Only the separate essential blocks will be described in detail. The vertical and horizontal waveform generators merely differ in certain component values due to the frequency difference.

Next, the circuitry will be described in detail as to how each "block" works and the function of each component.

First, the sync splitter will be described. The sync splitter consists (for channel No. 1) of sync separator IC1 and dual monostable multivibrators IC2 and IC3.

Referring to Fig. 4-34, video is fed through C1 and filter R1-C2 (to remove excess noise) to sync separator IC1, an LM1881N. Composite horizontal sync (negative going pulses) appears at pin 5. This is fed to horizontal delay multivibrator IC2A. R3 and C5 are power supply bypass capacitors. R5, R6, and C6 determine the period of the pulse generated by IC2A. This multivibrator produces a pulse triggered by the leading edge of the sync pulse. It has a width of about 8 microseconds. It is used to initiate a pulse generated by IC2B, which is active only during the line scan portion (video) of the video waveform. It will be used to split the video only component from the composite video waveform. R7, R8, and C7 determine the width of this pulse (53 microseconds).

IC3A and B perform a similar function on the vertical sync pulses. IC3A is a delay and IC3B generates a 16 millisecond pulse which is active during individual fields, of the TV signal. The delay of IC3A is set by R9, R10, and C8. During vertical retrace intervals it is desirable not to gate on the composite video so horizontal multivibrator IC2B is locked out during the vertical blanking interval, when pin 10 is low.

At pin 9 of IC2B is a positive going pulse during horizontal and vertical sync intervals. This pulse is fed directly to analog switch IC17A, a SPDT switch that is used to separate video from sync and also to sync selector IC4.

IC14, 15, and 16 perform the same functions for CH2 and their operation is exactly the same as IC1, 2, and 3 and therefore will not be discussed. Components and adjustments are identical to the circuitry consisting of IC1, 2, and 3.

Figure 4-35 shows the sync selector and PLL circuit. Sync from source 1 (Pin 9 IC2B) or source 2 (Pin 9 IC15) is selected by gates of IC4, a 7400. When sync select (pin 2) of IC4 is high, SYNC 1 is selected. When pin 2 IC4 is low, SYNC 2 is selected. Inverted logic level is taken from pins 10 and 11 to selectors IC17B and IC19B. Sync from pin 4 of IC4 is fed to filter network R16, R17, and C10, C11, and C12 to PLL IC LM1880 (IC5). C13, C14, R19, and R18 are loop parameter determining components. R21, R20, L1, C22, C23, and C24 are for the internal oscillator of IC5 operating at 503 kHz. C19, C20, R22, and R23 are feedback components.

R24 and C16 are vertical timing components necessary for correct operation of IC5. R25, C17 and C18 are supply decoupling components. A signal at

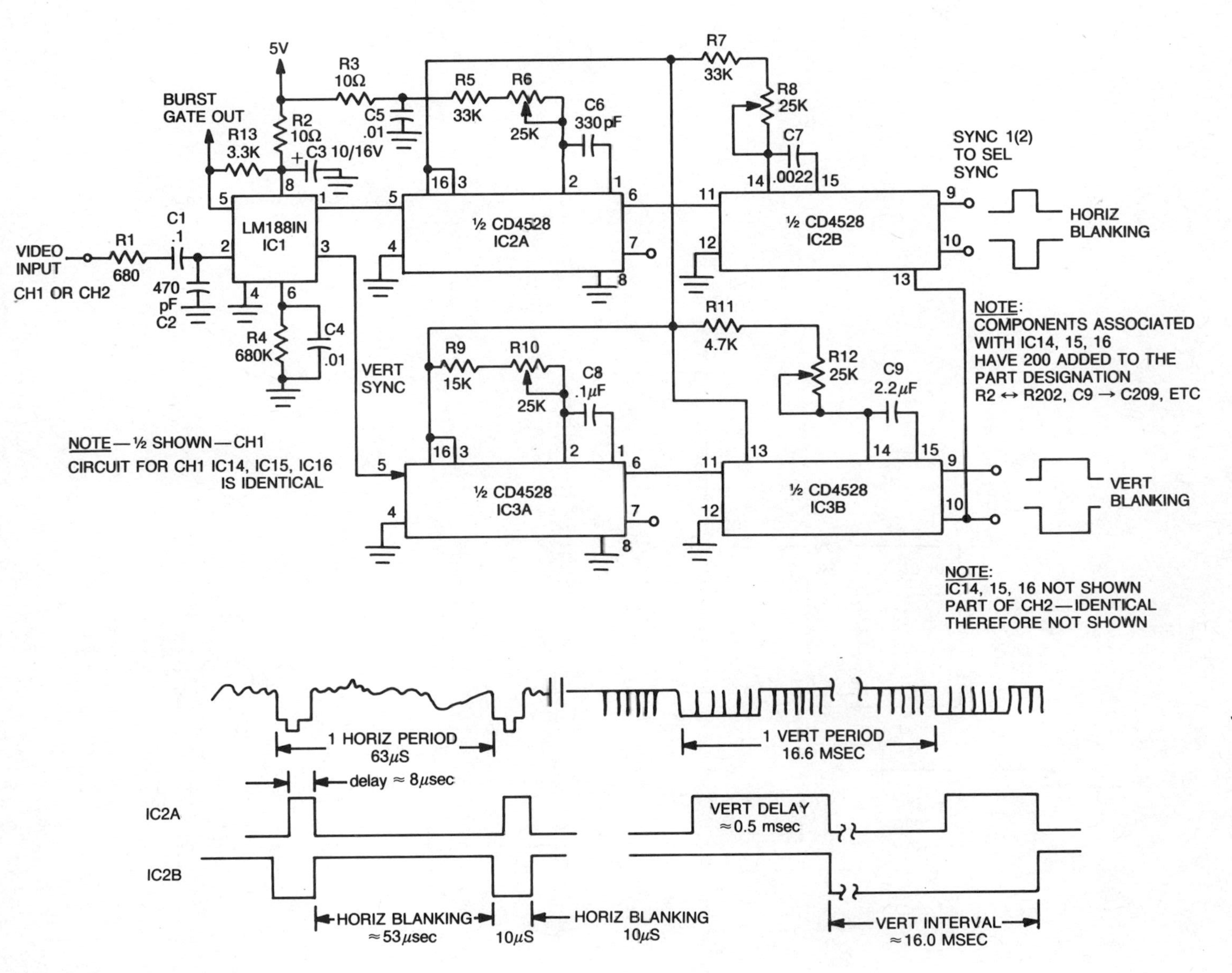

Fig. 4-34. A and B, sync—video splitter circuitry.

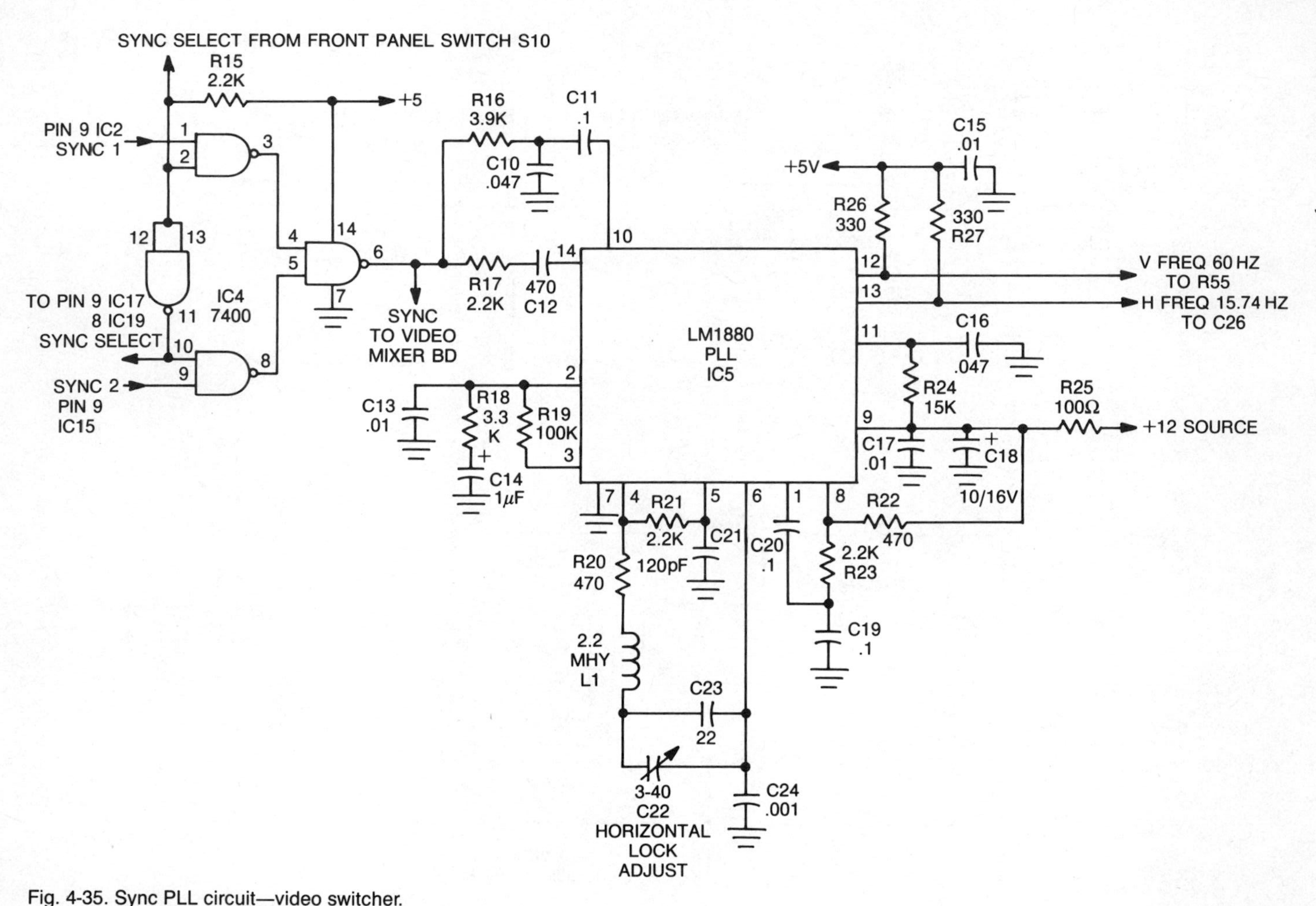

Fig. 4-35. Sync PLL circuit—video switcher.

the horizontal frequency appears across R26. C22 is adjusted for lockup with the SYNC 1 or SYNC 2 input. The outputs (from pins 12 and 13) are fed to the PLL sawtooth generators operating at vertical and horizontal frequencies, respectively. These sawtooth generators are used for generating the keying waveforms.

The keying circuits are shown in Fig. 4-36. There are four circuits—two for horizontal and two for vertical. First the horizontal and vertical rate sawtooth generators will be discussed.

Output from pin 12 IC5 produces horizontal frequency square wave pulses at junction C25 and C26 (positive going). These are differentiated by C25 and R28. Therefore Q1 is momentarily forward biased during sync intervals. C33 is thus discharged through R29. When Q1 is cut off, C33 charges toward +5V through R30 until discharged again at the next sync pulse. Q2 and R31 form an emitter follower to interface this waveform, which is a sawtooth of about 1–2 volts at horizontal frequency, to horizontal pattern select switch S1.

Vertical sync pulses are directly integrated by R60 and C42. D1 provides a discharge path for the integrator. The vertical sync pulse is a very short negative going pulse. Emitter follower Q6 and R61 feed the sawtooth waveform to one pole of S2, the vertical pattern selector switch.

The triangle waves at 8X horizontal and vertical frequencies required to produce certain keying waveforms are obtained from phase-lock loop circuits IC6, 7, 8 (for horizontal) and IC9, 10, 11 (for vertical). They are similar except for component values, so only the horizontal frequency circuit will be discussed, the vertical circuit being identical in operation.

Horizontal sync is fed to an LM565 PLL IC via C26. R32 and R33 are bias resistors. C27 and C30 are supply bypasses for the +5 and −5V lines. C28 is a loop filter capacitor and C29 suppresses spurious responses. VCO frequency out of pin 8 is nominally 126 kHz (480 Hz for the vertical circuit). It is set by R34, R35, and C31. The VCO output at pin 4 of IC6 is fed to the input of IC7 (pin 8) a 74C93 four-stage counter. Only three stages are used to obtain division by eight. The ÷ 8 output (pin 12 IC7) is fed back to pin 5, the phase detector input, of IC6. Therefore, under lock conditions, the VCO frequency at pin 9 will be 126 kHz (8 × 15.74) and will be a triangle wave. IC8 is a buffer amplifier and delivers this triangle wave to S1.

R49 is a mixer control that combines two of four available waveforms V, 8V, sawtooth, and H and 8H sawtooth. The exact proportion can be varied to achieve various key patterns. The waveforms resulting is fed to comparator IC13 through R50.

IC13 is biased to a threshhold by a dc voltage from S3, the auto-manual key selector switch, and voltage divider R46, R47 (variable) and R48, or by a slowly varying dc voltage from pin 6 of IC12, as selected by S3. The output of comparator IC13 feeds Q5 via R52 and R53. The output of Q5 is a square wave whose duty cycle depends on the dc levels from S3 and R49 or R45. It is used to drive the keying switches in the video mixer circuit to be described.

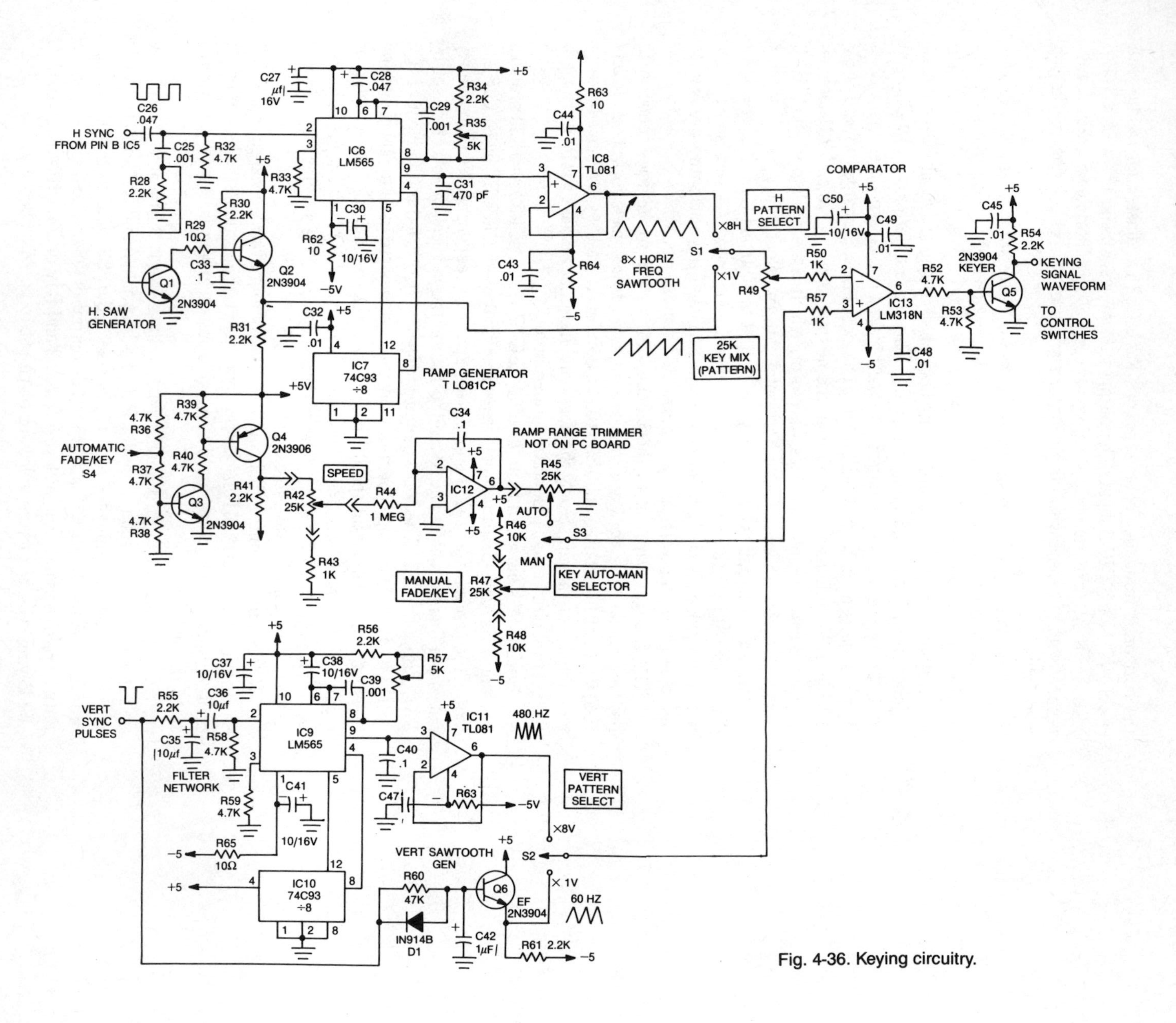

Fig. 4-36. Keying circuitry.

Ramp generator IC12 is used to generate a slowly varying dc voltage for slow fades, wipes or key-ins. It is fed either positive or negative signals through R44. The speed (rate) of the ramp depends on the setting of R42, the speed control. By varying R42, either a slow or fast key transition can be obtained. R47 is used where manual control of key transition is desired. Q3 and A4 feed either +5 or −5Vdc to R42, depending on the logic level at the junction of R37 and R36. R45 is used to set the voltage level in automotive operation.

R52, R53, R54, bypass capacitor C45, and Q5 square up the comparator output waveform and produce the final keying waveform at the junction of collector Q5 and R54.

Figure 4-37 shows the video switching circuits IC17 through IC20 are analog SPDT switches. Each has three sections that can be switched at over a 1 MHz rate and that can handle signals up to 5 MHz with good (50dB) isolation. They are CMOS CD4053 and controlled by a logic level at the input. All switches are in "up" positions (N.C.) when logic level is zero, "down" (N.O.) when logic level is high.

In this system, it is necessary to provide in several instances, a low drive (source) impedance (less than 20 ohm) from a high impedance (greater than 2k ohm) source.While a buffer amplifier may be used, if unity gain is adequate a simple emitter follower is the 0.6 volt dc offset they produce. This arises from the fact that about 0.6 volt dc must be present between base (input) and emitter (output) of the emitter followers. Referring to Fig. 4-37, examine the circuit of EF1 through EF6. Q103 used a NPN transistor. This places the emitter at −0.6 volts if the base (input) is grounded. However, a second stage Q104 using a PNP transistor is fed by Q103. Since the bias polarity of a PNP is opposite that of the NPN, the −0.6V level at the base of Q104 causes a 0 volt (ground) level at the emitter of Q104. What we have is a double emitter follower with effectively zero dc offset. Tests on this circuit indicate flat frequency response to over 5 MHz and essentially unity gain. It is used in six places in the video switching circuit, designated EF1 through EF6. All are identical and act as high impedance to low impedance interfaces.

Now the operations of the video switching network will be described in details. It consists of two identical channels and associated analog switches and emitter followers.

Video 1 (only one channel will be described) is inputted at pin 15 of IC17. IC1 is fed from this point as well. IC17 splits the video from sync. It is driven by IC2 in the keying section EFI (an emitter follower) and EF3 make both sync and video separately available at J2 and J7. (EF = emitter follower see Fig. 4-37). IC18A selects either input video or effects video derived from video 1 externally. IC18 selects either CH1 or a dc level (fade) used in fadeout. This dc level comes from R115 and is about −0.5 to +1.5 volts. It is blanked during sync intervals so as to not upset sync levels. Q100 through Q102 and D100, and R112 through R118 generate the required waveform. IC18C and IC20C are configured as a DPDT switch (reversing switch) to switch between CH1 and 2

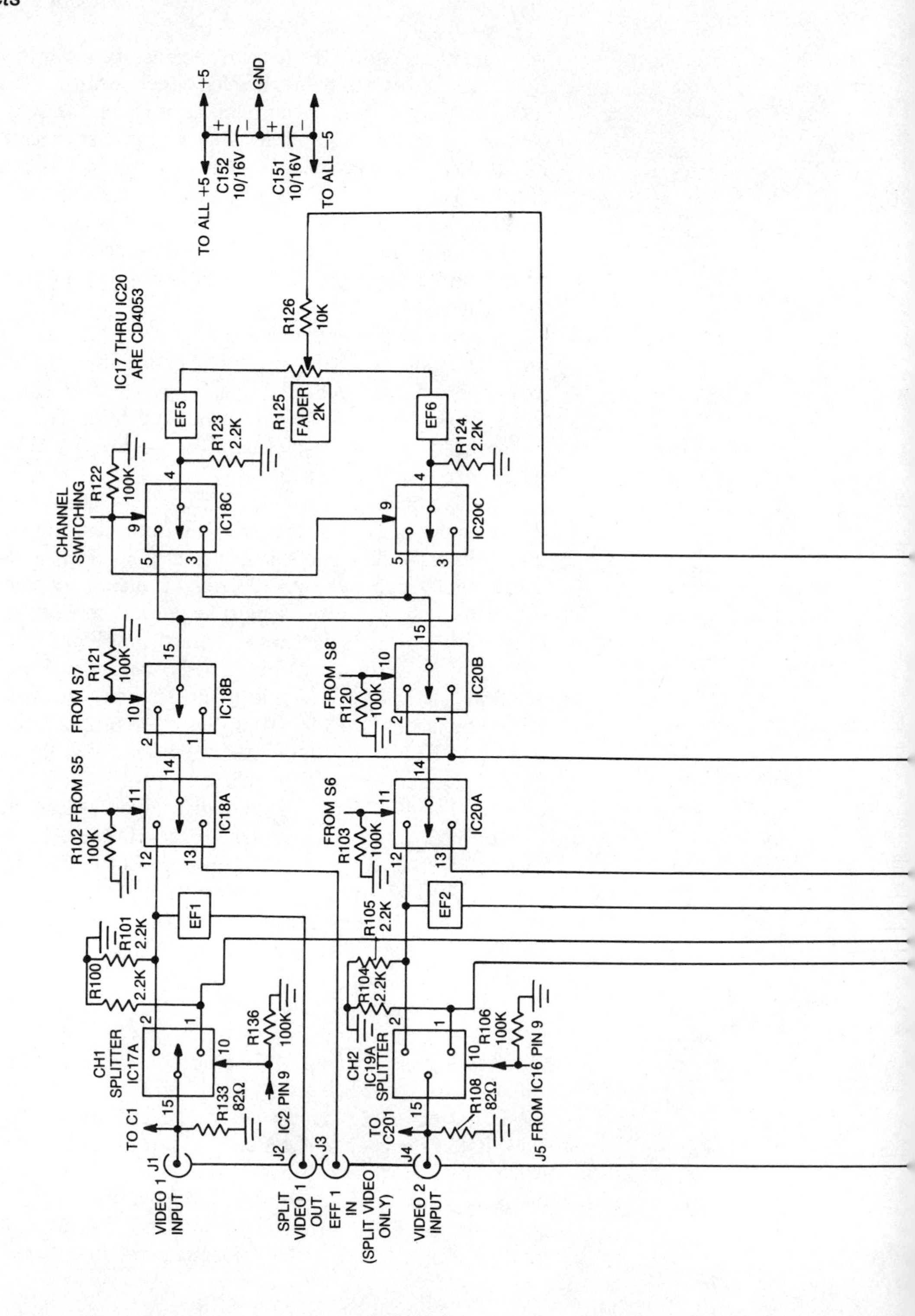
TO ALL +5
C152 10/16V
+5
GND
C151 10/16V
TO ALL −5
IC17 THRU IC20 ARE CD4053
R126 10K
R125 FADER 2K
EF5
EF6
R123 2.2K
R124 2.2K
CHANNEL SWITCHING
R122 100K
IC18C
IC20C
FROM S7
R121 100K
IC18B
FROM S8
R120 100K
IC20B
R102 FROM S5 100K
IC18A
FROM S6
R103 100K
IC20A
EF1
EF2
R101 2.2K
R105 2.2K
R100 2.2K
R104 2.2K
CH1 SPLITTER IC17A
R136 100K
IC2 PIN 9
R133 82Ω
TO C1
CH2 IC19A SPLITTER
R106 100K
R108 82Ω
J5 FROM IC16 PIN 9
TO C201
J1
J2
J3
J4
VIDEO 1 INPUT
SPLIT VIDEO 1 OUT
EFF 1 IN (SPLIT VIDEO ONLY)
VIDEO 2 INPUT

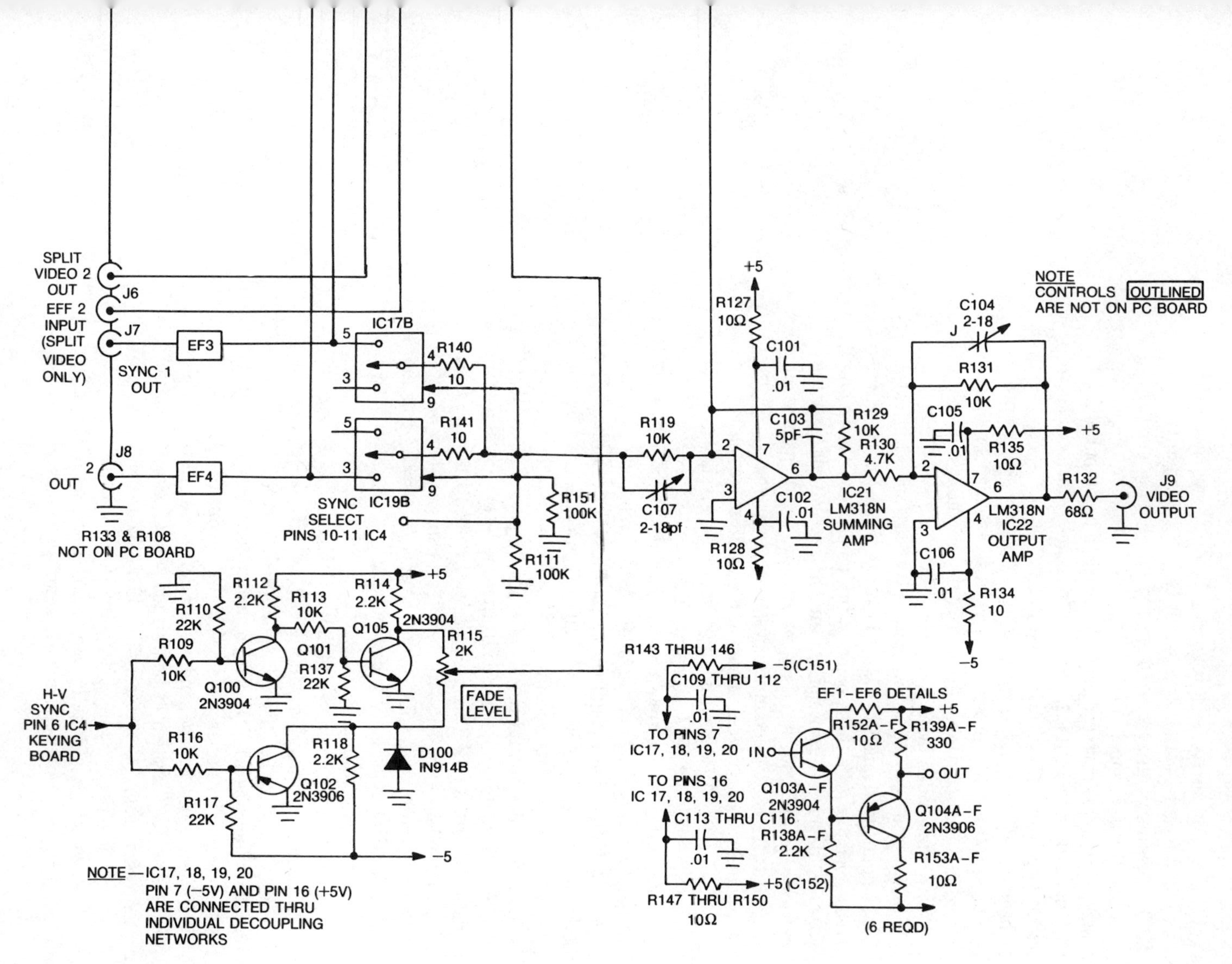

Fig. 4-37. Video switching board circuitry.

for direct fades, wipes, or key-ins (genlock sources required). Switched video from both channels is fed to fader R125. The output of R125 is taken to summing amplifier IC21, together with sync from sync switch IC17B and IC19B (sync selected for appropriate channel in use). R119 and C107 are frequency compensation components to maintain correct burst phase. The output of IC21 is a complete video signal (inverted). It is fed to IC22 for reinversion and then to J9 through termination resistor R132. This is the systems' output.

Now that we have described a video control system, there is still the problem of switching and controls for the various functions. A set of manual switches and controls will, with a little practice on the part of the operator, enable the implementation of quite a few video fades and keyed effects.

Figure 4-38 shows a basic set of control switches for use with the video control system. They connect to pins 9, 10, and 11 (control pins) of IC18 and 20. If desired, you can also interface these points with a computer or microprocessor setup of your own design. The possibilities are endless. We will leave the system largely up to you, as everyone has their own creativity and imagination. However, the basic hardware is provided in this system to do as you wish with. It is suggested you use the basic system shown in Fig. 4-38 to get an idea of what the unit can do, as a start. All leads from the seven switches shown in Fig. 4-38 are logic levels and are CMOS/TTL compatable.

As previously discussed, in the system there are two video channels, CH1 and CH2. Each channel feeds one side of the fader control. What the switches (S5 through S9) do is to determine exactly what signal each side of the fader control has applied to it. For example, suppose a fade to black is desired. In this case S7, which is the fade selector, would be set so that CH1 video passes directly to one side of the fader control R125. S8 would be placed in the fixed position, which applies a fixed dc level, selected by the fade level control, to the opposite side of fader control R125. This fixed dc level comes from the fade level generator circuitry and is set by R115 the fade level control. (This voltage is removed during sync intervals so as not to upset the sync levels.) By rotating R125, therefore a mix of CH1 video and the dc fade level is sent to the output amplifier and manual fading is accomplished.

If a fade from CH1 to CH2 video is desired, both CH1 and CH2 fade selectors are placed in the normal position. If a fade from CH2 to CH1 is desired, the switches S7 and S8 are placed in the ''fixed'' and normal position.

Mode switch S9 functions as a DPDT switch that swaps CH1 and CH2. It basically reverses the connections to each side of the fader control. If the fader control is set so that it is at one extreme, and say CH1 is coming through, then moving S9 to the ''Reverse'' position instantly routes CH2 into the output amplifier. It functions as a channel selector switch. If the fader control is centered, S9 has very little effect.

Since actual video switching is done by the analog switches IC17, IC18, IC19, and IC20, S5 through S9 merely apply dc control voltages and do not directly handle video signals.

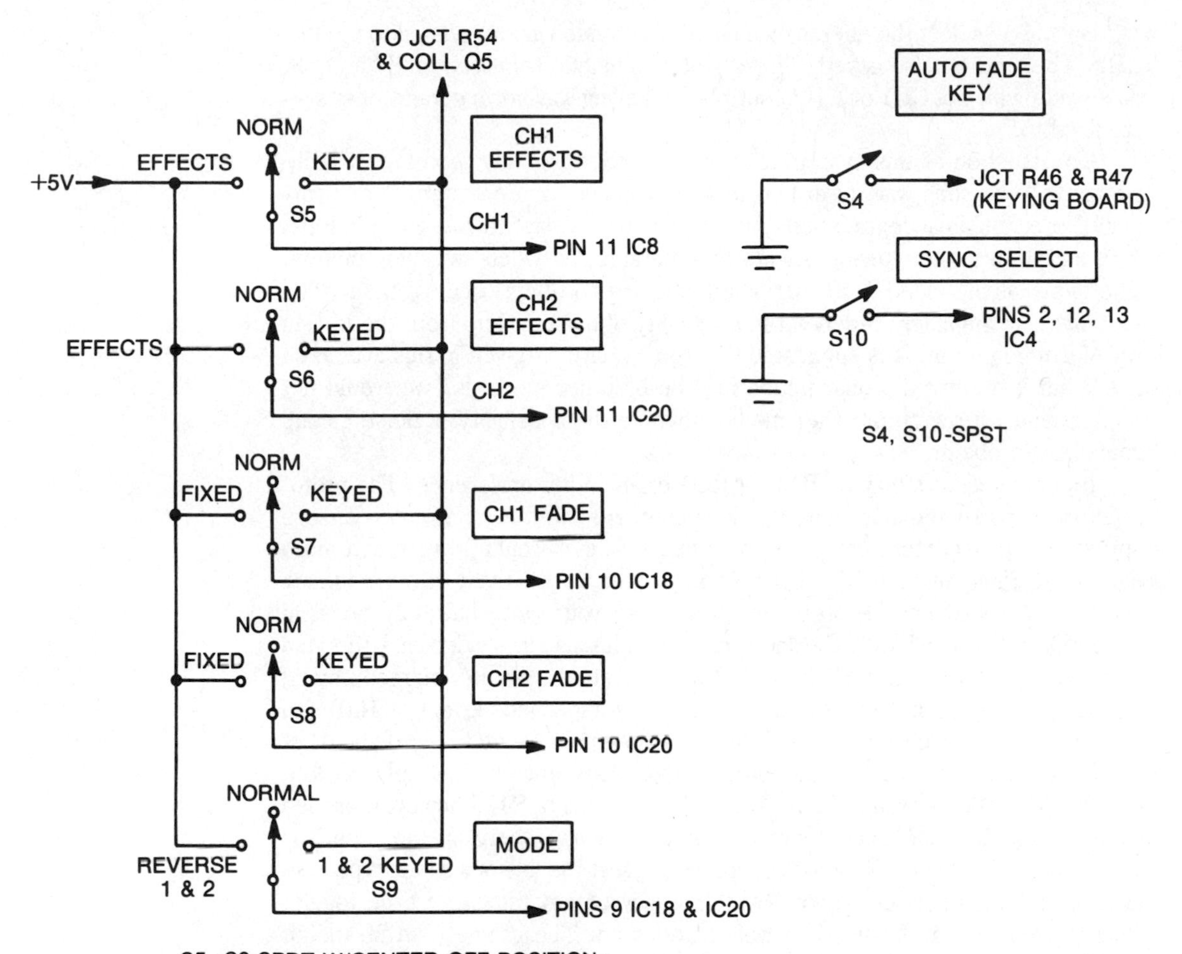

Fig. 4-38. Basic control panel for the video switching systems.

In the "keyed" positions S5 through S9 apply a waveform to electronically switch the video the "keyed" position is used to obtain keyed wipes and transitions, as well as keyed fades. The pattern selector switches S1 and S2 and the setting of key mix control R49 determine the particular pattern. S3 selects the manual fade/key mode whereby R47 manually manipulates the effect, or the auto key mode where a slowly varying voltage from the ramp generator produces the effect. S4 initiates the transition or effect. S4 has no effect in the "manual" position of S3.

The best way to get the "hang" of how the controls work is to spend some time experimenting with them, using a VCR or other video source, and watching a monitor to observe them directly.

S5 and S6 select the effects channel or other video inputs synced to CH1 or CH2. They can be also keyed. These switches function similarly to S7 and S8. S4 selects which of CH1 or CH2 controls the keying waveform generator is syncronized to.

Construction of this system is relatively noncritical but good layout techniques will go a long way toward noise reduction and assure system "transparency" (i.e., minimal degradation of video signals). PC layouts are given in Figs. 4-39 and 4-40 for the keying waveform generator and video switching boards. The fact that the video switching board contains no other circuitry than video switches and amplifiers reduces the possibility of noise pickup from the keying waveform generator. It is suggested that the PC layouts given in this article be used, but if preferred, construction may be by other methods. We would not recommend wire wrap or other methods where the possibility of noise pickup could be a problem.

Input connectors may be BNC or RCA or any other preference. The prototype was constructed in a Radio Shack cabinet. However, in many cases a more spread out arrangement using a sloping panel cabinet would prove much more desirable. Since most of the controls handle dc or relatively low frequency waveforms, layout can be pretty much any way you want. The only possible exception is around R125, the fader control. This handles video and the lead between R125 wiper and R126 should be kept short. We do not recommend shielding this lead, since it is at a relatively high impedance level ($\approx$ 1kΩ) and the stray capacitance to ground should be no more than about 25 pF. If you must shield it, remember that typical 50 ohm coax cable runs about 29 pF per foot and 75 ohm cable is around 22 pF per foot. The ends of R125 however, are fed from a very low impedance source so for these leads, there are no problems with stray capacitance. For a nice effect we suggest the use of a slider type control for R125. This is up to you. Switches may be any miniature type toggle. Rotary switches can be used as well if preferred. The cabinet can be metal, wood, plastic or any other preferred material.

The power supply may be any source of $\pm$5V at about 250 mA. Also, +12V at 50 mA is necessary for the operation of IC5. A simple supply is shown in Fig. 4-39 and the parts can be obtained at your local Radio Shack or other such store. Keep the power supply away from the video circuits to avoid 60 Hz hum pickup. If desired, a 12V wall plug-in transformer may be used for T1 and therefore only 12Vac must be brought into the main chassis.

External video effects units may be used with this unit. The Video Palette is excellent for use with this system. It can produce posterized images, partial or complete tonal reversal, and manipulate the video signal luminance and chroma for many interesting effects. It contains its own sync splitter so that it can simply be connected to the CH2 input of the video control system, with no other interface considerations. If preferred, just the effects board of the video palette can be connected between the split video (J2 or J5 and J3 or J6) if either or both channels, since the necessary stripped video for the operations of this

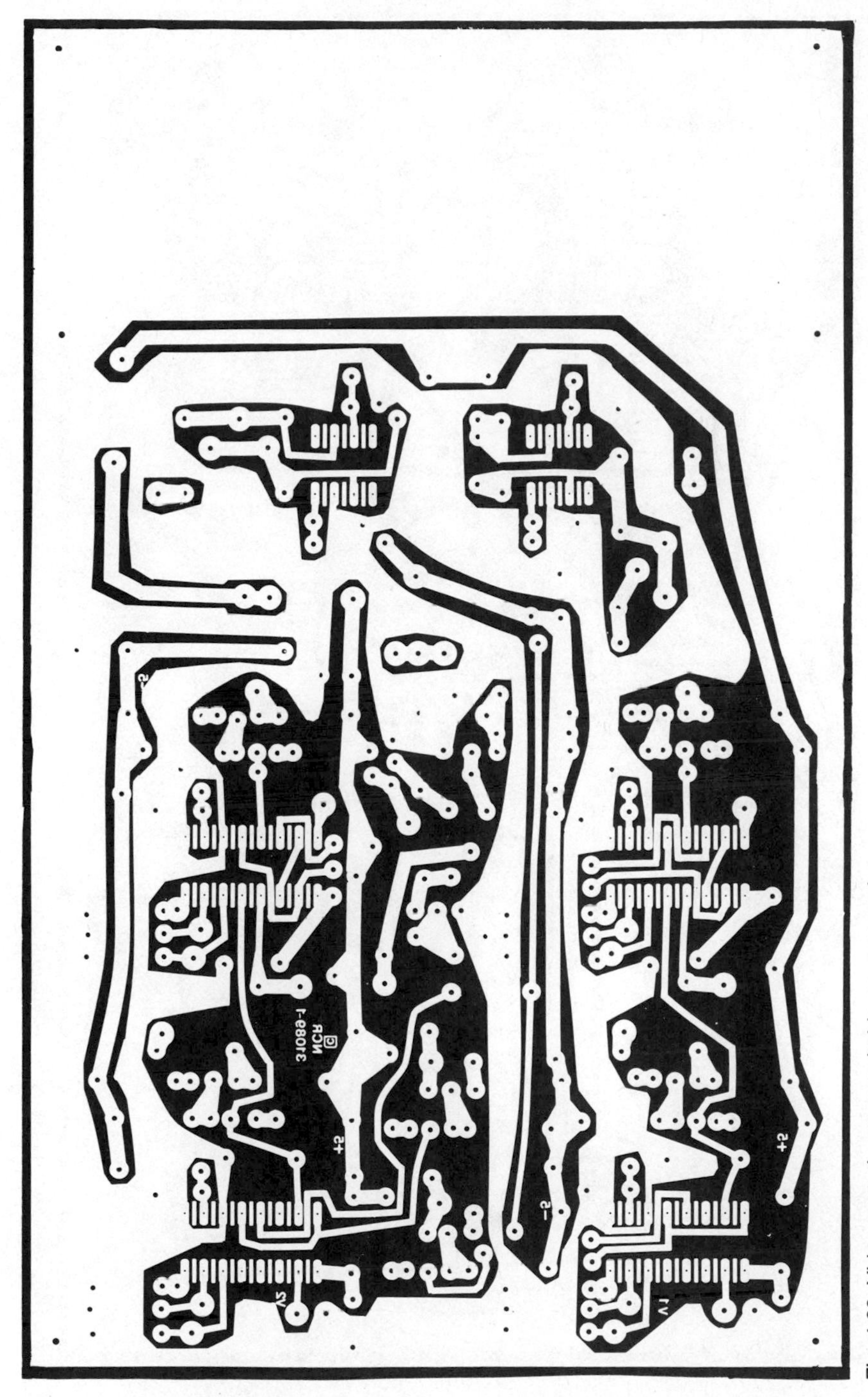

Fig. 4-39. Video control system suitable power supply.

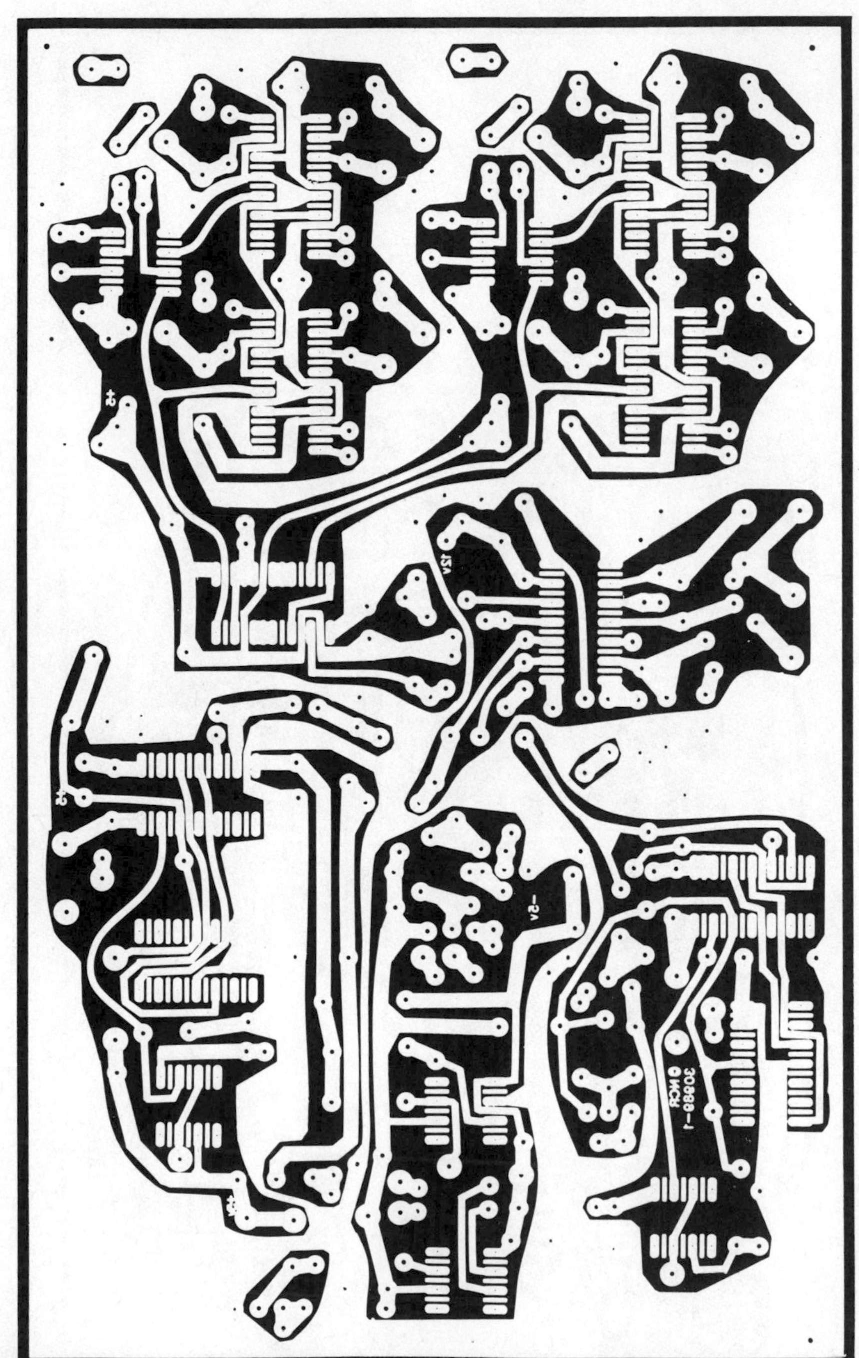

Fig. 4-40. Video keying generator board pattern.

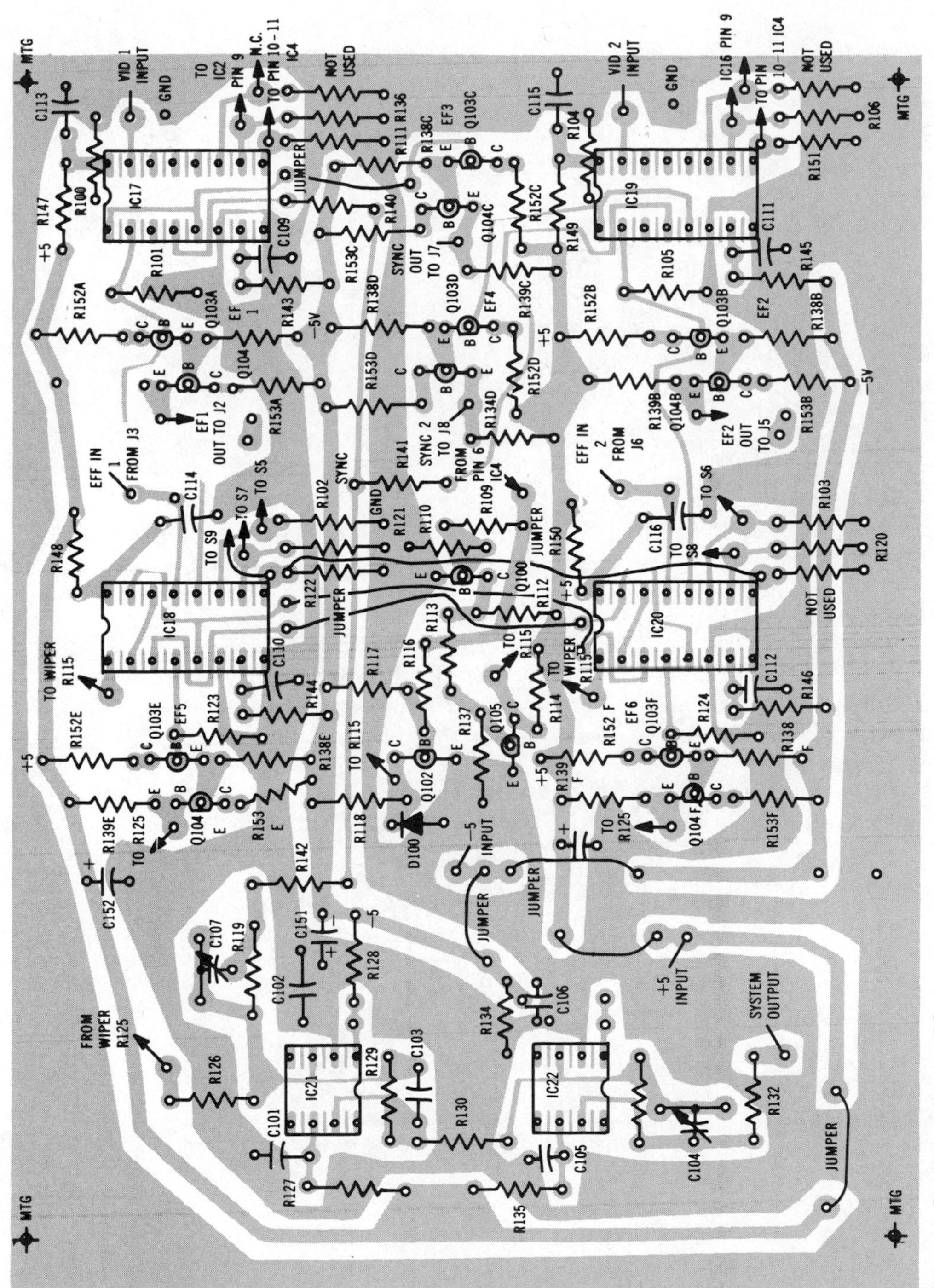

Fig. 4-41. Component layout (PC artwork 4-39)

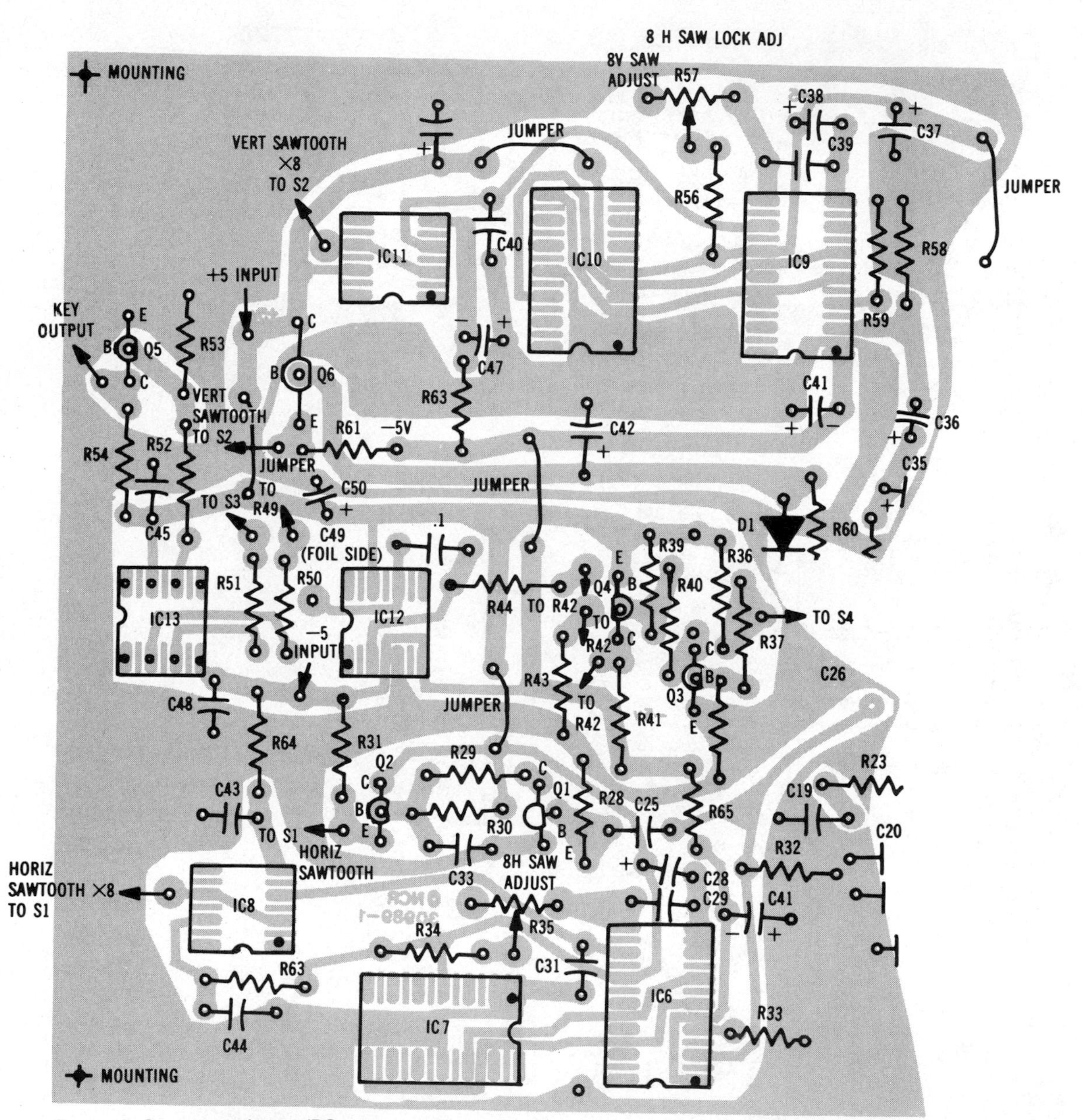

Fig. 4-42. Component layout (PC artwork 4-40).

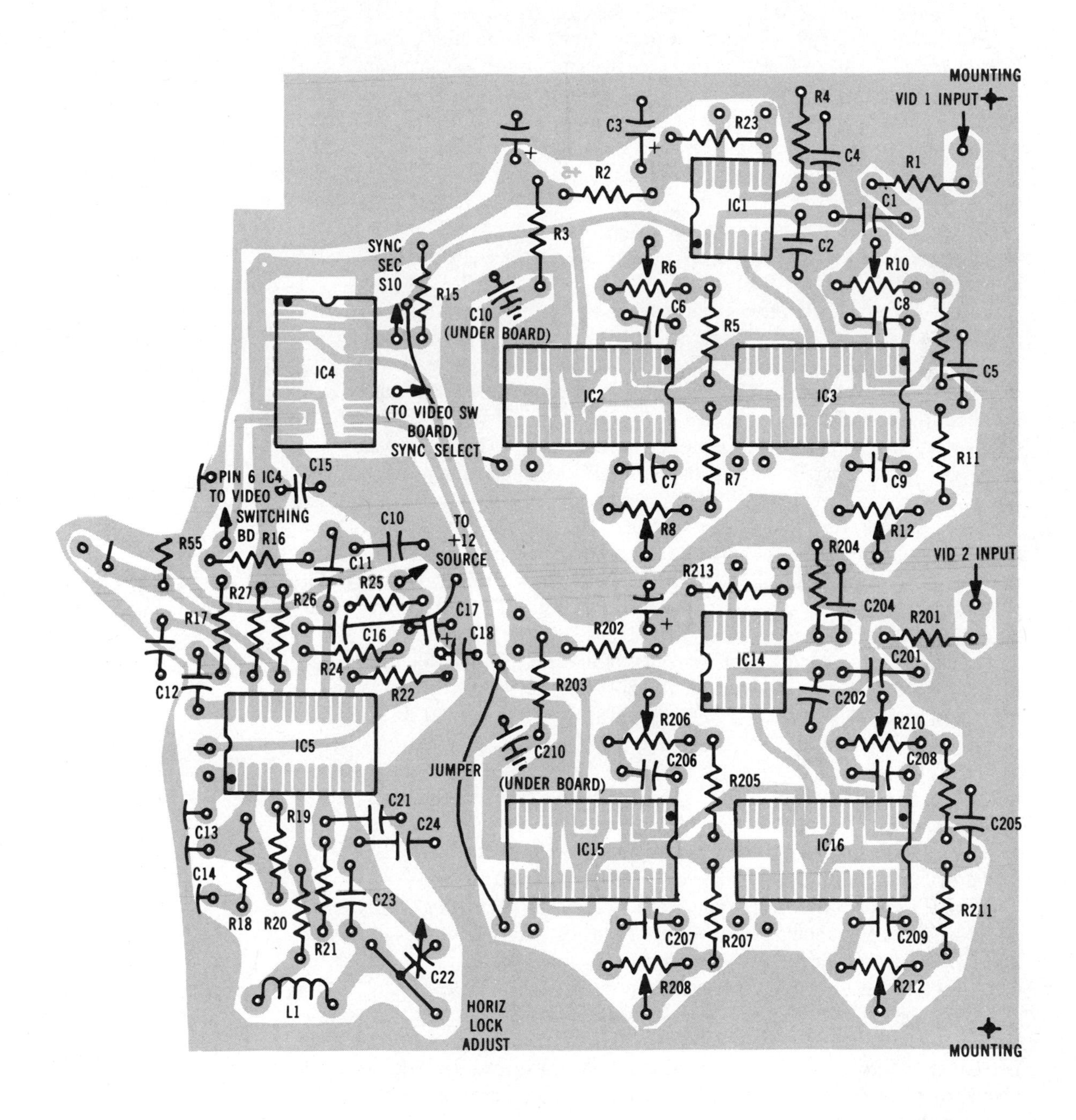
MOUNTING
VID 1 INPUT
R4
C3
R23
C4
R1
R2
IC1
C1
C2
R3
SYNC
SEC
S10
R15
R6
R10
C6
C8
C10
(UNDER BOARD)
R5
IC4
C5
IC2
IC3
(TO VIDEO SW
BOARD)
SYNC SELECT
R11
PIN 6 IC4
TO VIDEO
SWITCHING
BD
C15
C7
R7
C9
C10
TO
+12
SOURCE
R8
R12
R55
R16
C11
R204
VID 2 INPUT
R25
R213
R27
R26
R17
C17
C204
C16
C18
R202
R201
C12
R24
R22
R203
IC14
C201
C202
R206
R210
IC5
C210
C206
C208
JUMPER
(UNDER BOARD)
R205
C21
R19
C13
C24
C205
IC15
IC16
C14
C23
R18
R20
R21
R211
C207
R207
C209
C22
R208
R212
L1
HORIZ
LOCK
ADJUST
MOUNTING

board is available at J2 and J5. This will be discussed at another time. We strongly recommend you refer to the previously covered Video Palette.

Next, we will briefly describe how to set up the system after construction. First check all +5 and −5 pins on the IC devices for proper voltages (pins 7 and 4 of the LM318, pins 7, 8, and 16 of the CD4053, etc.). Verify +12V at the junction of R25 and the +12V supply. Also note that *no* IC device should get hot. Slight warmth may be noted on IC5. Make sure, of course, that you have checked all PC boards for solder bridges, splashes, or other foreign materials that could cause shorts.

An oscilloscope will be invaluable for the following checks. If you do not own a scope, try to obtain the use of one. We will describe setup for CH1—CH2 is identical.

Apply 1V p-p NTSC video negative sync, to J1. Verify negative sync pulses of about 5V p-p at IC1 pin 1. Next, adjust R6 so that IC2A pin 6 shows an 8μs pulse. Next, adjust R8 so that IC2B produces a 53μs wide pulse.

Verify a vertical sync pulse at pin 3 of IC1. It should be 60 Hz. Adjust R1Q for about 0.5 to 0.6 millisecond pulses at pin 6 IC3A. Next, set R12 for a 16 millisecond pulse at pin 9 IC3B. Note that if R12 is set too high in resistance no pulse will appear at pin 9. So start out with R12 at its minimum resistance setting.

Next, making sure that the SYNC select (S10) switch is in the CH1 position, check for sync pulses at pin 6 IC4. Connect the scope to pin 13 of IC5 and adjust C22 (use a nonmetallic tool) so that the pulses are syncronized to the video signal. They will "pop-in" at the correct setting. This setting should not be critical. Next, check for 60 Hz (vertical pulses at pin 12IC5.

Finally, place the scope probe on pin 4 of IC8 and adjust R35 so that a 126 kHz sawtooth waveform is obtained. It will "lock" in at the correct setting of R35. Now, place the scope on pin 6 IC11 and adjust R57 for a 480 Hz sawtooth. Again, it will lock in at the correct setting of R57.

Verify a horizontal frequency sawtooth at the junction of R31 and the emitter of Q2 and a vertical frequency sawtooth across C42. Note that these waveforms are at 15.7 kHz and 60 Hz.

Check for waveforms at the wiper of R49. It will be a mixture of two of the four previous waveforms depending on settings of S1, S2, and R49.

Check for ±2.5 volts at the wiper of R47 (depends on setting). Check voltage at pin 6 IC12. It should be between +4 and −4 volts. Activate S4—the voltage should slowly change. R42, should vary the rate of change. Set R45 at the center of its range. Set S3 to the manual position.

Place the scope at the collector of Q5 and rotate R49, you should see the keying waveform. At extremes of R49, the waveform will disappear and you will get either a zero or +5V level, depending on which extreme you are at. It should be possible to get solid zero, solid +5, or switching of any duty cycle between them (0 to 100%).

RESISTORS 1/4W 10%

SYMBOL	VALUE	SYMBOL	VALUE
R1, R201	680Ω	R26, R27, R139A – F	330Ω
R202, R203, R140, R141, R152A – F, R153A – F, R2, R3, R29, R127, R128, R134, R135, R143 thru R150, R62, R63, R64	10Ω	R35, R57	5k pot
R4, R204	680kΩ	R43, R50, R51	1k
R5, R7, R205, R207	33k	R44	1 Meg.
R45, R212, R206, R208, R210, R6, R8, R10, R12, R42, R47, R49	25k pot	R109, R46, R48, R113, R116, R119, R126, R129, R131, R142	10k
R130, R11, R32, R33, R36, R37, R38, R39, R40, R52, R53, R58, R59	4.7k	R60	47k
R15, R17, R21, R23, R28, R30, R31, R34, R41, R54, R55, R56, R61, R100, R101, R104, R105, R112, R114, R118, R123, R124, R138A – F	2.2k	R132	68Ω
R16	3.9k	R108, R133	82Ω
R18, R13, R213	3.3k	R115, R125	2k pot
R19, R102, R103, R106, R111, R120, R121, R122, R136, R151	100k	R117, R137, R110	22k
		R9, R24, R209	15kΩ
R20, R22	470Ω		
R25	100Ω		

CAPACITORS
10% tolerance (if not specified)
.01 is GMV, electrolytics – 50, + 10%

SYMBOL	VALUE	SYMBOL	VALUE
C1, C8, C11, C101, C208 C19, C20, C33, C34,C40	.1 Mylar	C22	3 – 40 trimmer
C2, C12, C202	470pF disc	C23	22pF NPO
C3, C18, C27, C30, C35, C36, C37, C38, C41, C47, C50, C151, C152, C203	10μF/16V electrolytic	C24, C25, C29, C39	.001 Mylar
C43, C44, C48, C49, C45, C109 thru C116, C4, C5, C15, C17, C32, C13, C101, C102, C105, C106, C204, C205	.01 disc		
C6, C206	330pF NPO	C31	470pF NPO
C7, C207	.0022 Mylar	C103	5pF NPO
C9, C209	2.2μF Tantalum	C104, C107	2 – 18pF trimmer
C10, C16, C26, C28	.047 Mylar		
C14, C42	1μF/35V electrolytic		
C21	120pF ±15% NPO		

Fig. 4-43. Parts list.

SEMICONDUCTORS

SYMBOL	TYPE	SYMBOL	TYPE
Q1, Q2, Q3, Q5, Q6, Q100, Q101, Q103A – F, Q105	2N3904	IC13, IC21, IC22	LM318N
Q4, Q102, Q104A – F	2N3906	IC17, IC18, IC19, IC20	CD4053B
D1, D100	1N914B		
IC1, IC14	LM1881N	**COILS**	
IC2, IC3, IC15, IC16	CD4528B	L1	80189-2R2
IC4	7400N		
IC5	LM1800N		
IC6, IC9	LM565N		
IC7, IC10	74C93		
IC8, IC11, IC12	TL081		

Fig. 4-43. Cont.

Next, place S5, S6, S7, S8, and S9 in the normal position. You should get video out of J9, the outputs, check it with a monitor. Adjust trimmer capacitor C104 for optimum sharpness. Next, adjust C107 for correct burst phase, as indicated by proper flesh tones on a video image. Next, place S8 in the ''fixed'' position. Vary R125—you should be able to fade to a level set by R115. Now try all the keying functions. This checks out the unit. You will spend a lot of time playing with this unit to see what it can do. There are too many combinations to describe them all. It is best to learn to use this system by playing with it—it will be fun. See Fig. 4-43 for the parts list.

A kit of parts consisting of the printed circuit boards and all parts that mount on the boards is available from:

North Country Radio
P.O. Box 53, Wykagyl Station
New Rochelle, NY 10804

Price: $137.50 plus $2.50 postage and handling
Both PC boards only $27.50 plus $2.50 shipping and handling

New York residents must include sales tax.

Part 5
Receiver Projects

This section discusses Subsidiary Communications Authorization (SCA) and FM reception, what SCA is, a review of basic FM theory, and how SCA is received. The method of detection is discussed in detail, and contruction details for a complete FM stereo receiver with SCA reception capability are given. Schematics and PC board layouts are covered, and a complete parts list is furnished. The receiver uses modern MOSFET/IC techniques, and is electronically tuned, using varactor diodes. Alignment, being very simple, requires no test equipment. A pre-etched and drilled PC board and a list of parts is available for those who wish to construct this receiver and/or experiment with modern FM receiver circuitry. It is rare that you see broadcast receiver construction articles any more. With modern IC devices, ceramic filters, and PC boards, construction and alignment is greatly simplified. This receiver project is well within the capability of the experimenter with some experience in audio or in TV servicing, and requires no critical alignment. Our receiver, during development, worked the first time it was tried. Most of the work involved the improvement of circuitry, and simplification of construction. You can learn much from building your own FM receiver. Unlike most hobby "gadgets," it will be used long after it is constructed. It is well worth the effort.

PROJECT 10: MODERN SCA AND MPX FM BROADCAST RECEIVER

Subsidiary Communications Authorization has been around for a number of years. It has been used for various purposes, such as background music transmission without commercials, medical news, second language programming, and radio services to the visually handicapped.

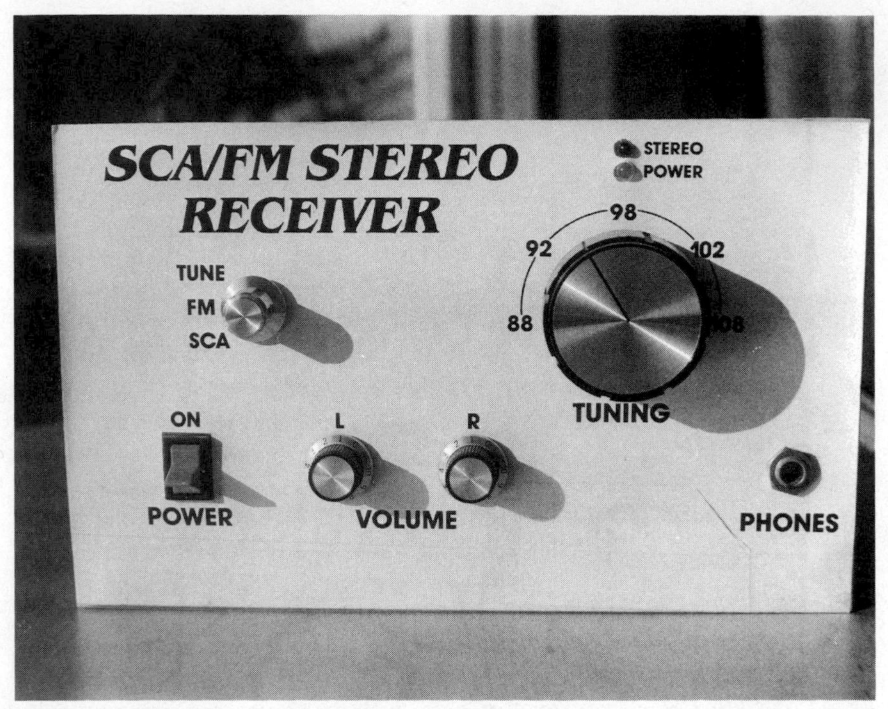

Fig. 5-1. SCA/FM stereo receiver mounted on wooden base.

SCA makes use of subcarriers in the audio baseband of an FM broadcast signal. Figure 5-2 shows the FM baseband spectrum. Subcarrier frequencies of 41, 57, 67, and 92 kHz are commonly used. The 41 kHz subcarrier cannot be used if FM stereo is being broadcast. A fairly new service called Automobile Radio Information (ARI) uses 57 kHz. By a wide margin, 67 kHz is the most popular subcarrier (80 – 90%) with 92 kHz and 57 kHz accounting for the balance.

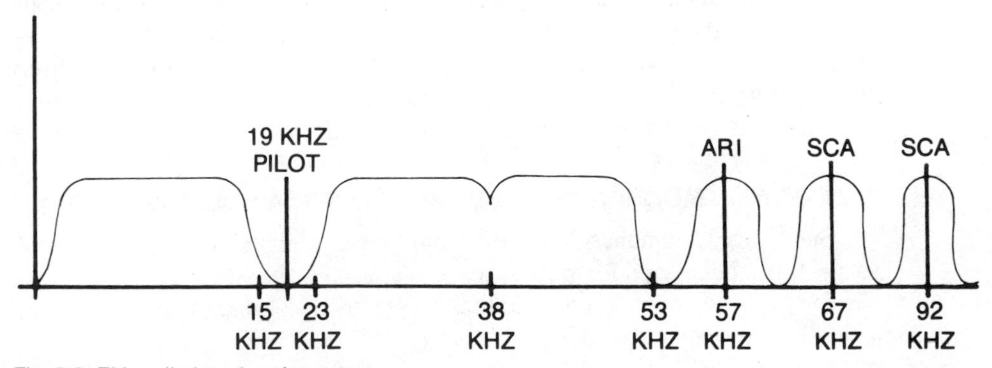

Fig. 5-2. FM audio baseband spectrum.

Before we go too deeply into details, let us digress briefly for a discussion of FM theory, for the benefit of those readers not familiar with FM.

An FM (abbreviation for Frequency Modulation) signal is simply any rf signal that is modulated instantaneously changing the frequency of the signal in accordance with the modulation. (In our case, the FM baseband signal.) Since frequency is measured in cycles per second (Hz), and cycles per second can be thought of as the number of times a sine wave goes through a complete cycle (0 to 360 degrees) in one second, you might consider frequency as a rate of change of phase (degrees) per unit time (second). Frequency modulation is related to phase modulation. Phase modulation is that modulation in which the relative phase of a carrier wave, with respect to a reference, is varied in accordance with a modulating signal. In order to change the phase of a signal, there has to be an instantaneous change in frequency. As an example, if we want to change the phase of a carrier from 0° to 180°, we must "reverse" one cycle (see Fig. 5-3).

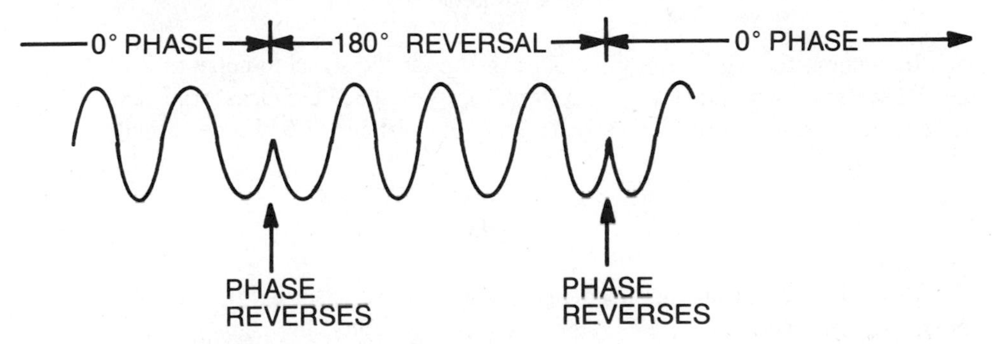

Fig. 5-3. Phase reversals—phase modulation.

If we want to vary the frequency of a signal, the instantaneous phase of the signal must continuously change. For example, suppose we want to sweep from 100 MHz to 101 MHz in one second. This implies a frequency change of 1 MHz per second, (or 1 Hz per microsecond). In one microsecond, the 100 MHz carrier will go through 100 cycles. In this brief period, 1 cycle must be added so that after one second, one million cycles per second has been added. This can be done by advancing the phase of the carrier 1/100 cycle, or 3.6 degrees, every cycle, so that after 1 microsecond, 101 cycles will have been gone through, and the new *instantaneous* frequency will be 100,000,001 cycles (100 MHz + 1 Hz) per second. Notice that a *constant rate of change of phase has been introduced to produce a "sliding" frequency*. If this rate of change of phase is kept up indefinitely, the frequency would continue to sweep upward by 1 MHz per second, so that at $t=1$ second the carrier frequency would be 101 MHz, at $t=2$ seconds, 102 MHz, etc. If we retarded the phase by the same 3.6°, we would have produced a *decrease* in frequency of 1 MHz per second (in this example).

If you did not follow this, or cannot understand the concept, simply remember that a change in frequency implies a phase change, and vice versa.

The deviation of an FM signal is the component of change in carrier frequency that the modulating signal produces. Note that, in general, the *frequency* of the modulating signal is *not* the same as the deviation. In the USA, FM broadcast stations are permitted $\pm$ 75 kHz deviation, which corresponds to 100% modulation. Both a 20 Hz audio signal and a 20 kHz audio signal can produce 75 kHz deviation. The amplitude of the modulating signal (program audio) produces the deviation. If one volt of audio produces $\pm$75 kHz deviation, then one tenth of a volt would produce $\pm$7.5 kHz deviation. Deviation and modulation frequency are independent variables.

The ratio of deviation to modulation frequency is called the *modulation index*. The greek letter β (beta) is usually used to represent this quantity.

$$\beta = \frac{\text{deviation}}{\text{modulation frequency}}$$

In general, the higher the value of β, the better the signal to noise ratio of an FM system compared to an AM system. In a typical FM broadcast transmitter situation, with a 1 kHz audio signal at 50% modulation (37.5 kHz deviation).

$$\beta = \frac{37.5 \text{ kHz}}{1 \text{ kHz}}$$

Note that, for constant modulation signal level, β increases at low audio frequencies, and decreases at high audio frequencies. Thus, the signal-to-noise ratio of an FM system is poorer for high audio frequencies than for low frequencies. This is the reason *pre-emphasis* is used in FM broadcasting, so as to boost the modulation index at the high end of the audio spectrum.

It turns out that, additionally, the noise spectrum of a detected output, in an FM receiver, has a rising noise-vs-frequency characteristic. If we employ *de-emphasis* at the receiver, we not only correct for the pre-emphasis at the FM broadcast transmitter (which would make the audio sound "tinny"), but also pick up *additional* signal-to-noise improvement at the receiver as well. Pre-emphasis and de-emphasis are used in both broadcasting and commercial FM two-way radio (which usually uses $\pm$5 kHz deviation) to materially increase signal-to-noise ratio. This gives FM its excellent quality and "clean" sound. Since most static is AM in nature, very little static gets into the system. An FM receiver deliberately limits the incoming signal to a constant level before detection further reducing AM static.

The occupied bandwidth of an FM signal, at first glance, appears to be simply the peak-to-peak deviation. However, this is not always true. A 75 kHz deviation FM broadcast signal rquires somewhat more bandwidth. It is necessary to know this required bandwidth for various reasons, among them channel spacing, necessary receiver bandwidth, and signal-to-noise ratio considerations.

For very high modulation indices, the necessary bandwidth is very close to the peak-to-peak deviation. As an example, this would be true for, say, a 100% modulated FM signal (75 kHz in commercial broadcasting) with a low audio frequency modulation (i.e., 20 Hz). However, the situation changes for low modulation indices. At a modulation index of 10 the bandwidth required would be about 2.8 times 75 kHz, or 210 kHz. At a modulation index of 5 (for example 15 kHz audio, with 75 kHz deviation) about 3.3 times 75 kHz, or 247 kHz bandwidth, would be required. (See Fig. 5-4.)

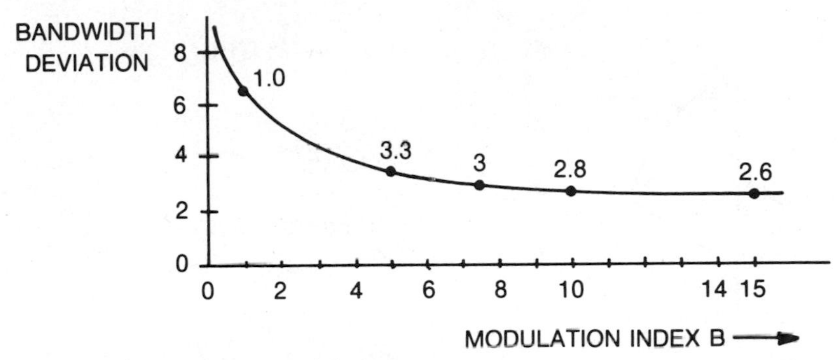

Fig. 5-4. Occupied bandwidth vs. modulation index.

This bandwidth change is due to the sidebands generated in FM. The sidebands, as in the AM case, are separated by the modulation frequency from the carrier. However, depending on the modulation index, the sidebands vary in amplitude. They appear, reach a maximum, then, at higher modulation indices, some sidebands disappear. In fact, the carrier disappears at a modulation index of 2.4. This means, if we apply a tone of about 31 kHz to an FM transmitter and adjust the level of this tone to produce a deviation of 75 kHz, the carrier will actually null out. Of course, the FM signal has not disappeared—all of its energy is now contained in sidebands spaced 31 kHz from the carrier—at ±31 kHz, ±62 kHz, ±93 kHz, etc. (This is similar to what happens in an SSB signal, where all the energy is in one sideband, with the carrier and other redundant sidebands being suppressed.) As a matter of fact, the null of the carrier at a modulation index of 2.4 can be used to measure the deviation of an FM transmitter, using a spectrum analyzer.

The sidebands vary in amplitude in a pattern that can be expressed mathematically by Bessel functions of the first kind. A given sideband N, where N is the order (carrier N = 0, first sidebands N = 1, second N = 2 etc.) of the Bessel function. We will not discuss this further, except to state that Fig. 5-5 is an approximate illustration of how these functions appear. Further discussion would require the reader to have a background in analytic geometry and integral calculus. We cannot assume this and instead refer the interested reader to a more advanced text on modulation theory, such as Schwartz's "*Information Transmission, Modulation and Noise*" (McGraw-Hill 1959).

As a rule of thumb that works out in practice, the required bandwidth is *approximately* twice the deviation plus the highest modulating frequency. This figures out to about 240 kHz for an FM-Stereo/SCA receiver, and bandwidths of 230 to 280 kHz are used in practice. Note that the FM detector must be *very* linear over this bandwidth to reduce distortion on the recovered audio. As another example, commercial 2-way FM radio used for police, fire, taxicab, etc. and also amateur radio 2 meter FM use ±5 kHz deviation with audio restricted to 3 kHz (300 Hz). These receivers use 13 kHz bandwidth i-f filters. This is

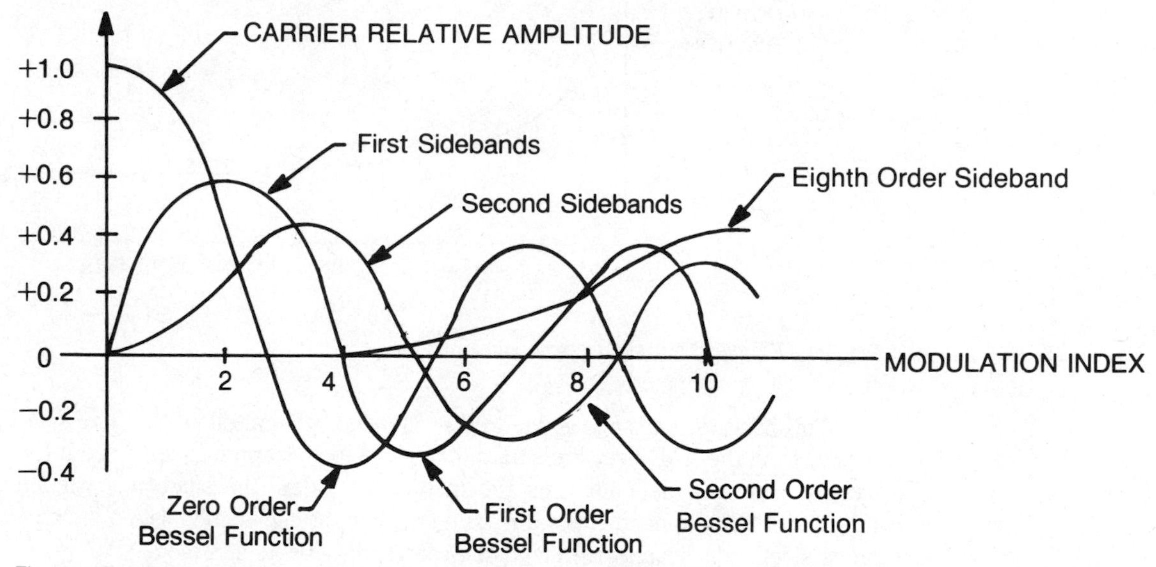

Fig. 5-5. Relative amplitudes of carrier and sidebands vs. modulation index.

twice the deviation plus the highest modulation frequency, as is seen from the figures given.

Referring back to Fig. 5-2, the various components of an FM broadcast (stereo) signal are as follows:

1. Audio baseband 0–15 kHz. This is comprised of the *sum* of left and right audio channels, and is the basic program audio a monophonic FM radio would receive.
2. Stereo baseband (19 kHz and 23–53 kHz). This consists of a pilot carrier at 19 kHz, and a DSB (double sideband) suppressed carrier AM signal centered at 38 kHz. The 38 kHz carrier is suppressed, and a low level pilot carrier at 19 kHz is used to regenerate the 38 kHz suppressed carrier. In this way the 38 kHz DSB AM signal is recovered and

detected. It has the *difference* of left and right channel audio. This difference signal is *zero*.

3. ARI (Automobile Radio Information) subcarrier. This is 57 kHz (3 times 19 kHz) and is a narrow-band (voice frequencies) channel used for traffic bulletins. Originated in Europe, this service has been instituted in America and may become popular in the future. It is currently used on a trial basis in major metropolitan areas such as New York City.
4. SCA subcarrier at 67 kHz. (Also at 92 kHz.) This subcarrier is used for "hidden" radio programs, background music and digital data transmission. It is frequency modulated with a ±7.5 kHz deviation *maximum*. It is not a high fidelity service—it is similar in AM radio with its bandwidth capability of up to about 5000 Hz.

Our immediate interest is in this SCA channel. It is normally used as an auxiliary, income-producing service by the operators of FM broadcast stations. However, we do not get something for nothing. It requires about 10% max of the total 75 kHz deviation (100% modulation). This reduces the available energy in the modulation baseband for the main carrier. In practice, this costs about 1dB. Normally, it would not be missed. The use of both 67 kHz and 92 kHz would cost about 2dB. In certain areas of the country, where "loudness wars" are commonplace, such as in the New York City area rock music market, this 2dB would be somewhat noticeable. However, additional revenues to the FM station from the leasing of these subcarriers to various services are another source of income.

Our technical problem encountered is in the "crosstalk" between the SCA channels and the main stereo program. However, the modern phase-locked loop type of stereo demodulator used in almost all of the FM receivers manufactured since about 1975 have about 40dB or better rejection of the SCA channels, and crosstalk is not a problem. However, the problem of multipath reception of FM signals can cause distortion and unwanted crosstalk of main channel audio into the SCA channels. Transmitter distortion can also contribute to this, although the FM broadcast operator will take precautions to reduce this. SCA reception tends to be somewhat noisy, but a good FM antenna system, a sensitive receiver with wide bandwidth (needed anyway for stereo) and adequate signal levels at the receiver, will generally yield good reception. However, do not expect SCA to be as good as main-channel audio—you do not have the bandwidth available. It is comparable to AM broadcast reception, which is adequate for voice, background music, and data transmission.

The SCA subcarrier can be and is being used for data services such as stock quotes, telemetry, and has even been used for slow scan color TV signals.

How do we receive these signals? *First a word of warning*. SCA, at present, is a private radio service *not* intended for general use by the public. Some services (such as the Physician's Radio Network) rent decoders for this service to medical personnel. However, *at present* there is no law against an experimenter constructing a receiver, and listening to these signals for *his own purposes. Do not record these signals, divulge them, listen to them in public, or use them for any commercial purposes. You may otherwise be subject to prosecution for theft of services, etc.*, with appropriate penalties.

Figure 5-6 is a block diagram of the SCA and FM stereo receiver. It uses a MOSFET rf amplifier, with both input and output (mixer) circuits tuned by a varactor diode, which can be thought of as a voltage-variable tuning capacitor. The dc tuning voltage is variable from about 1.5 to 8 volts dc. The local oscillator operates at the tuned signal frequency − 10.7 MHz. The oscillator is also tuned by means of a varactor diode (varicap is also used as a name for this component). The three varactors are biased by a common dc bias line, so as to simultaneously tune the rf amp, mixer, and oscillator circuits. The mixer output circuit is tuned to 10.7 MHz and feeds an i-f preamplifier that has about 30dB gain. This preamplifier uses two transistors and three fixed-tuned ceramic i-f filters centered at 10.7 MHz. No alignment is necessary and the bandwidth is about 250 kHz. This eliminates complex sweep alignment and allows a novice builder to automatically get the good i-f bandpass response necessary for FM stereo/SCA reception.

An RCA CA3189E (or National LM3189N) IC performs limiting and quadrature detection of the FM signal, and recovers the original audio baseband. This chip has high gain, good limiting, very low distortion detection. It also provides AFC voltage to correct drift in the local oscillator and to aid in tuning a selected station. Due to the very high gain, layout is *very* critical and we strongly recommend the PC layout herein. Otherwise you are on your own with a possibly very difficult rf stability problem.

Quadrature alignment is very easy. Simply adjust the quadrature network for maximum audio recovery. No alignment equipment is necessary—the ear does a fine job.

The audio output of the CA3189E (or LM3189N—they are interchangeable) is fed to a 2N3565 audio amplifier for an output level of about 3V peak-to-peak. This baseband audio is used to feed the SCA detector (NE565) and MPX detector (MC1310).

A high pass and twin tee L-C filter designed to reject frequencies below 50 kHz passes the SCA carrier to the NE565 PLL. The output of the PLL is the VCO control voltage, which follows instantaneous frequency variations of the 67 or 92 kHz subcarrier. This output (about 50 to 100 millivolts peak-to-peak) is the SCA audio. It is passed through a low pass R-C network to remove noise above about 5 kHz and improve the signal to noise ratio. An SCA audio amp (2N3565) amplifies this to about 500 millivolts peak-to-peak, sufficient signal to fully drive the audio power amplifiers. Tune-up is simple and noncritical—adjust the VCO potentiometer for clearest audio. Signals from 57 to 92 kHz can be received.

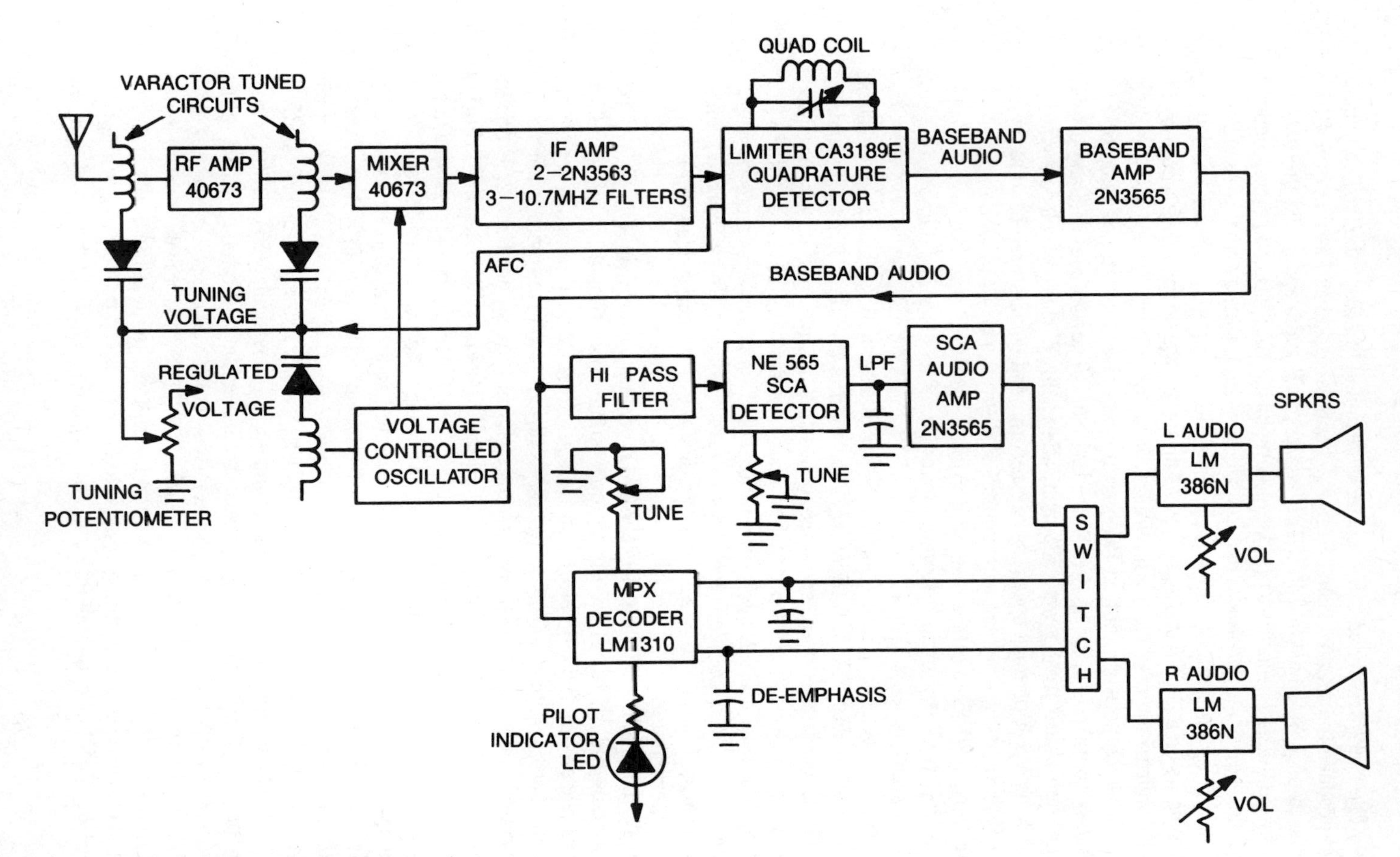

Fig. 5-6. SCA/FM stereo receiver block diagram.

The multiplex decoder MC1310P is designed to accept the baseband audio and produce the original L and R audio channels. Baseband audio of the 2 – 3V peak-to-peak is fed to the MC1310P and L and R audio appear on the outputs. Shunt connected capacitors provide de-emphasis. An LED can be connected to the chip so as to provide a visual indication when a stereo signal is received.

Tune-up consists of simply adjusting the tune potentiometer so that the LED lights on a known stereo station. The stereo will be readily detected by ear as a confirmation of correct adjustment. It is not critical.

A switch network (not on PC board) connects either stereo audio or SCA audio (or both) to the dual power amps LM386N. They are identical and provide about one watt of total power (500 milliwatts each). While they will do a good job of driving a pair of small speakers, we recommend using stereo head phones of the Walkman® variety for best sound. If desired, the LM386N amps can be omitted and the outputs of the switching network fed into the line inputs or tuner inputs of any audio system. About 500 millivolts of audio into a 10k ohm load is available.

Sensitivity is excellent—about 2μV for 30dB quieting. This is comparable to a good quality FM hi-fi receiver. Distortion is low (less than 1%) and the receiver, as stated before, sounds excellent. With a pair of 8 or 16 ohms 6×9 speakers, a wood cabinet and a few knobs, this makes an excellent FM receiver, especially for fringe areas where high sensitivity is necessary. SCA reception is also excellent, within its inherent technical limits. Further refinements would include tone controls and digital PLL tuning, which will be left to the experimenter.

A schematic of an FM-MPX/SCA receiver is shown in Fig. 5-7. Upon examining the PC layout, you may notice some unusual methods of layout and parts placement. This receiver is a high gain, very high frequency device and layout is *very* critical in order to avoid various problems with stability and unwanted coupling of signals into circuits where they do not belong. Frequencies as high as 120 MHz are present in the LO (local oscillator). Over 110dB of gain is present at the i-f frequency of 10.7 MHz. Since electronic tuning is used, any stray signals on the tuning voltage will cause the LO to be FM modulated, and appear in the receiver output as unwanted audio. Peak FM deviation is 75 kHz. A 75 Hz uncertainty in oscillator frequency will cause an FM modulation 60dB below peak audio level to appear on a received signal. A 750 Hz shift will result in 40dB below peak. While 60dB below peak is not too bad, 40dB is awful. It shows up as residual background noise. In particular, 60 or 120 Hz hum may be one of these undesired signals. An inspection of the tuning curve of the receiver reveals that one volt change in the tuning voltage will cause a 2 or 3 MHz shift in the receiver tuning voltage. A little arithmetic will show that one millivolt will cause a 2 to 3 kHz shift, and therefore, to meet the 60dB or better signal to noise ratio, the tuning voltage to the LO should be ''clean'' to the level of about 25μV peak-to-peak or less. The mixer and antenna tuning voltages can tolerate more noise, since the tuned circuits in these stages do not materially affect the signal, (although theoretically they could introduce phase modulation).

In order to bring a 2μV received signal up to several volts peak-to-peak, over 120dB gain is required. An FM receiver should limit on its own internally generated noise. Adequate sensitivity (better than 3μV) should be obtained over its tuning range of 87 – 109 MHz, with good spurious signal rejection. The detector should not introduce any unnecessary distortion and it should be possible to get better than 1% THD (total harmonic distortion) overall. Stereo separation should be at least 30dB between L and R channels.

In addition, the receiver should have some form of AFC (unless digital phase-lock loop tuning is used) to prevent inevitable oscillator drift from detuning the receiver. And, although it may seem like a tall order, the receiver should be capable of being aligned with only a VOM or even "by ear," with absolutely no critical "tweaking" requiring test equipment and considerable experience few of us possess.

This layout was designed based on much experience with receivers and the particular IC devices and transistors used in this design. The PC board is single sided without plated through holes. This gives the hobbyist with limited resources the chance to reproduce the PC board at home, and also keep costs down. It is also a lot easier to work with single sided PC boards, since you can see through them with a strong light source. In addition, unwanted stray capacitances, a problem at FM broadcast frequencies, are minimized. However, a single sided board is also much more difficult to lay out and get good rf grounding and rf stability. We *strongly* urge that you use our layout if you are contemplating building this receiver or any portion of it. In particular, the CA3189E/LM3189N can be extremely *nasty* to stabilize in a poorly laid out board.

In order to allay any doubts you may have, this layout has proved to be stable, reproduceable, and, as aforementioned, worked the first time without stability problems. However, much experience is behind it and it always helps if you know the answers to problems in advance. A discussion of the receiver circuitry and construction hints follows. Either the PC board or a complete kit of parts is available to the experimenter—see the end of the article.

Referring to Fig. 5-7, incoming FM signals from the antenna are applied between the tap on L1 and ground. L1 is the antenna coil and is tuned by tuner C1, and varactor diode CR1 to the signal frequency. CR1 is nominally about 22 pF at 2 volts bias. C1 will normally be set at about 3 to 6 pF (plates about 1/3 meshed) and the rest of the circuit stray capacitance, about 6 pF, which is made up of PC board strays and the input capacitance of Q1 (about 4 pF). This gives a total capacitance of about 32 pF (at 100 MHz) to tune antenna coil L1. CR1 has a variable back bias of from 1.5 to 8 volts dc across it. It varies in capacitance from about 15 pF at 8V to about 30 pF at 1.5V. This swing in capacitance is adequate to tune L1 over the range of 87 – 109 MHz, which is the complete FM

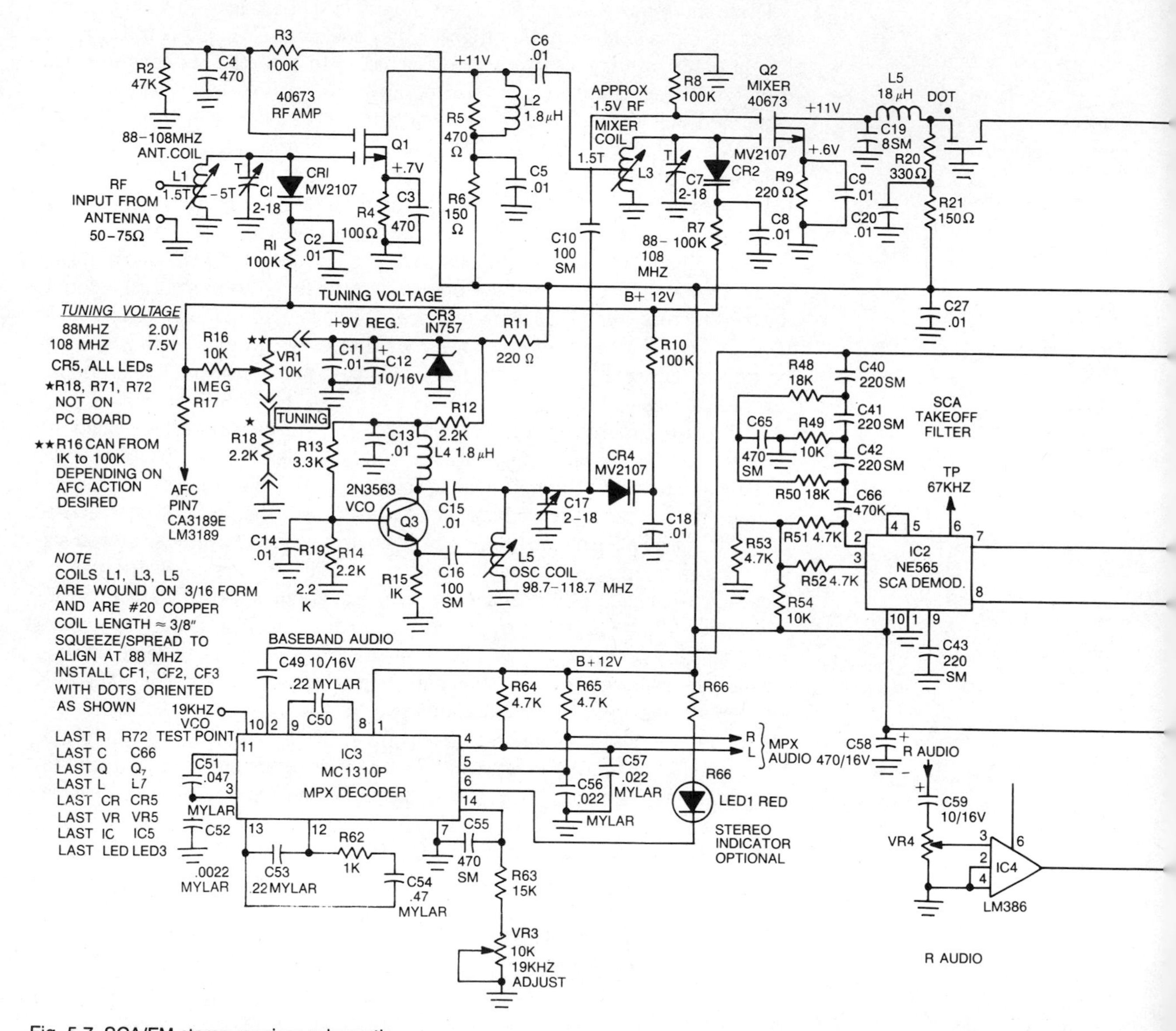

Fig. 5-7. SCA/FM stereo receiver schematic.

broadcast band plus some overlap. The tap on L1 is selected so that the gate of Q1 (gate No. 1) sees a source impedance of about 500 to 1,000 ohms. This is the 70 ohm antenna impedance stepped up by a factor of about 10. This step-up ratio gives an optimum trade-off between noise figure, strong signal handling capability, and stability of the rf amplifier stage.

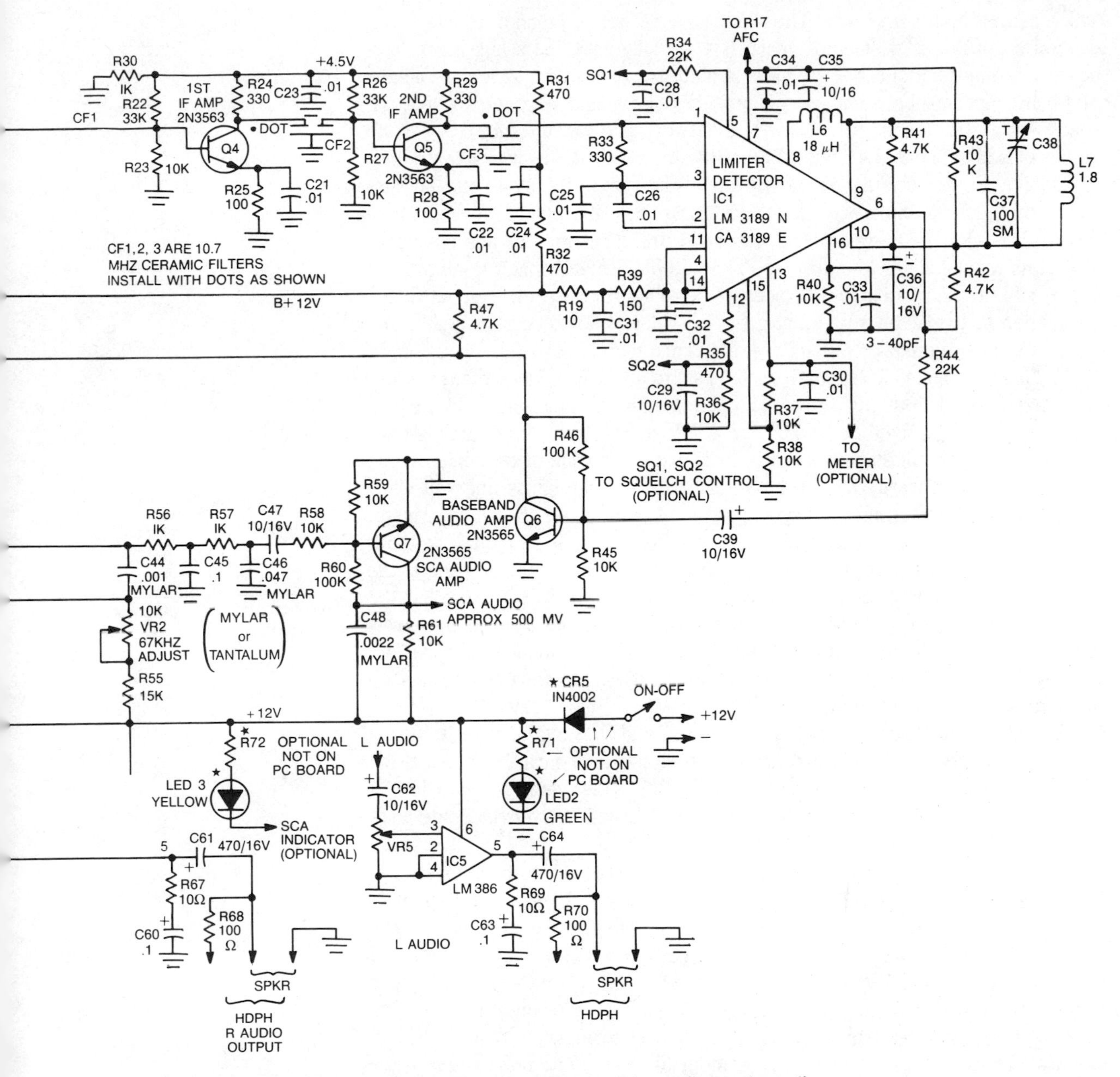

C2 provides an rf ground and allows dc bias from the tuning voltage line to be supplied through R1. It also cleans up any noise present on the tuning voltage line. No dc current flows in R1—hence there is no voltage drop.

Q1 is a 40673 MOSFET device which gives about 15dB power gain in this circuit and a noise figure of 4dB or less (typically 2 – 3dB at FM frequencies).

This ensures high sensitivity. There is no base-emitter junction to cause unwanted rectification of strong signals. (Q1 actually acts like a high gain, low-noise pentode vacuum tube, and its equivalent (electrical) circuit is not unlike a 6DC6 or similar tube.) R4 and C3 provide biasing and rf grounding for the source of Q1 (analogous to cathode bias network in a vacuum tube). A second gate is biased at about +4V with R2 and R3, and C4 bypasses gate 2 to ground. (Analogous to the screen grid if a pentode.) The gain of the stage may be controlled by reducing this bias from +4V (full gain) to −2V (cut-off) for AGC purposes. However, AGC was not necessary in this receiver, and was not used. The drain (analogous to the plate of a pentode) is biased through R6 and L2 to about +11Vdc. Drain current (exactly equal to source current) is about 6 to 8 milliamperes. It may be noted that gate No. 2 draws *no* dc current, unlike the vacuum tube screen grid, which does draw current.

L2 is an rf choke. R5 limits the stage gain to about 6×, since the transconductance (about 12,000 micromhos or, microsiemens if you prefer) and the load is about 450 ohms, fixed by R5 and circuit losses. This amount of gain is optimum to ensure circuit stability, adequate enough to override mixer noise, not so high as to unnecessarily overload the mixer on strong signals and will allow about a 3dB margin for mistracking and errors in alignment of the tuned circuits. C6 couples the rf signal to L3, which is used as a mixer tuned circuit. The tap is chosen at about one third of the way up. The overall voltage set-up seen at the mixer gate is about 36 times, (referred to the antenna). Remember that the gate of the mixer is high impedance (about 3,000 ohms) and that the antenna is low impedance (70 ohms) so that actually the equivalent voltage gain is about 6× and the power gain about 36×, or about 15dB. C5 is an rf bypass and R6 decouples the rf stage from the B+ line.

The mixer input tuned circuit is tuned by C7 and CR2, with strays. It ideally is actually exactly equal to the antenna circuit with regard to capacitances. However, the operating Q is a little higher (about 30). The overall rf bandwidth is about 2 to 3 MHz, which allows some mistracking and still provides adequate image rejection about −30dB or better.

The mixer is driven by a signal of about 3−4 volts peak-to-peak on gate 2. Since the transconductance of the 40673 is a function of the gate 2 voltage with respect to the source the LO signal modulates the transconductance of the 40673 mixer. This results in the 40673 acting as a mixer. R8 returns gate No. 2 to dc ground. R9 and C9 provide about 0.6 volt bias, which makes both gates about −0.6V with respect to the source. C19 and L5 provide about 15:1 impedance step-down, matching the 330 ohm i-f filter terminating resistor R20 to the drain of the 40673 mixer (which is about 5k ohm). The power gain of the mixer (ratio of i-f signal at 10.7 MHz to rf input signal) is about 12 to 15dB, depending on LO drive level.

The LO uses a 2N3563 transistor (Q3) operated at about 4 volts. R12, R13, R14, and R15 establish the operating point at about 4V at 1.5 milliamperes. The LO is actually a VCO, or voltage controlled oscillator. It operates in common base mode for rf. C14 grounds the base for rf. L4 is an rf choke to feed dc voltage to the collector of Q3. C15 couples the tank circuit L5, C17, and varactor diode CR4. The tank circuit determines the oscillator frequency, which should be 10.7 MHz above (or below) the signal frequency. In this receiver, the LO operates 10.7 MHz above the incoming signal. Therefore, it must tune from about 98 to 120 MHz. The spacing should be 10.7 MHz over the entire tuning range of 87 – 109 MHz. R16 and R17 are used to couple AFC correction voltage to the tuning line, eliminating the need for a separate AFC tuning diode. R16 can be anything from 1k to 100k depending on how much AFC is desired. We used 10k ohm.

L5 and C19 as before mentioned, match the mixer to ceramic filter CF1. L5 and C19 also help prevent unwanted VHF components from leaking into the i-f stages, which could cause spurious responses. CF1 is a piezo electric device that is the equivalent of an i-f transformer. It has a 1dB bandwidth of 250 kHz, centered at 10.7 MHz. It acts as a double tuned circuit. Insertion loss is about 6dB, and termination impedance is specified at 330 ohms. R20 terminates the input and the input impedance of 1st i-f amp. Q4 terminates the output of CF1. Q4 is biased with R22, R23, R24, and R25 to about 2 milliamperes at a VCC = 4 volts. CF2 couples Q4 to Q5 which is biased identically to Q4, using R26, R27, R28, and R29. C21 and C22 bypass the emitters of Q4 and Q5 respectively. R32, R31, C24, and C23 provide decoupling of the i-f stages Q4 and Q5 from the power supply line. R30 is a ballast resistor used to determine the operating points of Q4 and Q5. It results in a +4.5V supply to these stages, forming a voltage divider with R31 and R32. CF3 couples i-f signal to the limiter detector stage (IC1 and peripheral components). CF1, CF2, and CF3 shape the i-f bandpass of this receiver. They are fixed tuned and no alignment is required.

The adjustment of C17 and L5 can accomplish the oscillator tracking very closely. L5 is the oscillator coil. A tap at about 1.5 turns provides feedback to the emitter of Q3. C16 is used to couple rf to the emitter of Q3. It is 100 pF in value. Actually, C15 could be a lot smaller than the 0.01μF used, but here is the possibility of spurious rf resonance with L4 if it is made too small. The resonant point of the 1.8μH choke and 0.01μF capacitor is ground at 1.1 MHz—too low to have any appreciable Q and cause unwanted parasitic oscillations. As in the mixer circuit and the antenna circuit, C18 a .01 capacitor and R10, a 100k resistor, couple dc tuning voltage from the tuning voltage line.

The oscillator derives its dc power from a regulated +9V source from zener diode CR3, filter capacitors C11 and C12, and 220 ohm ballast resistor R11. The tuning voltage is derived from potentiometer VR1 connected across the 9V supply. Since the tuning voltage does not go below 1.5V a resistor R18 of about 2.2k can be connected in series with the low end of the pot (not necessary but limits the low frequency coverage of the receiver). If the tuning line is grounded to zero volts, the receiver will tune down to 75 MHz to establish the low band limit of 87 MHz. Gain of Q4 and Q5 is about 26 to 30dB. This gives a total gain so far, from the antenna, of about 55 to 60dB, ensuring that the front end noise will cause limiting in IC1. This noise floor is about 0.3 to 0.4 microvolts, referred to the input. Therefore, we would have about 100 to 300 microvolts of noise present at the output of CF3. This will rise to several millivolts in the presence of a received signal of 3 microvolts or so. The maximum output of Q5 is about 0.25 volts i-f signal, which is the saturation point, no matter how strong a signal is received. IC1 can easily handle this without distortion. No AGC was found necessary in this receiver.

IC1 provides all the functions of an FM i-f system. It includes a 3-stage limiter, signal level detectors, a quadrature detector and an audio amplifier with optional muting circuit (squelch). It has its own internal regulators for dc voltages, and can drive an external tuning meter. All of this is a single 16 pin DIP package configuration. For further details, consult the manufacturer data sheet. IC1 is either a CA3189E (RCA) or LM3189N (National Semiconductor) and either type will work well in this circuit, and for our purpose here they are directly interchangeable.

IC1 will limit on input signals as low as 25μV. It has extremely high gain and great care is needed in layout of the PC board. Therefore, unless you are experienced with rf engineering and PC layout, *do not* change our layout—it works very well. It is not the same as recommended by the manufacturer, which did not fit our layout. However, we have had lots of experience with this IC and we know what works. Do *not* use a socket or spacer pad, DIP header, etc. The chip *must* be grounded at pins 4 and 14 with virtually zero lead length. C38 and L7 must be kept as far away as possible from CF3 to prevent feedback.

Input signal from CF3 is applied to pin 1 of IC1. R33 is a bias resistor and also terminates CF1. C24 and C26 are rf bypass capacitors. R19, C31, and R39, and C32 are used to feed B+ voltage to pin 11 of IC1, and provide rf decoupling as well. R34, C28, C29, R35, and R36 provide termination for squelch (mute) circuits in the chip. We did not use the mute feature. (If you want to use it, connect a 10k pot from SQ2 to ground, and connect the wiper to SQ1 junction C28 and R34, remove R36.) R37 and R38 provide a load for the tuning meter circuit connected from pin 13 to ground. We did not use a meter, but pin 13 provides a handy test point for aligning the front end tuned circuits. R40 and C33 terminate the AGC circuit at pin 16, which we did not need.

The components L7, C38, and C37 form a tuned circuit at 10.7 MHz. R41 acts as a swamping resistor to obtain a wide bandwidth of the quadrature circuit C37 – C38 and L7, L6 provides drive voltage from pin 8 (i-f out) to pin 9 (quadrature detector input). It is somewhat critical for proper squelch circuit operation. It should be about 20μH at 10.7 MHz. (18 or 22μH works, we used 18.) R43 provides a load for the AFC circuit, and R42 biases the audio circuit in the

IC. C33, C34, C35, and C36 are bypass capacitors for rf and audio. C38 is used to tune the quadrature circuit to 10.7 MHz. It is adjusted for best received audio and zero dc voltage across R43, which should ideally occur at the same setting of C38. Recovered baseband audio is present at pin 6. It contains the FM baseband and SCA signal. It is taken off through R44 and C39.

Q6 (a 2N3565) is an audio amplifier stage. It is set up for a nominal gain of about 5 (the ratio of R44 to R46 is the approximate gain of this stage). R45 and R46 bias Q6 to about $V_{CE} = 6V$ $I_C = 1$ milliampere. R47 is a load resistor. About 2 volts of baseband audio is present at the collector of Q6.

Audio from Q6 is fed to two separate circuits. One circuit is an SCA demodulator. The other is a multiplex decoder. Referring to Fig. 5-7, C40, R48, R49, C41, C42, and R65 form a twin tee filter and signal attenuator. This attenuates the 38 kHz MPX and audio components below about 50 kHz. IC2 is a NE565 phase-locked loop. It contains a voltage controlled oscillator and phase detector comparator. If a signal of sufficient amplitude (about 100 millivolts) is fed into pin 2 or pin 3 and sufficiently close (say within ±30%) of the VCO frequency, the VCO will follow any changes in the frequency of the signal at pin 2 or 3. This control voltage for the VCO is present at pin 7 and is a linear function of the input signal frequency. Therefore, the NE565 can function as an FM detector, with no external inductive components required. At the SCA frequencies of 67 or 92 kHz, coils can become rather large and somewhat costly. The elimination of these coils and their alignment is advantageous. The NE565 is biased by external resistors R51, R52, R53, and R54. The VCO frequency is determined by C4 and the resistance of VR2 and R55. VR2 is adjusted so that the VCO frequency, which can be measured at pin 4, is near 67 kHz.

Adjustment of VR2 is not critical, and simply adjusting VR2 for clearest SCA reception is adequate. (If 92 kHz operation is desired, R55 should be changed to about 6.8k ohms.) C44 (a .001μF Mylar capacitor) is used as a loop filter for the phase-locked loop. Audio appears at pin 7 of the NE565. R56, C45, R57, and C46 suppress 67 kHz components and eliminate excess high frequency noise and provide some de-emphasis, SCA reception can be somewhat noisy unless adequate signal is available to the receiver. C47 couples audio (about 50 millivolts) to amplifier stage Q7, R58, R59, R60, R61, and C48. This stage brings up the detected audio level to about 500 millivolts, comparable to the MPX audio L and R channels from IC3. C48 further reduces high frequency noise. R59 and R60 are biasing resistors, and R61 is a load resistor for Q7, a 2N3565 audio transistor. SCA audio from Q7 goes to an external selector switch, where it can be routed to IC4 and IC5 (audio amplifiers) when SCA reception is desired. This switch is not shown on Fig. 5-6 since it is not on the PC board. Figure 5-8 shows a suggested switching network. A small DPDT toggle switch would be suitable, or a slide switch or rotary switch could be used. Other arrangements might be the use of a third switch position to simultaneously monitor one MPX channel and the SCA channel. This is left up to the constructors' individual requirements.

IC3, a MC1310P or LM1310N (interchangeable) is fed signal from the collector of Q6 through dc blocking capacitor C49. The MC1310P contains a VCO, a phase-lock loop for regenerating the 38 kHz stereo subcarrier, a lock detector used as a stereo indicator circuit, and a decoder circuit for deriving L and R channels. The internal VCO actually operates at 76 kHz and the 19 kHz and 38 kHz signals are derived from an internal frequency divider. No coils are required and alignment consists simply of adjusting VR3 for a 19 kHz signal at pin 10. In fact, the adjustment can be carried out by simply adjusting VR3 so that LED1 lights when receiving a marginally weak stereo station. The stereo (and correct adjustment) will be pretty obvious to anyone that has ever heard what stereo audio sounds like. C53, R62, and C54 form a compensating network for the internal phase-lock loop (PLL). C50 is also a filter capacitor used in the internal PLL circuit. C55, R63, and VR3 control the center frequency of the internal VCO, which should be 76 kHz. The 19 kHz pilot signal (derived from an internal divider) is available at pin 7 for test purposes. Audio output appears at pin 4 (left) and pin 5 (right). R64 and R65 serve as load resistors for the internal audio amplifiers. C56 and C57 provide de-emphasis for the FM audio. The R and L audio from pins 4 and 5 is fed to the selector switch (not shown on the schematic). See Fig. 5-8.

IC4 and IC5 are LM386N audio amplifiers, with about 0.5 watt output, adequate for driving two 8 ohm speakers. Do not use speakers with less than 8 ohms. Both IC circuits are identical. We recommend that, if more audio is desired, that an external audio amp be connected to this receiver. Audio can be taken from the tops of volume controls VR4 and VR5 and fed to the tuner, line, or auxiliary inputs of an external amplifier. The receiver will sound excellent if a good amplifier/speaker combination is used. The LM386N were meant to drive two small speakers or to be used as drivers for a pair of stereo earphones. The receiver will sound excellent when used with a good pair of stereo earphones. These audio amplifiers have no tone control circuits either. They should be thought of as amplifiers for general purpose listening not as hi-fi power amplifiers. Measured frequency response of the entire receiver is ±1dB from 15 Hz

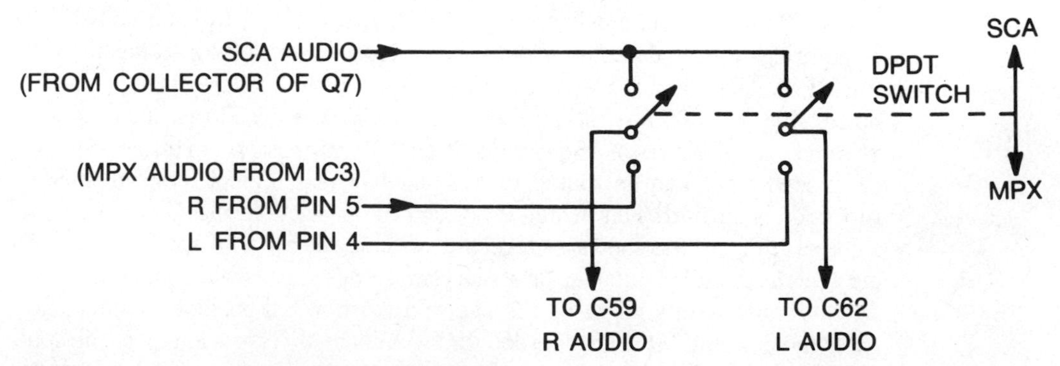

Fig. 5-8. Suggested audio switching for the receiver.

to over 15 kHz (allowing for de-emphasis, the response dips 6dB per octave above about 2,100 Hz, which is normal). (This measurement was made with 32 ohm ''Walkman'' style™ headphones.)

The receiver can be built without IC4 and IC5 if desired to save cost. The inclusion of a more powerful amplifier with tone controls would have made this project more costly, therefore, the use of the LM386N devices.

Audio from the switching network is coupled via C59 (and C62) to volume controls VR4 (and VR5) IC4 and IC5 have a voltage gain of about 20, which means that about 1 volt peak-to-peak audio will drive them to full output. C58 is a +12V line bypass to maintain a low supply source impedance for IC4 and IC5. R67, C60, and R69 and C63 suppress any tendency to high frequency instability in the audio amplifier. C61 and C64 are blocking capacitors to keep dc from the speakers. R68 and R70 are 100 ohm resistors used to limit the drive to 32 ohm stereo headphones to about 120 milliwatts, to protect the earphones (and not break the eardrums of the listener).

The entire receiver draws about 125 milliamps at +12V (recommended supply voltage). The supply should be ''stiff'' and have good no load to full load regulation (250 millivolts or better) about 250 milliamperes at full output will be needed. Noise should be low—less than about 50 millivolts. This should be no problem with a 12V storage battery, or an LM3407-12 regulator IC. Figure 5-9 shows a suitable supply of this type.

Construction and Alignment

Construction of this receiver consists mainly of inserting components in the PC board. All capacitors should be inserted with as close to zero lead length as possible. Make sure to use values specified in the parts list. C15, C19, C37, C43, and C55 should be silver dipped mica or NPO ceramic. Resistors are all 1/4 watt ±10% tolerance types. Make sure that the components are flush with

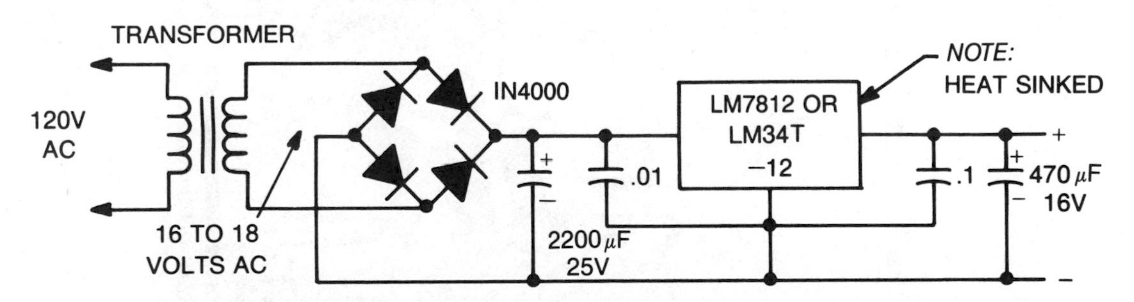

DIODES— IN 1000 SERIES, ECG 125, OR ANY SUITABLE 50 PIV 1A DIODE
TRANSFORMER— 16 TO 18VAC @ ≥ 500 MA

(POWER SUPPLY PARTS ARE AVAILABLE AT RADIO SHACK)

Fig. 5-9A. Suggested power supply for the receiver.

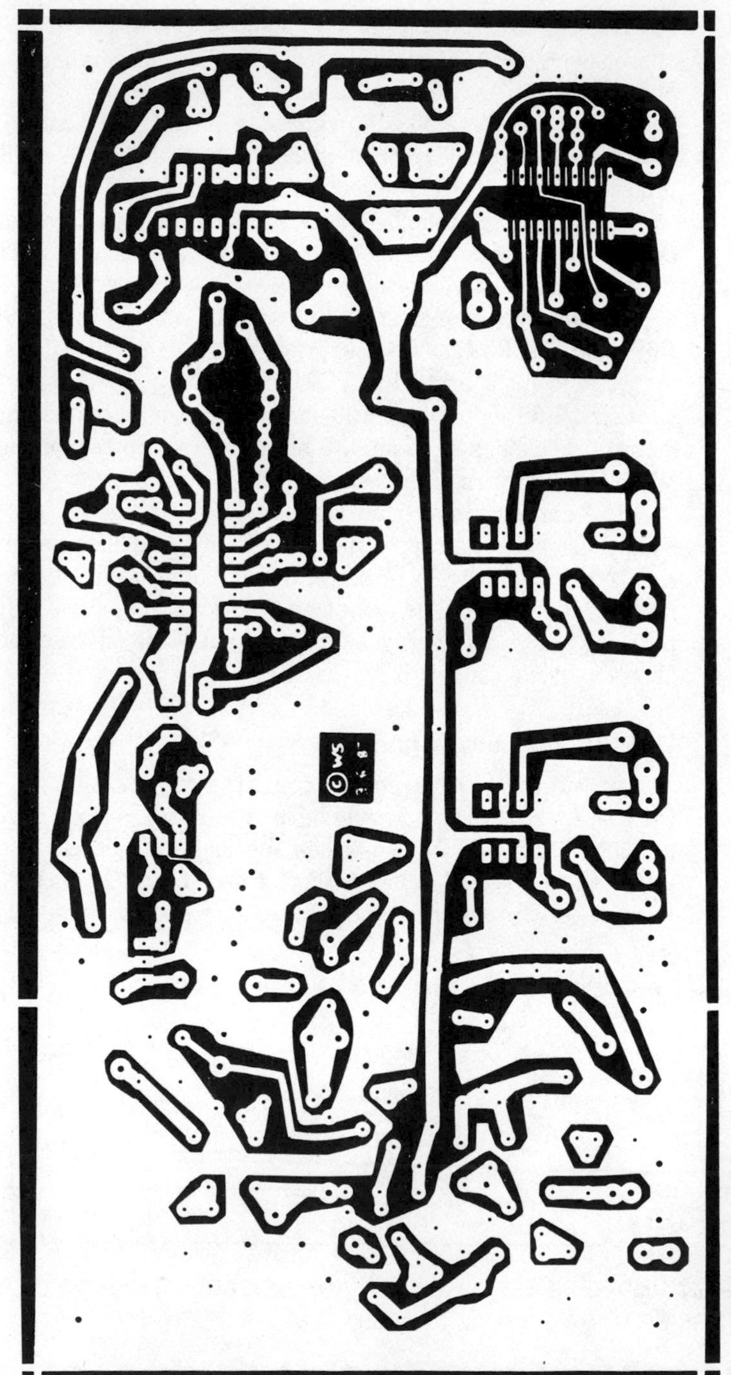

Fig. 5-9B. Project 10 PC layout.

the PC board. Insert components, bend over the leads, leave about 1/16" length past the bend, and cut off the rest. Use a 25 or 40 watt iron. It is better to use a hot iron and solder quickly than use a too small iron and prolong the time that a component has to be heated. Watch polarity on all electrolytic capacitors. Tuning diodes CR1, CR2, and CR4 and zener diode CR3 must be correctly oriented otherwise the circuit will be inoperative. IC devices must be correctly inserted, otherwise they may be damaged instantly upon application of power to the receiver. Be careful with CF1, CF2, and CF3, they are somewhat delicate and easily broken. VR1, the tuning potentiometer, while it can be mounted on the PC board, is better mounted separately, and the use of a high quality (multiturn if possible) 5 to 20k pot is recommended. If the kit of parts listed at the end of this article is purchased, the 10k pot supplied for VR1, while satisfactory for testing and experimenting would best be replaced by a good quality multiturn pot for greatest ease of tuning. Mount the PC board at *least* 1/2" away from any metal or glossy surfaces such as plastic or wood. We used 1/2" metal spacers. Do not forget the AFC jumper between C34 and R17. It is run above the PC board, diagonally, in a straight line between the two points. Should AFC not be needed, simply eliminate this wire.

The completed board should be mounted preferably in a chassis, with a front panel made to satisfy the constructor's needs. Placement of controls is not critical and VR4 and VR5 may be mounted off the PC board, using short wires, or if desired (best for long runs) shielded wire. Shielded wire for VR1 would also be a good idea. Number 24 gauge is adequate. The rf input connection should be with coaxial cable and a suitable connector, such as a UHF, BNC, N., or F type. The headphone jack can be a standard 1/4" or 1/8" stereo jack as required. If an ac power supply is mounted near the PC board, keep 120Vac leads far away from VR1, VR4, and R5. The leads of the controls in this case, should be shielded to prevent ac hum pickup.

Coils for the front end are wound from No. 20 tinned copper wire. They should be wound on a 3/16" mandrel. A 3/16" drill is ideal for this purpose. L1 and L3 are 5 turns and are tapped at 1.5 turns. In order to tap the coil, it is easiest to wind the coil, install it in the PC board, and then, using a length of No. 22 or 24 wire (a clipped-off lead of a resistor or capacitor is useful for this purpose) install the tap on the coil—simply solder it on. See Fig. 5-10. Also, install test points at pin 10 of IC3 and pin 6 of IC2, and the collector of Q6 (see PC board layout). It might be handy to install them at IC1 pin 13, pin 7, and pin 10 (optional). See Fig. 5-11 and Fig. 5-12.

The coil L5 is like L1 or L3 except only four turns are used. Mount the coils about 1/8 to 3/16" away from the PC board. Tap on L5 is at 1.5 turns also.

Alignment is very simple. First, check that *all components are inserted correctly*. Check for solder bridges and unsoldered or poorly soldered joints. When you are sure everything is OK, proceed as follows:

1. Set all pots at the middle of their range.

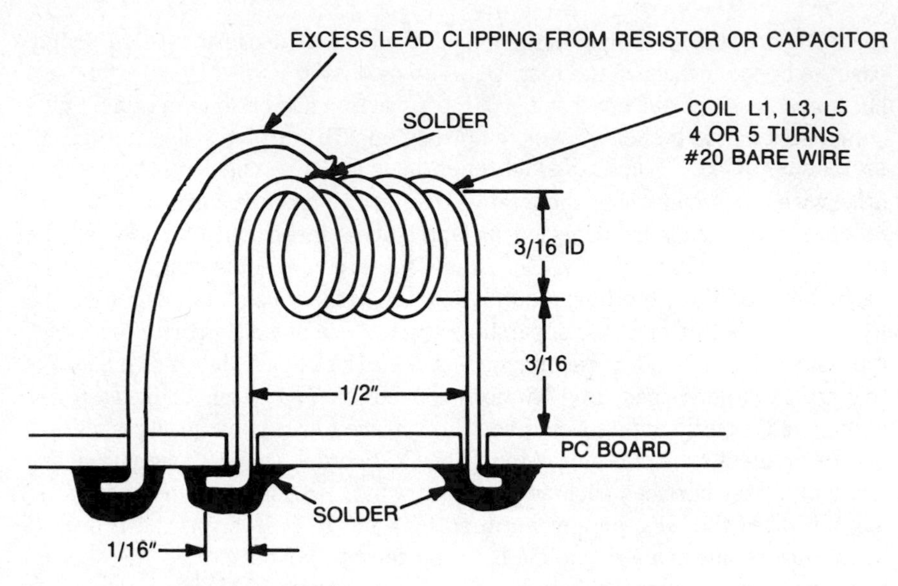

Fig. 5-10. Coil construction details.

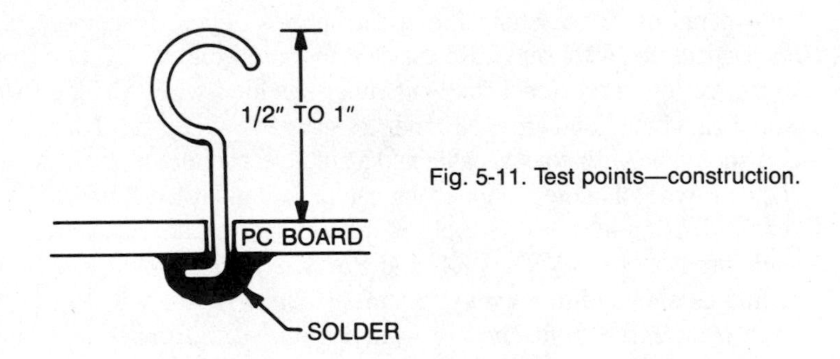

Fig. 5-11. Test points—construction.

2. Set C38 so it is 1/2 meshed (quadrature adjust).
3. Set C1, C7, and C17 (antenna, mixer, and VCO) so they are about 1/4 meshed.
4. Connect two 8 ohm speakers to the audio outputs, or a pair of 32 ohm headphones between R68 – R70 and ground.
5. Connect about 6 feet of hookup wire to the rf input (tap on L1) to serve as a temporary antenna.
6. If you have not done so, wire up the switching network shown in Fig. 5-8, (or else simply use jumper leads) and connect to the PC board.
7. Measure dc resistance, using the positive meter probe on B+ line and the negative probe on ground foil. You should read above 100 ohms. If not, something may be wrong. Note that on some model multimeters, the red probe *may be negative*. If you get a low reading, try reversing the ohmmeter leads.

Table 5-1. Tuning Curve of the SCA/FM Stereo Receiver.

TUNED FREQ. MHz	TUNING VOLTS*
87	1.6
88	1.8
90	2.1
93	2.7
96	3.4
98	3.8
100	4.4
102	5.0
105	6.1
108	7.7
109	8.3

*Average values — some deviations may be expected in any given unit.

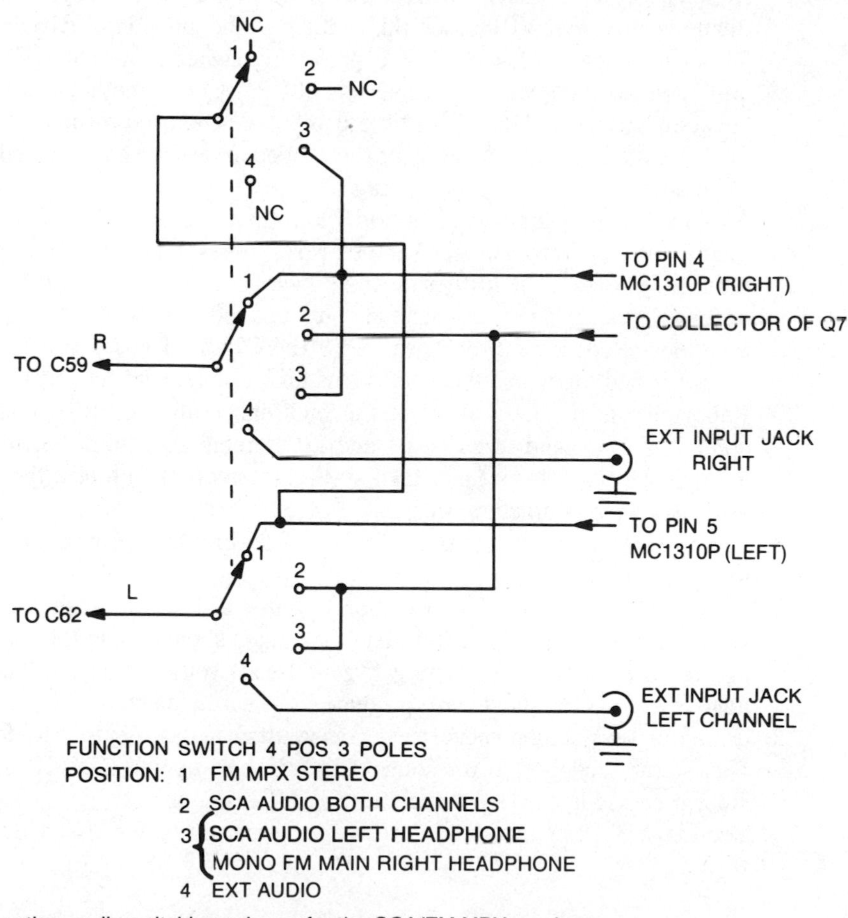

Fig. 5-12. Another audio switching scheme for the SCA/FM-MPX receiver.

8. Apply +12Vdc to the B+ line. Current drain should be about 125–150 mA at 12V.
9. You should hear a rushing noise in both speakers (or headphones). If not, check wiring, and make sure everything is in place. If one channel is dead, check IC4, IC5, or IC3 as required.
10. Rotate VR1—In most areas of the US you should hear a few FM stations. The SCA-MPX switch (Fig. 5-8) should be in MPX positions. At first, stations may sound distorted—Step 11 will correct this.
11. Adjust C38, using a *nonmetallic* tool for clearest audio (lowest distortion). Adjust VR4 and VR5 for a comfortable level.
12. Find a weak station at the high end of the FM broadcast band (106–108 MHz). You may have to adjust C17 if the receiver will not tune this high. Use your FM tuner or FM portable radio to see where the station is located if you do not recognize it. Then when you find the known stations, get its frequency, and look up in Table 5-1 the corresponding tuning voltage. Set VR1 to get this voltage at the junction of R16 and R17. Use at least a 20k/volt VOM, preferably higher. Now adjust C17 until the station comes in at this setting of VR1. Check another station between 88 and 91 MHz. This time compress or expand turns on L5 until this station comes in. Repeat the procedure at high and low ends to be sure of complete band coverage.
13. Adjust C1 and C7 for best reception. Note their setting.
14. Find a weak station between 88 and 91 MHz. Adjust C1 for best reception. If C1 needs to be further meshed (increased capacitance), return to the C1 setting in (13) and instead compress the turns on L1 for best reception. Repeat the procedure on L3 and C7. If C1 or C7 needs to be decreased (unmeshed) expand turns on L1 or L3 respectively.
15. Repeatedly adjust C1 and C7 at the high end (106–108 MHz) and compress or expand turns on L1 and L3 as required until no further improvement is noted. It may take several attempts to optimize these settings, since interaction occurs.
16. You may have to repeat steps 12 through 15 several times for best performance.
17. Readjust C38 so that when a station is tuned in "on the nose," the voltage across R43 is zero. Detuning the station should cause this voltage to go either positive or negative (±1 to 2 volts is typical). This adjustment of C38 should also produce the clearest audio.
18. Tune in a weak station known to be transmitting stereo. Adjust VR3 for best stereo reception. If you connect an LED between pin 6 of IC3 and R66, it should light. This completes FM MPX alignment.
19. See Fig. 5-8. Place the switch (or change jumper leads) for SCA operation.

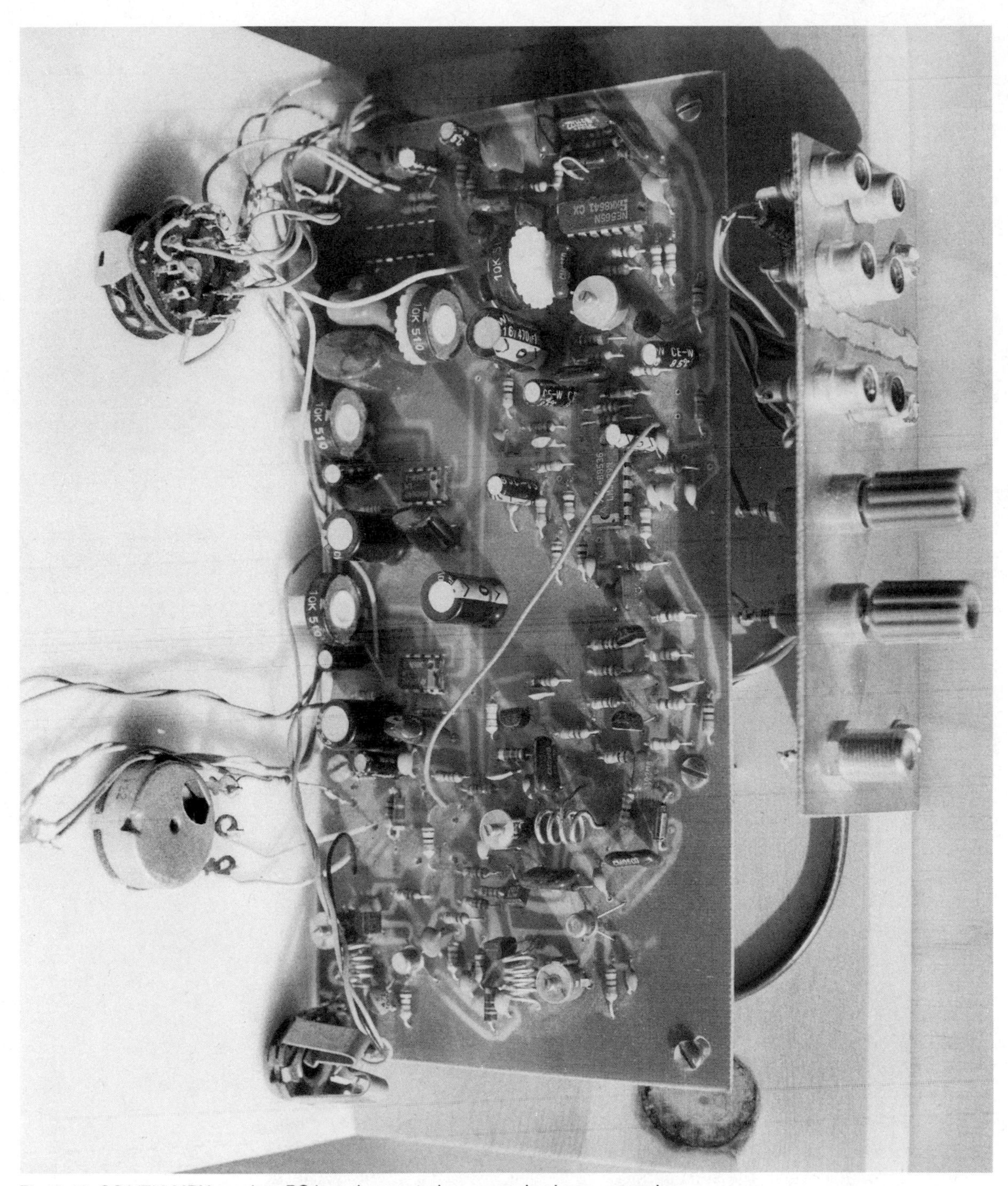

Fig. 5-13. SCA/FM-MPX receiver PC board, mounted on a wooden base—rear view.

20. *Slowly* tune over the FM broadcast band. You may hear several SCA stations. Adjust VR2 for best reception. If no SCA signals are heard, adjust VR2 for a 67 kHz signal (square wave) at pin 6 of IC2.

This completes alignment. If any difficulty is encountered, check your wiring. Most of the adjustments are broad and noncritical and the receiver should work with the initial adjustments given in Steps 1, 2, and 3. C38 is somewhat "sharp" in its adjustment, as may be C17. However, if anything is so critical that "breathing on it" causes problems, something is wrong.

The completed receiver should exhibit a 20dB quieting sensitivity of better than 3 microvolts, and you will find that it is probably much better than this. Extra time spent in alignment will pay dividends here. Take your time and no problems should be encountered duplicating this receiver. One of the authors (Mr. Sheets) lives in a rural area 50 miles NE of Albany, NY, in the hills near the New York-Vermont border. The receiver, at this location, picks up stations from Plattsburg, NY, 120 miles away, Albany, NY, Binghamton, NY (170 miles away), Burlington, VT, and many others in Vermont, New Hampshire and Massachusetts. And, this with only a 2-foot clip lead for an antenna! Oh yes, about five or six SCA subcarriers, as well. In a metropolitan area you will do even better, since more SCA carriers are on the air.

A kit of parts consisting of the printed circuit board and all parts that mount on the board is available from:

North Country Radio
P.O. Box 53, Wykagyl Station
New Rochelle, NY 10804

Price: $75.00 plus $2.50 postage and handling.
New York residents must include sales tax.

Note that cabinets, speakers, headphone jack, and switches are *not* included—only PC board mounted components since individual builders will want to tailor this project to their own requirements.

PROJECT 11: CARRIER CURRENT RECEIVER

Another section of this book described a carrier current transmitter. This chapter will describe two different receivers for use with the transmitter. The receivers differ in the fact that one is a simple AM receiver that may be tuned from about 220 to 300 kHz. The other receiver is set at nominally 270 kHz, for wideband FM use. Both receivers include an audio power amplifier for loudspeaker monitoring of receiver output. For monophonic audio (voice, music, etc.) applications they can be used just as is. For nonaudio uses, such as data or control applications, or for stereophonic audio use, external circuitry can be interfaced to these receivers. The AM receiver is simpler and will be described first. The AM receiver will doubtless find more application in noncritical speech audio and control/data application where some noise can be tolerated, or in environments that are relatively quiet from an electrical noise consideration.

Figure 5-14 shows the schematic of a simple AM receiver using two IC devices and one transistor. It is a TRF (tuned radio frequency) type of receiver, since tuning is generally not required once the receiver has been installed in its intended application. It has a sensitivity of about 1 millivolt (1,000μV) at the input for a reasonable audio output ($^1/_2$ watt). While this may seem low, it is more than adequate since line noise and pickup of static are the limited factors, and usually are at least this level, even in quiet locations. C22 couples signals from the power line to the PC board. R10 serves as a static drain and discharges C22 and C1. It also will cause F1, a 1 amp fuse, to blow if C22 shorts. C22 *must* be a capacitor rated to withstand 250Vac line voltage (safety factor, for 220Vac line use a 600Vac rated cap) and no other type is satisfactory. A potential shock hazard can exist otherwise. Also, the chassis *must* be grounded. If an older two-wire system is used, the operation of this receiver must *only be done with the chassis grounded to a cold water pipe*.

Signal from C22 and R10 goes to a tuned network having about a 20 kHz bandwidth. C1, C2, C3, C4, C5, and L1 and L2 make up this network. It allows only the desired signal to pass and rejects other signals. C2 and C4 are variable, and are simply adjusted for maximum signal level. A jumper between line (antenna) and the input network can be removed for reception of very strong signals if distortion (overload) becomes a problem. With the components shown, about 220 to 300 kHz can be covered.

IC1 is a "gain block" i-f chip, used for TV i-f applications, but perfectly useful at low frequencies as well. It has AGC capability and has approximately 60dB of gain. C6 and C7 provide rf by passing C8, C9 and L3 are placed across the output of IC1 and are broadly resonant around 280 kHz. C11 provides rf bypassing, R4 supplies B+ to IC1, and C10 couples rf energy to detector diode D1. If operation much removed from 280 kHz is desired, change either C8 or C9 from 330 pF to 220 pF for 300 kHz, or 470 pF for 250 kHz. If operation around 220 kHz is desired use a 470 pF (NPO or silver mica) capacitor at both C8 and C9. However, C8 and C9 can be left as 330 pF without much loss in receiver gain.

Diode D1 acts as an envelope detector L4 keeps rf energy from entering the audio system, and C14 provides rf bypassing. The cathode of D1 is returned to a variable resistor R5. A voltage of about 4 to 6 volts appears on the cathode of D1 even with no received signal. When a signal is received, the dc voltage at the anode of D1 increases (goes positive) due to recertification and detection of the signal. This dc voltage appears across C14 (detector output). It is fed thru R3 and C13, which remove audio components, to the base of Q1, an emitter follower. The voltage at the emitter of Q1 (across R1) is fed to pin 5 of IC1. A positive going voltage tends to reduce the gain of IC1. This in turn reduces the

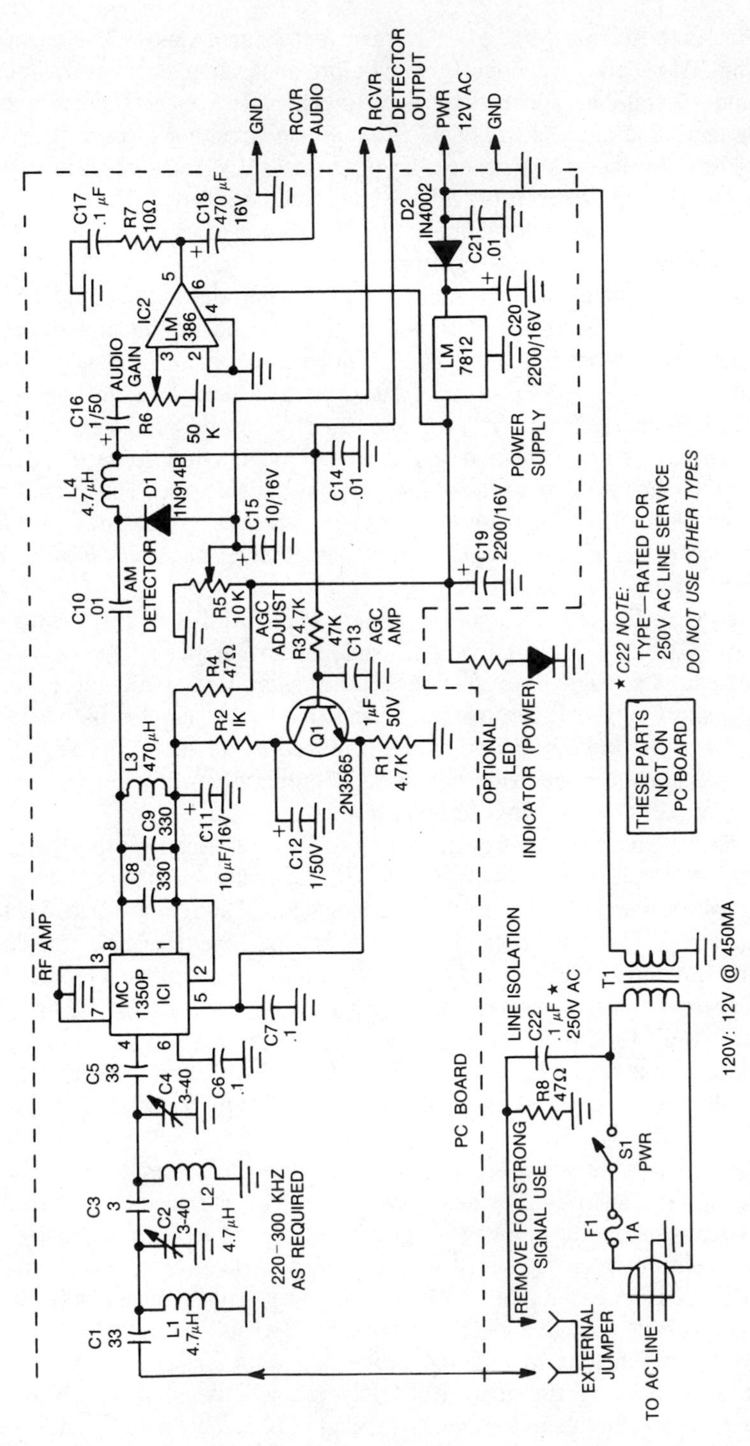

Fig. 5-14. AM carrier current receiver—schematic.

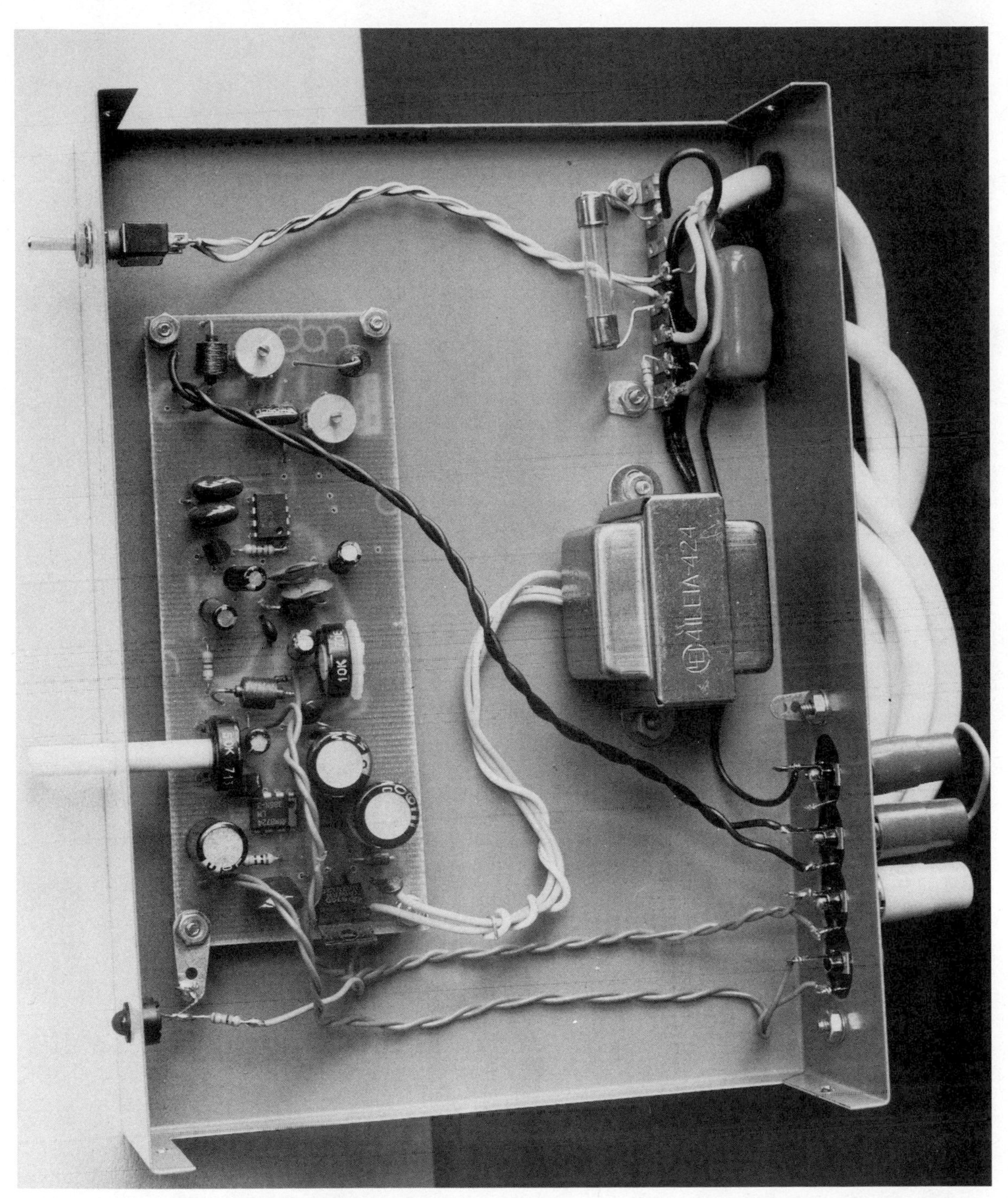

Fig. 5-15. Internal view—AM carrier current receiver.

signal fed to D1 and subsequently the dc voltage across R1. In this manner AGC (automatic gain control) is obtained, which keeps the receiver output relatively constant despite received signal level changes. R5 is set for best AGC action.

Detector output is taken from C14. In this design, upper frequency limit is 10 kHz or so. By reducing the value of C14, higher frequency response can be obtained. Also, increasing the value of C3 to 5 pF and returning the input bandpass filter for wider bandwidth can be done as well. However, the FM approach to be described later would be more likely useful for higher modulation frequencies. The detector output is brought out to an external terminal, jack, etc.

IC2 is an audio amplifier for listening or monitor purposes. Audio components on the effected signal are fed to R6, an audio gain control, through C16 to IC2. R7 and C17 suppress a tendency for IC2 to oscillate under certain load conditions. C18 couples audio to an external speaker, up to 1/2 watt of audio is available.

Dc power for the receiver is obtained from a power supply consisting of D2, C20, IC3, and C19. About 12Vac at D2 gives +12Vdc across C19. T1 is a Radio Shack P/N 273-1365 or other 12V 450 mA to 1 amp transformer C21 suppresses noise generated by D2 from causing internal receiver noise.

The PC board layout is shown in Fig. 5-16 and the component placement in Fig. 5-18. The board can be mounted in a small cabinet, with T1 and a few parts obtainable at Radio Shack (see parts list). The original prototype was built this way, but you can do as you see fit in your particular application. Leads are attached to the PC board and brought out to a terminal strip for interfacing with other equipment as required. We used a Radio Shack 270-272A cabinet and 274-322A phono jack assembly in our prototypes. However, you may use whatever you may require or have on hand.

The FM version of the carrier current receiver is shown in Fig. 5-15. It differs from the AM version as to detector circuitry and the use of a broadband filter at the input. Like the AM receiver, it provides audio output to a speaker for monitoring purposes, and a detector output for interfacing with external equipment. It operates from a 12Vdc supply. Since the detector is voltage sensitive, a regulated supply is recommended, such as a LM7805-12 regulator or similar. Figure 5-17 is the PC layout, Fig. 5-19 is the parts placement.

Referring to Fig. 5-15, input signals from the power line are coupled through C23 and R19 to the input filter network. As in the AM version, C23 *must* be a type of capacitor that is rated at 250Vac for ac line service. *Do not* substitute other types or else a shock hazard may develop. Again, only use a capacitor that is manufacturer recommended for this type of service. Signals above about 500 kHz are rejected by C1, which serves to reduce a tendency for the triple tuned filter network to "leak" signals at the frequencies far above the passband. C2 through C7, L1 through L3, R1, R20 (termination resistors) form a triple tuned bandpass filter having a passband from 220 to 340 kHz. In order to reduce distortion, it is important that this network has a relatively flat bandpass within ±1dB. Signals from the filter are applied to pin 4 of an MC1350P, a

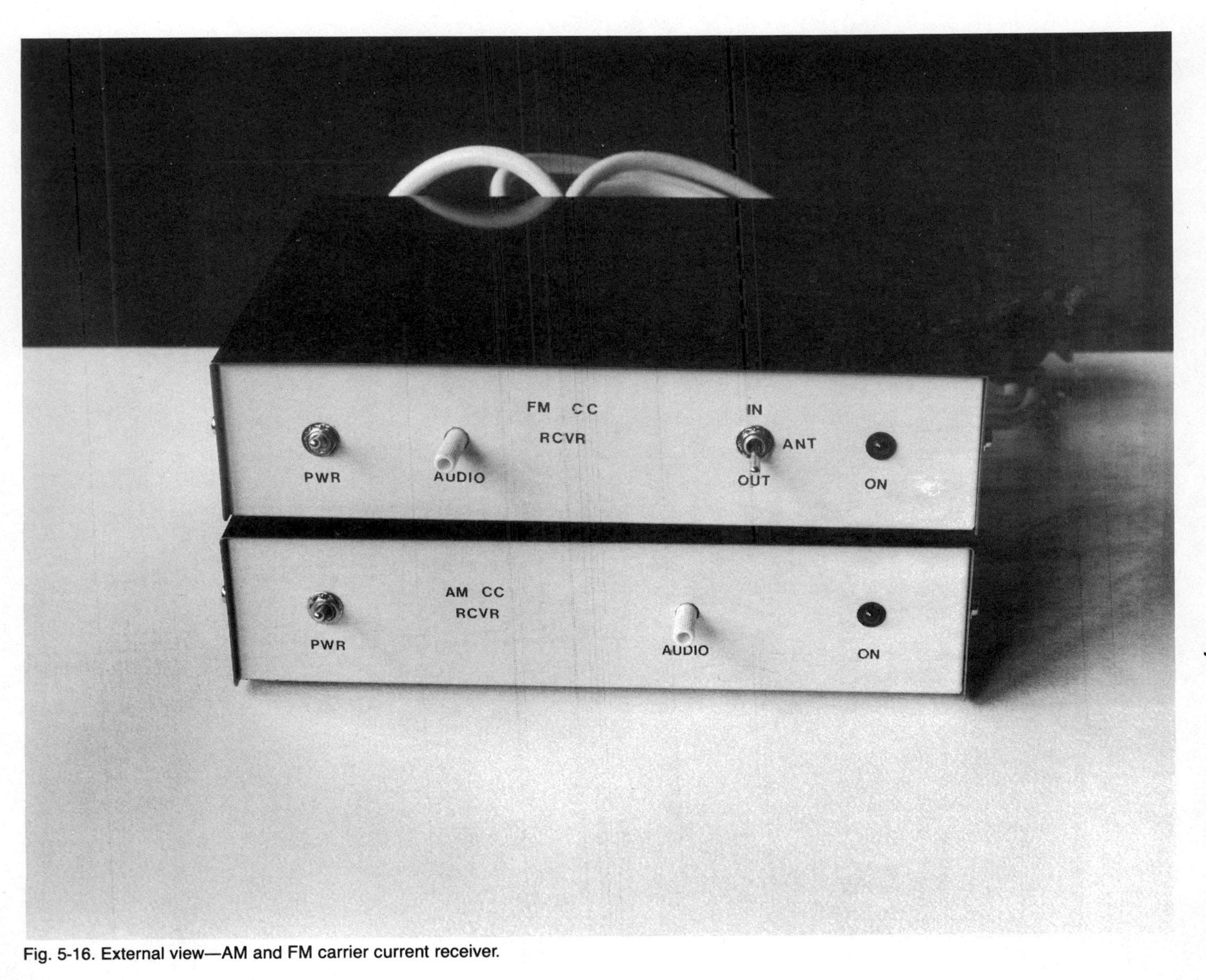

Fig. 5-16. External view—AM and FM carrier current receiver.

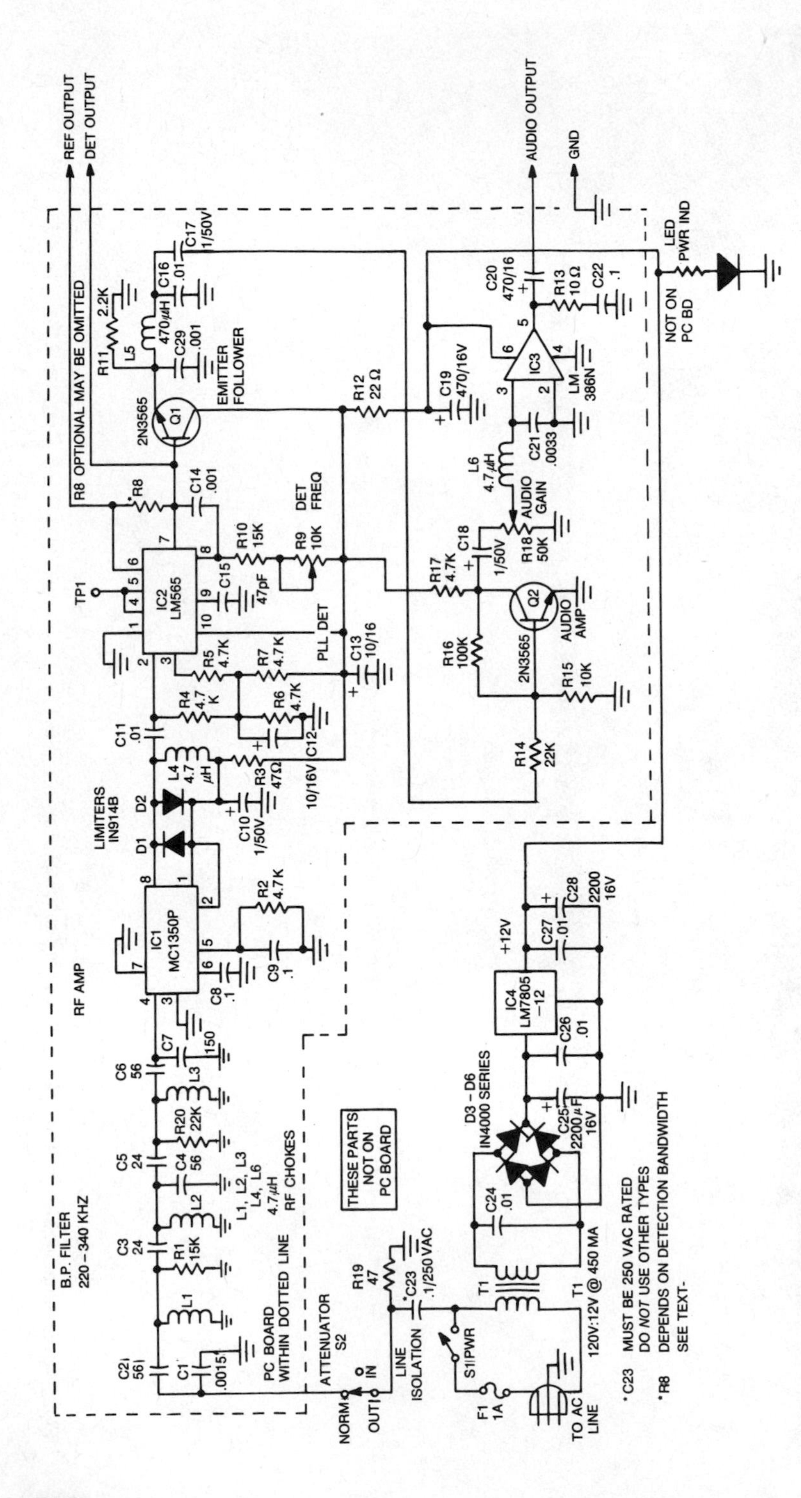

Fig. 5-17. FM carrier current receiver—schematic.

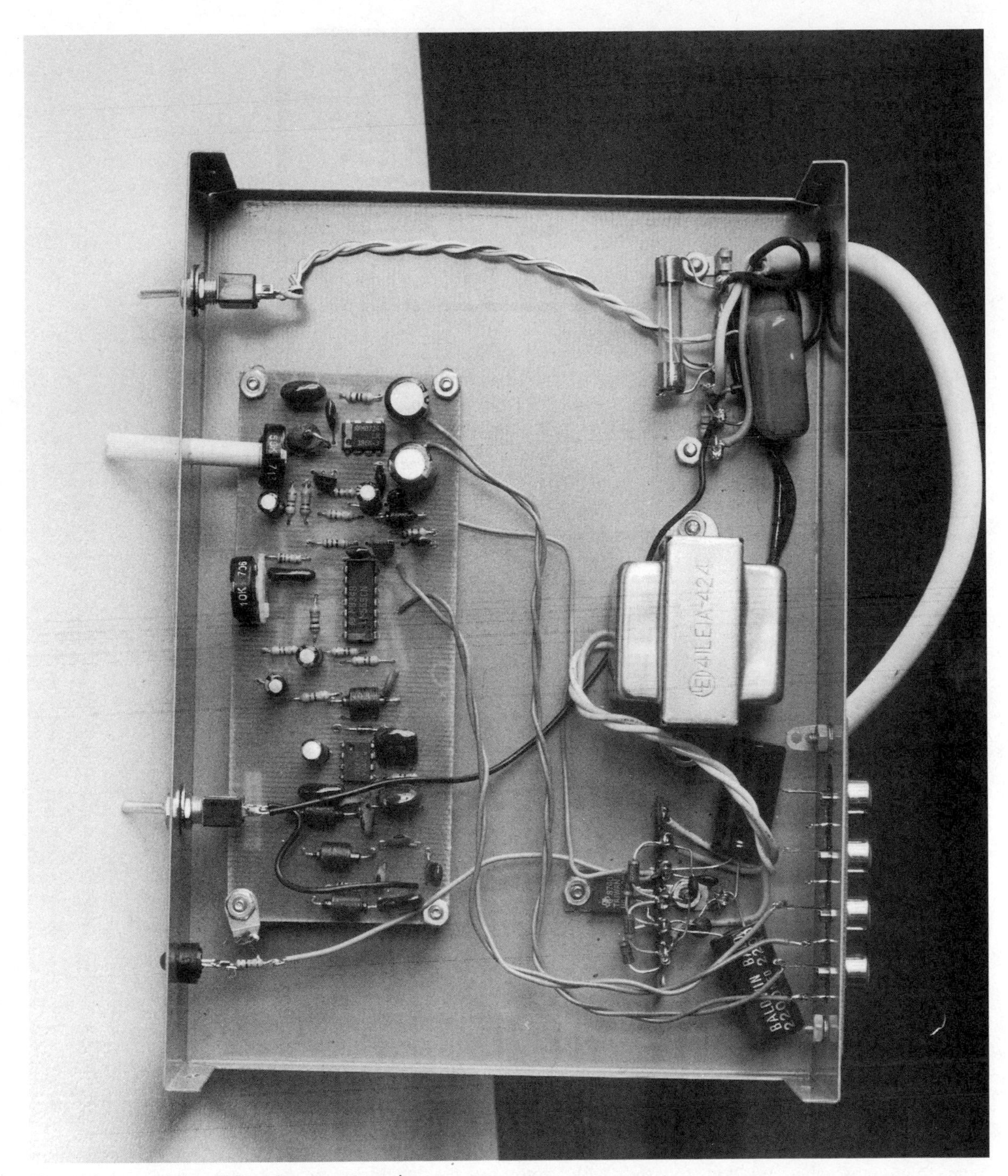

Fig. 5-18. Internal view FM carrier current receiver.

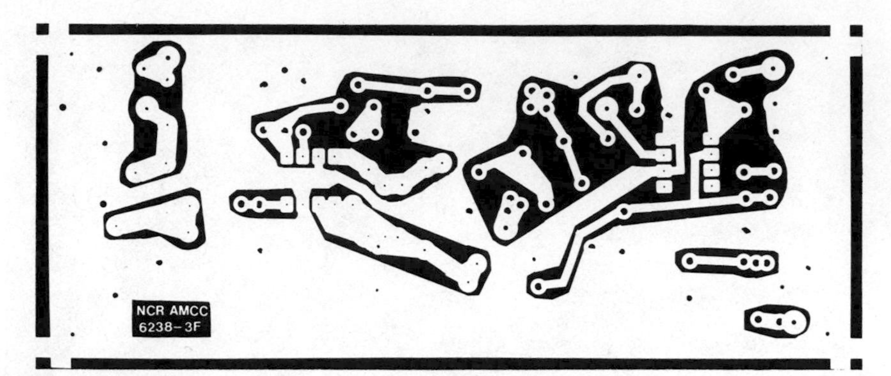

Fig. 5-19. AM carrier current receiver—PC artwork.

"gain block" IC. As in the AM receiver, it is used as a tuned rf amplifier. C8 and C9 provide internal bypassing for the chip. R2 returns pin 5 to ground and provides bias for pin 5 so that IC1 operates at maximum gain. In this receiver, no AGC is required, unlike the AM version. Amplified signal appears at pin 8 of IC1. L4 provides dc bias and high rf impedance to pin 8, and D1 and D2 provide amplitude limiting of the FM signal. C10 and R3 form a B+ decoupling network to prevent stray signal coupling via the power supply. C11 provides signal coupling and dc blocking between IC1 and detector IC2. About a 1.2 volt peak-to-peak signal exists between either side of C11 and ground (ac component only) and is normally this level under full limiting conditions. Also, D1 and D2 reduce power line noise and strong "spikes."

IC2 is a PLL detector, the NE/LM565 chip that has been around a number of years and has proven to be reliable and very versatile. It can follow better than ±50% variation in input frequency and makes an excellent FM demodulator. It typically has about 300 mV peak-to-peak output level at ±10% signal frequency deviation. The detector output pin 7 and reference pin 6 are brought out for interface purposes for use with external equipment.

C11 couples the signal to pin 2 of IC2. R7, R6, and C12 derive a nominal +3.9V for biasing the input pin 2 and the other (unused) input pin 3, via R4 and R5 respectively. Pins 8 and 9 connect to an internal voltage controlled oscillator, and C15 at pin 9, and R10 plus variable R9 connect to pin 8. These components set the VCO free running frequency. VCO signal and input signals (at pin 2) are compared in an internal phase detector. The output of the VCO appears at pin 4. Pin 4 couples to pin 5, the VCO input of the phase detector. The jumper between pins 4 and 5 provides a test point for setting up the phase-lock loop VCO frequency, which is normally 280 kHz in this receiver. The phase detector output drives an internal amplifier whose output (VCO control voltage) appears at pin 7 of the NE/LM565. This voltage follows frequency variations of the input signal and is a replica of the original modulation on the input FM signal. This output (pin 7) is therefore the recovered audio or data modulation. It is brought out to an external connection, together with a reference dc signal from pin 6 of

IC2. The detector output has about a +10.5Vdc component. This common mode voltage, (also on pin 6) can be eliminated and level shifted with an external operational amplifier.

C14 is used to suppress oscillations that may occur under certain conditions. Detector output at pin 7 is coupled to emitter follower Q1 and R11. C29, L5, and C16 form a low pass filter and are used to reduce 280 kHz components that are present in the output of IC2 from reaching the audio amplifier system Q2 and IC3. C17 couples audio through R14 to the base of Q2. Q2, in conjunction with R17, C18, R15, and R16 form an audio amplifier to bring the recovered audio up to around 1 volt peak-to-peak. This Q2 stage is only used to drive IC3 and is only required for audio program or speech application (as is true of IC3 also). R18 is a volume control, L6 and C21 suppress any remaining 280 kHz components from IC2, and the input from L6 and C27 is applied to IC3, an LM386N audio amplifier used to drive an external speaker. It has a voltage gain of 20× (26dB) and can deliver 1/2 watt into an 8 ohm speaker. R13 and C22 provide suppression of parasitic ultrasonic oscillation and C19 provides supply bypassing. C20 couples audio to an external speaker. Dc power is supplied by a power supply using an IC regulator (IC4) and a few external components. T1 supplies 12Vac to bridge rectifier D3–D6 (1N4000 series, ECG 125A, etc.). C25 provides filtering and supplies about +16V to IC4A, an LM7805-12 regulator. C26 and C27 are stabilizing capacitors for IC4. About +12V at up to 250 mA is supplied to the receiver. The supply requirements as to ripple and noise are more stringent for the FM receiver than the AM receiver.

Both receivers are constructed on a 5″×2″ PC board, and may be housed in whatever enclosure you desire. If purely data or control applications are required, the audio monitor amplifiers (LM386) may be omitted, with associated components. However, the amplifiers are inexpensive and serve as handy monitors and may just as well be included. A source of PC boards and parts is listed at the end of this article.

The FM receiver has a frequency response to over 50 kHz. Stereo audio from a multiplex generator can be transmitted over the transmitter previously described, and also data signals up to 50,000 bits per second as well as ultrasonic signals from TV remote control chips (excellent for use as a low cost control system) or other signals of a similar nature. Either direct FSK (frequency shift keying) signals, or tone-modulated carriers can be received. We have tried to keep each receiver as versatile as possible therefore external data demodulators or other circuitry for control uses in specific cases have been kept off the receiver boards.

In the case of stereo FM signal transmission, a suitable multiplex generator can be constructed using the PC board from the "Wireless FM Stereo Link" described in Project 6 on p. 104 of this book. In this case a complete MPX signal can be obtained from the wiper of R64 (see article) and fed to the transmitter. The 88 MHz VCO and amplifier circuitry (Q3, Q4) can be simply omitted if

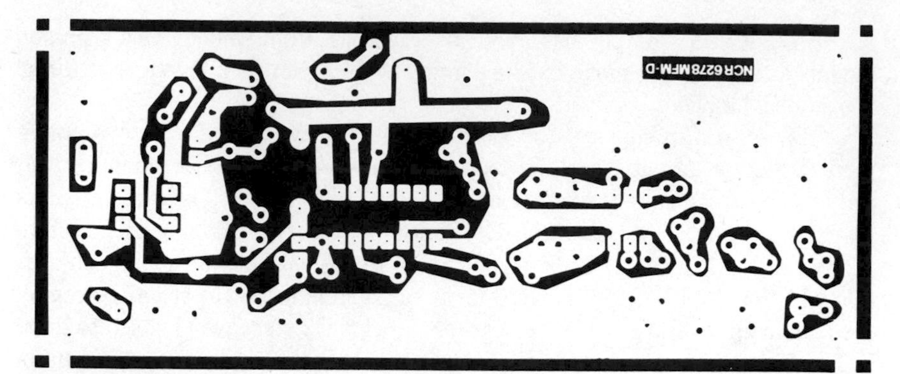

Fig. 5-20. FM carrier current receiver—PC artwork.

desired. See Fig. 5-23 and the RE article for details. A PC board and/or parts is available for this item from North Country Radio.

For control purposes, there are many options. A set of chips made by Motorola, the MC14457 transmitter and MC14458 receiver for IR remote control application can be used to make a simple multifunction controller. They use data signal frequencies of 38 and 41 kHz, well within the range of this receiver. Also, the MC14469 asynchronous receiver/transmitter, or the MC14497/MC3373 receiver could be interfaced to the system.

In cases where simple control or monitor functions are needed, chips such as the LM567 tone decoder, and the NE565/LM565 phase-lock loops can be used to construct tone detectors, tone demodulators or other such functions. The AM receiver can be used to advantage for simple control applications such as tone signalling, etc. A few applications will be discussed in a subsequent article. There are so many possibilities due to the availability of low cost digital IC devices and functions.

In order to construct these receivers, it is suggested that the PC layout provided be used. While the frequencies are fairly low and nothing is particularly critical, these layouts have been proven to work well and should be easy to duplicate. For those that prefer, ready made PC boards and parts kits are available. All other components with the exception of the chokes and coils can be obtained at a neighborhood Radio Shack or similar store.

The receivers are constructed by first inserting fixed resistors in the PC board, then taller components such as capacitors. Next, the coils and potentiometers are installed. Lastly, install the IC devices. It is recommended that all other parts are installed and soldered in place before installing the IC devices, to minimize any handling and potential static damage. Use only rosin core solder, *not* common acid core or paste fluxes and solid solder that is sold at hardware stores. Make sure that the solder is intended for radio-TV and electronic applications. We suggest that the PC board be carefully inspected for shorts, solder bridges, or missed solder joints before applying power and testing the

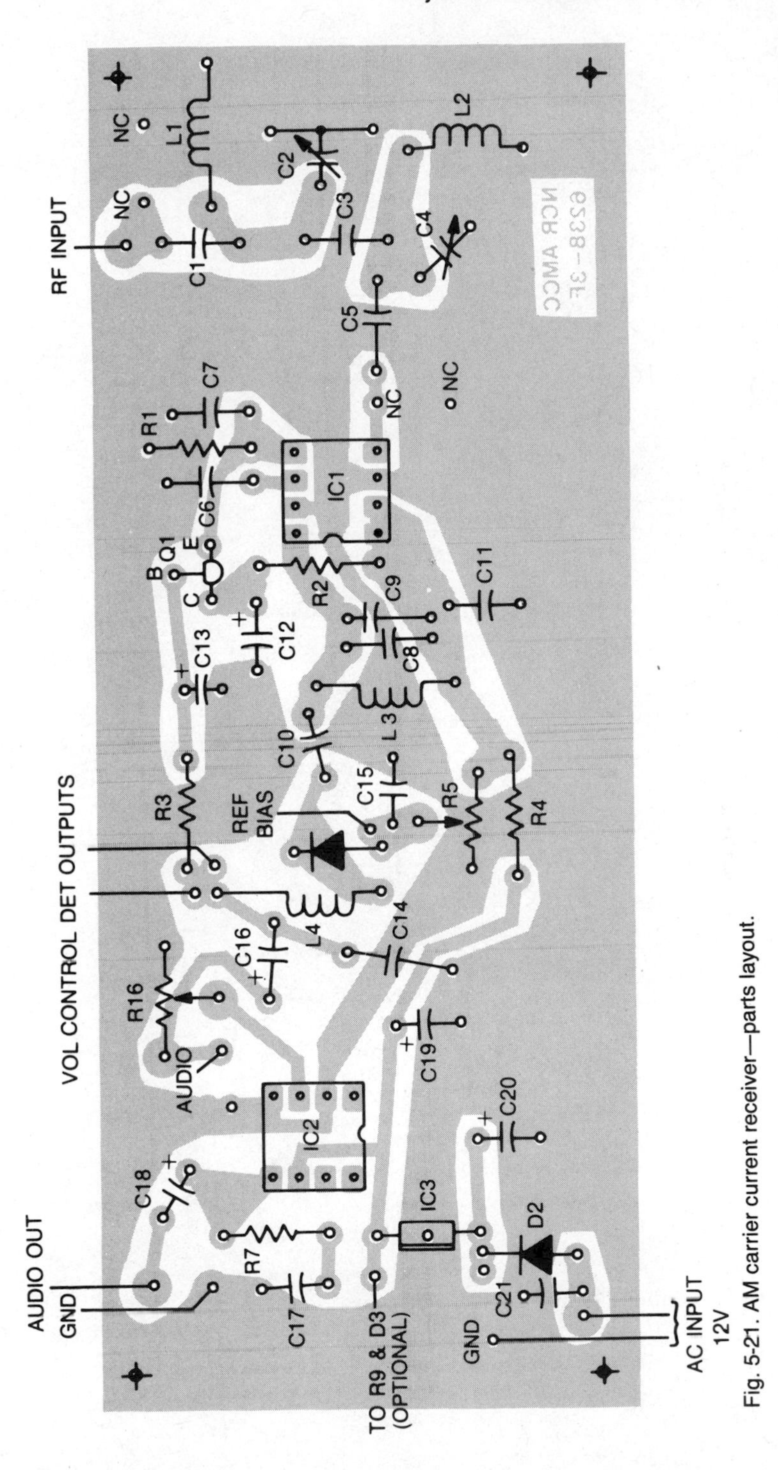

Fig. 5-21. AM carrier current receiver—parts layout.

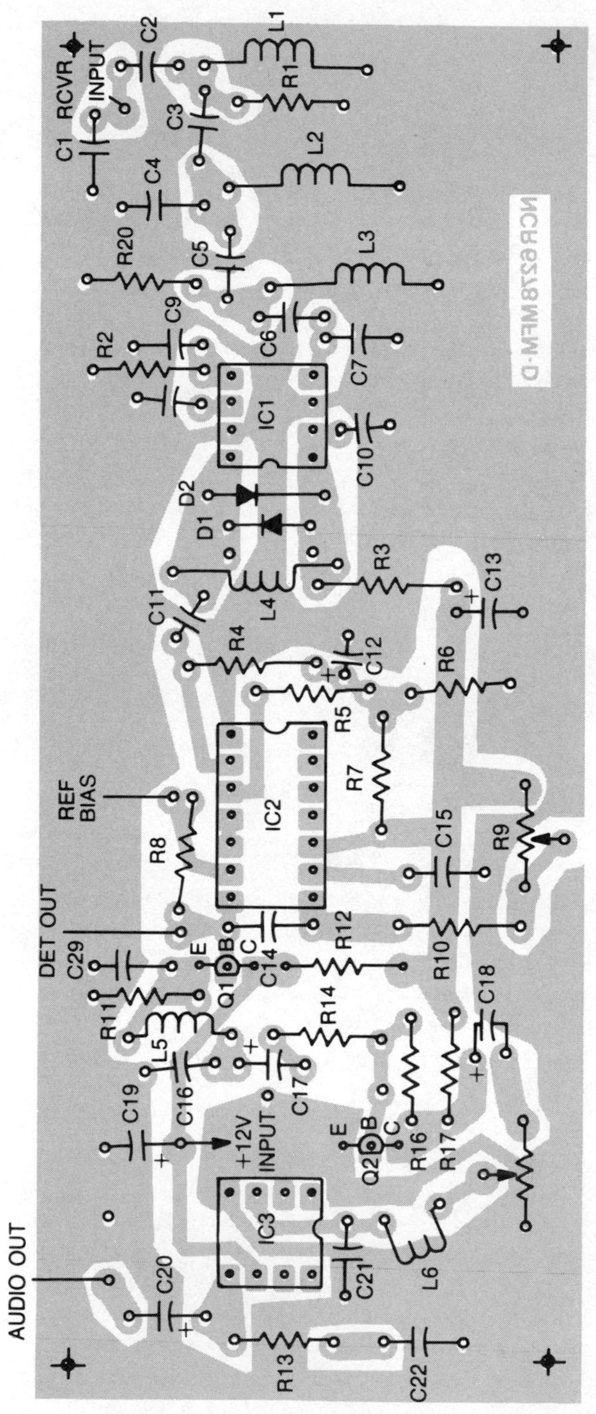

Fig. 5-22. FM carrier current receiver—parts layout.

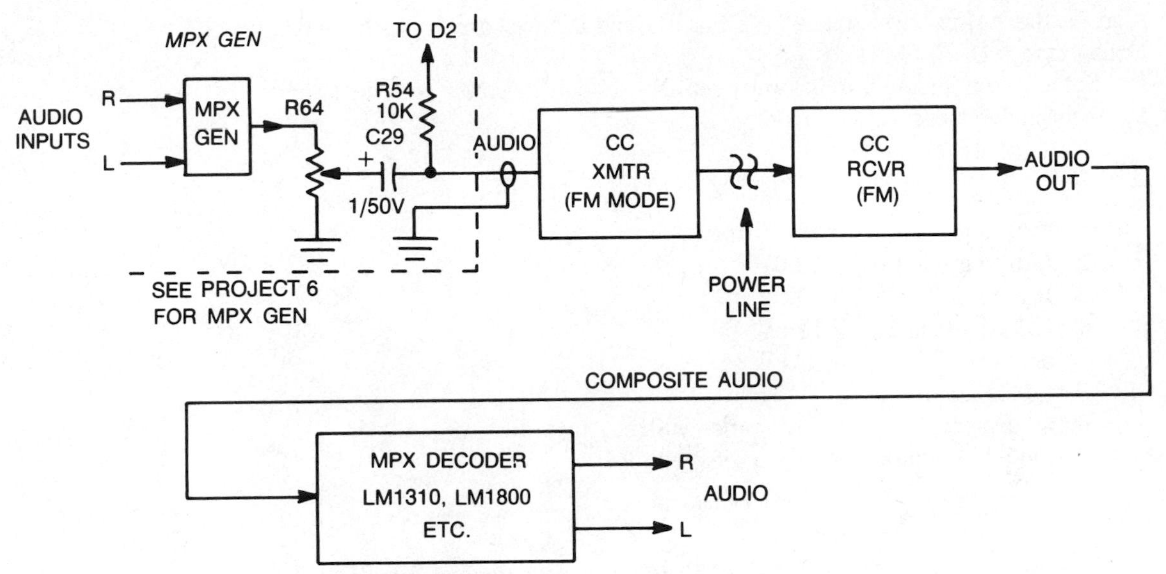

Fig. 5-23. Stereo transmission methods via carrier current.

receiver. Also, do not bend wires to rf chokes closer than about 4 – 5mm (3/16″) from the body.

The following checks should be made before power is applied. We used a digital VOM on the ohms ranges. See Figs. 5-14 and 5-15. Note: Positive (red lead of ohmmeter) may actually be *negative*.

AM Receiver

1. Check all coils for dc resistance
   L1, L2, and L4 = 48 ohms
   L3 = 22 ohms
2. Pin 6 IC2 to ground greater than 500 ohms (after 10 seconds)
3. Pin 5 IC2 to ground greater than 10 ohms
4. Pins 1, 2, 4, 5, 6, 8 of IC1 should not show shorts to ground (less than 500 ohms, for example).

FM Receiver

1. Check all coils for dc resistance
   L1, through L4 and L6 = 48 ohms
   L5 = 22 ohms
2. Pin 6 IC3 to ground greater than 500 ohms (after 10 seconds)
3. Pin 5 IC3 to ground greater than 10k ohms
4. Pins 1, 2, 4, 5, 6, 8 of IC1 and pins 2 through 10 of IC2 should not show shorts to ground (less than 500 ohms).

Also, in the AM receiver, make sure that D1 and D2 are correctly polarized. Doubly check D2.

Set all potentiometers to halfway open. Next, apply dc power and check for the following voltages:

AM Receiver

1.	C20 to ground	+16V (approx.)
2.	C19 to ground	+12.0V
3.	IC2 pin 5	+ 6.0V
4.	IC1 pin 1 and 2	+11.8V
	pin 8	+11.8V
5.	Q1 Collector	+11V
6.	Q1 Emitter	+ 6V (varies with setting of R5)
7.	Pin 5 IC1 approx.	+ 5V (varies with R5)

FM Receiver

1.	Voltage across C25 approx.	+16V (exact voltage depends on T1)
2.	Pin 6 IC3	+12V
3.	Pin 5 IC3	+ 6V
4.	Collector Q2	+ 7V
5.	Emitter Q1	+ 9 to +10V
6.	Pin 7 ICR	+10 to +11V
7.	Pin 8 IC2 approx.	+10.5V
8.	Pins 2 and 3 IC2 approx.	+ 4V
9.	Pins 1 and 2 IC1	+11V
10.	Pin 5 IC1 about	+ 4V

If any appreciable discrepancies are found, trouble may be indicated. Some variations may occur depending on instruments, but any significant (±20%) variations should be checked out before proceeding further.

Connect a monitor speaker to the output of either the AM or FM receiver. Apply power and in the case of the AM receiver, set R10 about two-thirds open (i.e., to get about 6 – 7 volts across C15) or in the FM receiver, set R9 at midpoint then, make the following tests:

AM Receiver. Apply about 1 millivolt 30% AM modulated signal at 280 kHz between C1 and ground. If no signal generator is available, connect a long (25′) piece of wire to this point and use either noise or a radio beacon signal, etc. Peak C2 and C4 for maximum audio output. Adjust R5 so that the audio output stays relatively constant over a range of inputs between 1 millivolt and 1 volt. R5 will affect receiver gain and if set too high will cut off IC1 and the receiver will be "dead." You should obtain adequate speaker volume at high settings of R6. Terminate the detector output terminals with a 1k resistor for this test.

FM Receiver. Apply about 1 millivolt of a modulated signal ±40 kHz deviation and adjust R9 for minimum distortion. If no generator is available, set R9 at midpoint. You should obtain a hiss in the speaker with no input signal. If possible, verify the frequency response of the input network by connecting a scope to pin 8 IC1 and applying a CW signal (unmodulated) to the junction of C1 and C2. The signal should be low enough so that no more than 0.5V peak-to-peak appears at pin 8 IC1, so that D1 and D2 do not conduct. Vary the signal frequency between 200 and 350 kHz and plot the response. (Keep the input *constant*.) You should get ±1dB flatness or better between 240 and 330 kHz. If not, try adjusting the value of C4, or C3 and C5 as required. If the recommended components and layout are used, this should not be a problem. A flat passband is necessary to reduce distortion of the wideband FM signal, which is sensitive to phase distortion. Ideally, the phase response should be a linear function of frequency (constant delay) and the frequency response should be flat.

The finished receiver boards are mounted in cabinets. Figure 5-24 shows a suggested layout using a small metal cabinet. If desired, the monitor speaker can be mounted inside the same cabinet, or conversely, a wooden speaker cabinet can house the receiver board and power supply. An inexpensive speaker sold for use with small stereo systems usually has a cabinet large enough for this purpose, with room to spare.

Interfacing these receivers to external equipment of circuitry generally involves level shifting and/or amplification. As an example, take a look at Fig. 5-25. This shows the AM receiver detector circuitry. The detector is biased above ground by several volts. An operational amplifier may be used to "level shift" the dc voltage that is present. The op amp is connected as a differential amplifier. The gain is set by the ratio of R_B to R_A. R_A would typically be 10k ohm. R_C and R_D adjust the dc operating point of IC1 and R_C would be about 10k ohm and R_D a pot of 250k. Either the dc level at the output or the ac signal or both can be fed to a tone decoder, level sensor, or TTL interface or the drive CMOS logic, etc. The FM detector output can be interfaced in a similar way. The detector output appears across IC2 pins 6 and 7. The level is up around +10.5Vdc. A resistor network with a balance pot in one leg is used to bring the dc level down to about 4–5 volts. (We are assuming the use of a 741 or similar op amp, operated from a single 12V supply in this case.) A gain control is provided to adjust the output level. The dc and audio signal components are taken from the op amp output.

Note that if only audio outputs are required, such as FSK tone keying or data, the audio output can be simply taken from the detectors with a coupling capacitor since the dc information present is not required. However, if simple on-off carrier keying or frequency shift between two frequencies is used, then the interfaces may be needed since dc information (levels) would be needed.

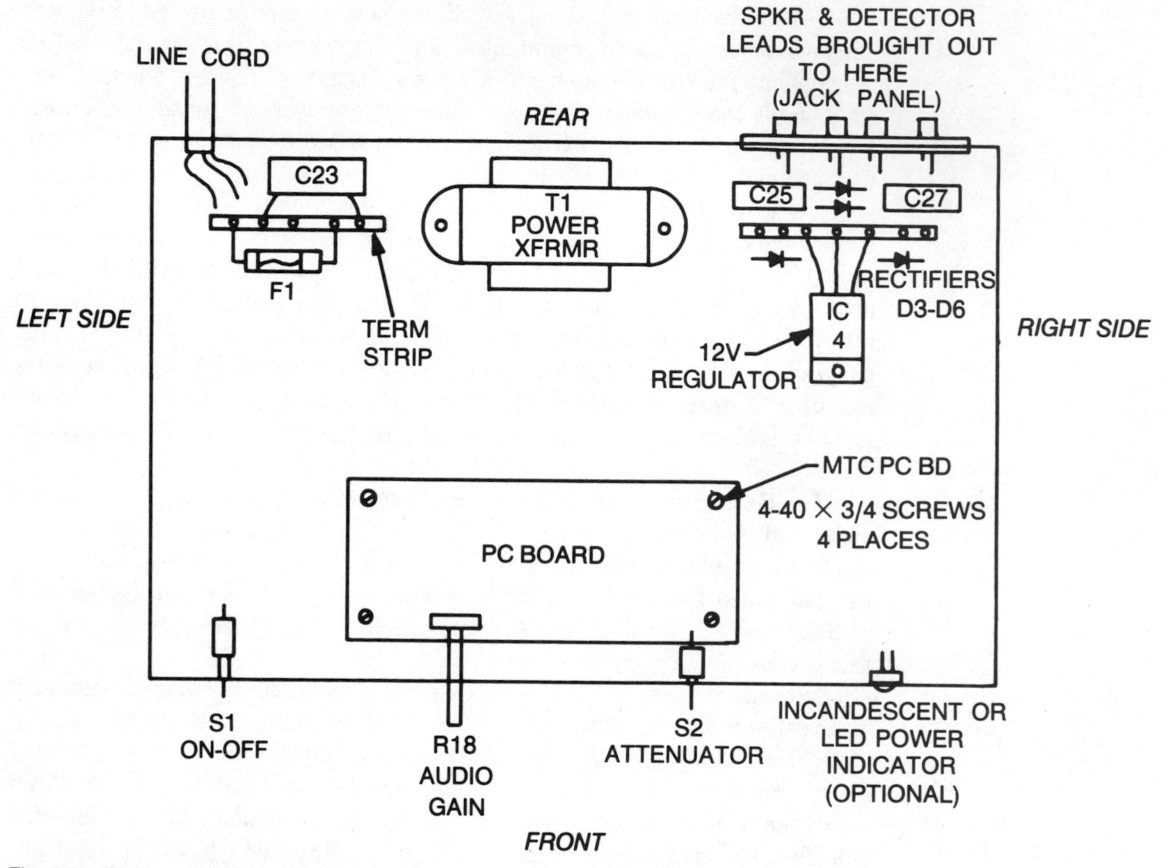

Fig. 5-24. Carrier current receiver—parts layout.

Data transmission can be done with suitable modems. Modems suitable for telephone line use (300, 600, 1200 Band, etc.) can be directly interfaced since their line inputs and outputs are simply audio tones in the 300 – 3000 Hz range. Slow scan video can also be sent in a similar manner.

Figure 5-26 shows a few simple circuits using the popular NE/LM565 PLL and NE567 Tone decoders for FSK demodulator and also remote control by the use of tone modulation. Figure 5-27 shows the use of a TV remote control chip normally meant for IR hand held controls, to effect a multicommand control system. By the use of filters, there is no reason why data and audio cannot be sent simultaneously over the same transmission link. Simply remember to not exceed the dynamic range of the system so there will be little intermodulation and crosstalk. See Fig. 5-29.

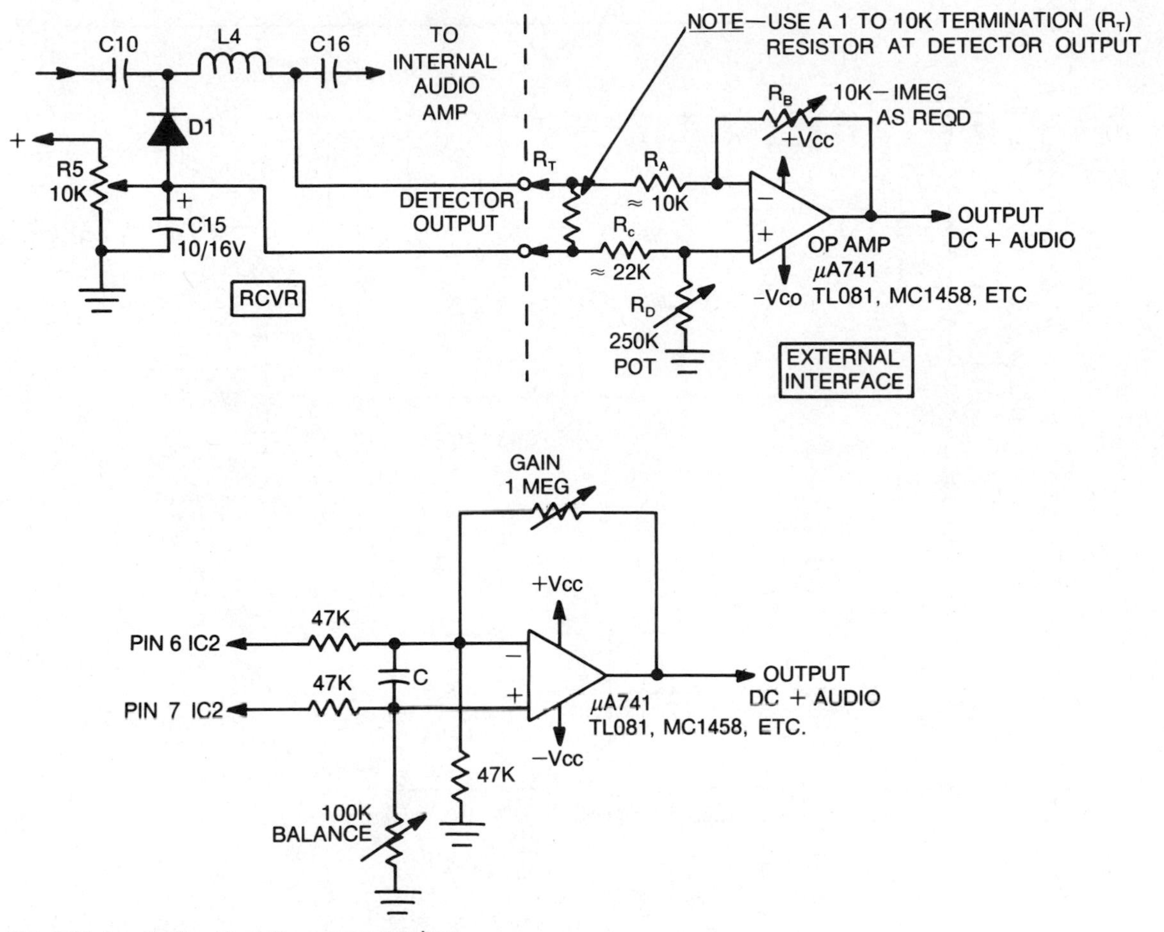

Fig. 5-25. Interfacing of carrier current receiver.

A kit of parts consisting of the printed circuit board and all parts that mount on the board is available from:

North Country Radio
P.O. Box 53E, Wykagyl Station
New Rochelle, NY 10804

Price: AM receiver list $28.50
FM receiver list $38.50
Carrier current transmitter kit $54.50

Plus $2.50 postage and handling.
New York residents must include sales tax.

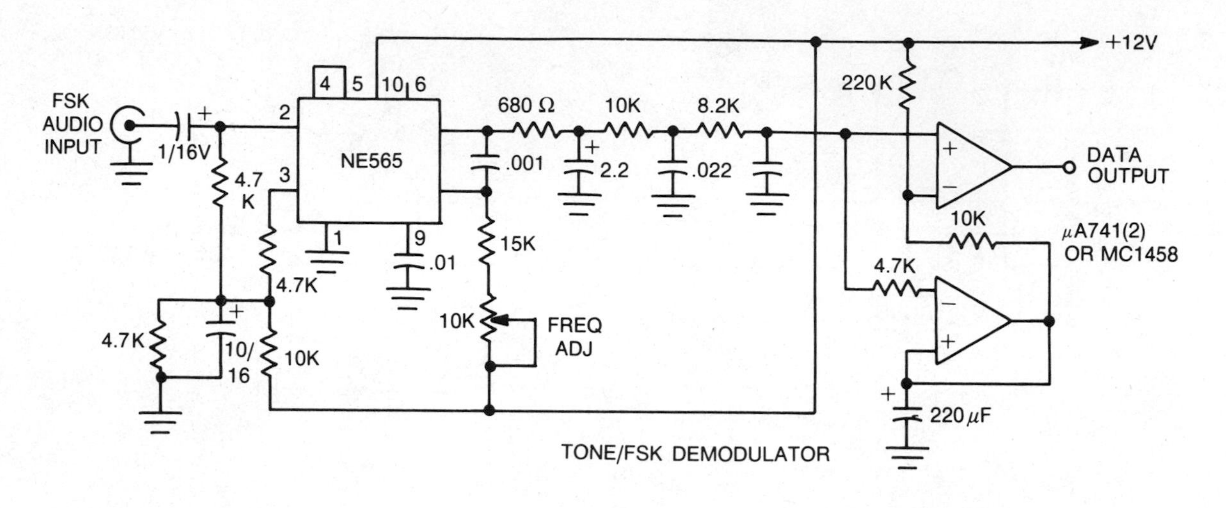

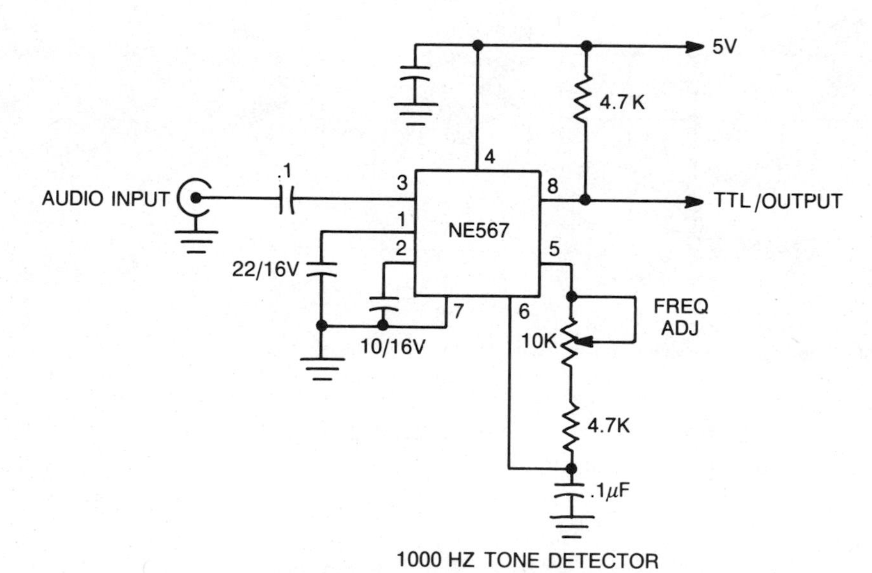

Fig. 5-26. Control circuitry for carrier current receiver use.

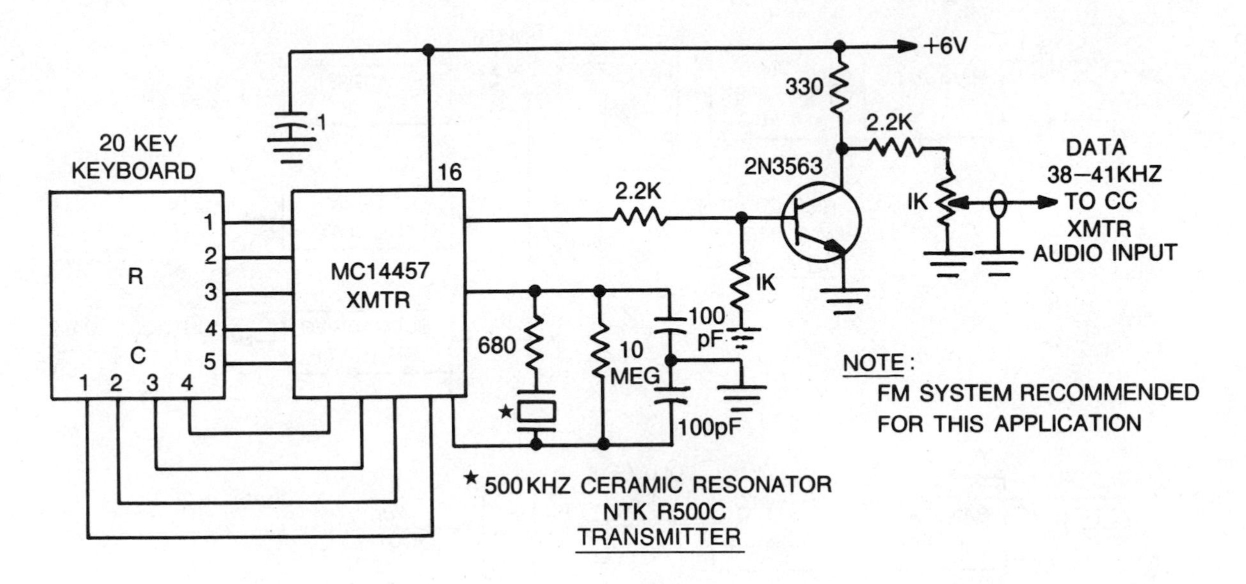

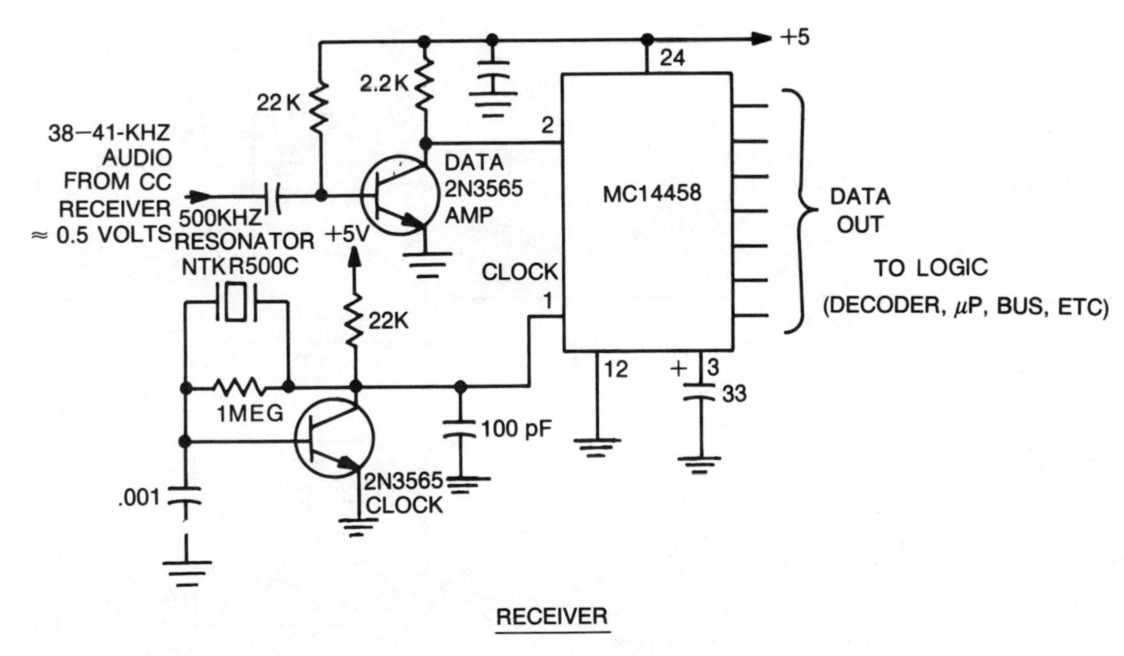

Fig. 5-27. Use of TV remote control chips with the carrier current system.

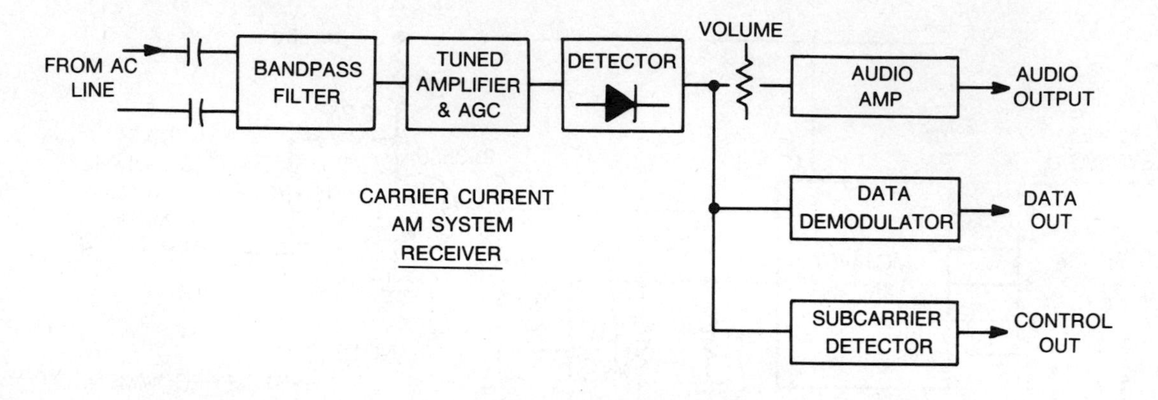

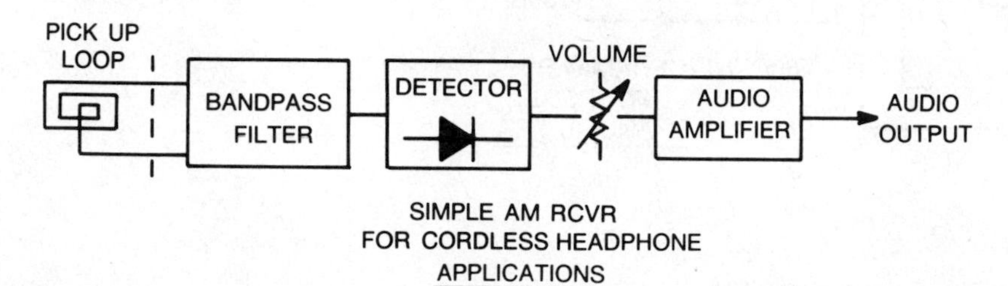

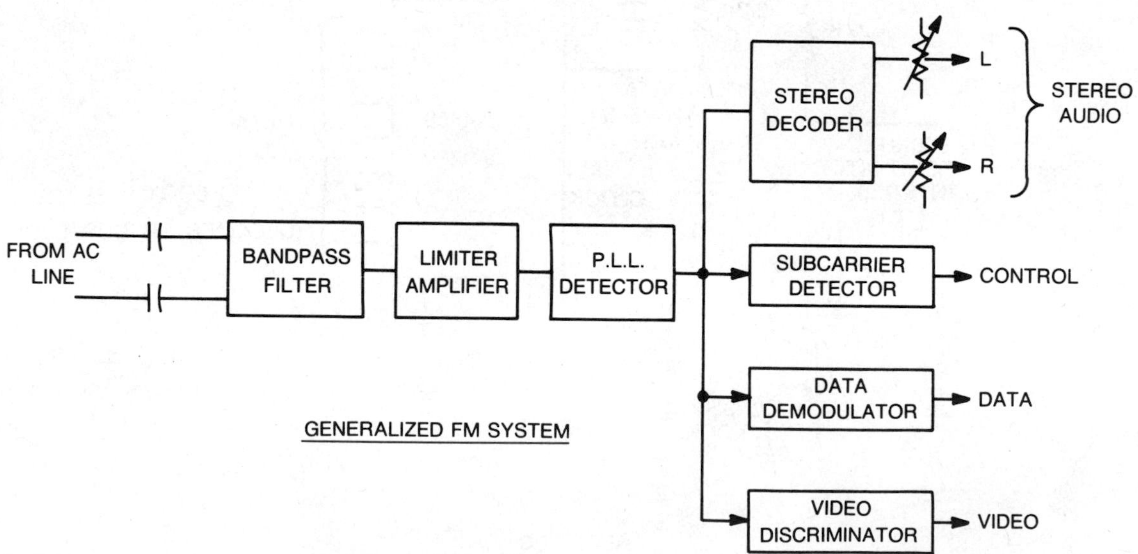

Fig. 5-28. Basic carrier current receiver block diagrams.

A.

Part	Value
R1	15k
R2	4.7k
R3	47 ohm
R4	4.7k
R5	4.7k
R6	4.7k
R7	4.7k
R8	10k
R9	10k trim pot
R10	15k
R11	2.2k
R12	22 ohm
R13	10 ohm
R14	22k
R15	10k
R16	100k
R17	4.7k
R18	50k pot w/shaft
R19	47 ohm
R20	22k
R21	2.2k (not required if LED omitted)
D1, D2	1N914B
D3-D6	1N4000 series diode 1A
D7	LED optional
Q1, Q2	2N3565

Part	Value
C1	.0015 Mylar 50V
C2	56pF NPO or SM
C3	24pF NPO or SM
C4	56pF NPO or SM
C5	24pF NPO or SM
C6	56pF NPO or SM
C7	150pF NPO or SM
C8	.1 Mylar 50V
C9	.1 Mylar 50V
C10	1μF/50V elec
C11	.01 disc
C12	10μF/16V elec
C13	10μF/16V elec
C14	.001 Mylar 50V
C15	47pF NPO or SM
C16	.01 Mylar
C17	1/50V elec
C18	1/50V elec
C19	470μF/16V elec
C20	470μF/16V elec
C21	.0033 Mylar 50V
C22	.1 Mylar 50V
C23	.1/250Vac rated
C24	.01 disc
C25	2200μF/16V elec
C26	.01 disc
C27	.01 disc
C28	2200μF/16V elec
C29	.001 Mylar 50V

Part	Value
IC1	MC1350P Amp
IC2	LM565 PLL Det
IC3	LM386N AF Amp
IC4	LM7805 – 12 12V neg
L1	4.7mH P/N 988 472
L2	4.7mH P/N 988 472
L3	4.7mH P/N 988 472
L4	4.7mH P/N 988 472
L5	470μH P/N 988 471
L6	4.7mH P/N 988 472

Ac line cord and plug

MISCELLANEOUS ITEMS*

Part	Value
T1	120V:12V 450 mA Radio Shack P/N 273-1365 or equivalent
S1	SPST switch
FI	1A fuse

RCA phono jacks as reqd.
PC board
Cabinet (Radio Shack)
Hardware as required
Ac line cord and plug 3-wire with ground

*Miscellaneous items are *not* included in kit. Purchase locally at Radio Shack.

B.

Part	Value
R1	4.7k
R2	1k
R3	47k
R4	47 ohm
R5	10k pot
R6	50k pot
R7	10 ohm
R8	47 ohm
R9	2.2k ohm
IC1	MC1350P
IC2	LM386
IC3	LM7812-12V regulator
Q1	2N3565
D1	1N914B
D2	1N4002 or 1N4007
D3	LED
L1	4.7mH P/N 988472
L2	4.7mH P/N 988472
L3	470μH P/N 988471
L4	4.7mH P/N 988472

Part	Value
C1	33pF NPO
C2	3 – 40 trimmer
C3	3pF NPO
C4	3 – 40 trimmer
C5	33pF NPO
C6	.1 Mylar 50V
C7	.1 Mylar 50V
C8	330pF NPO
C9	330pF NPO
C10	.01 disc
C11	10μF/16V elec
C12	1μF/50V elec
C13	1μF/50V elec
C14	.01 disc
C15	10μF/16V elec
C16	1μF/50 elec
C17	.1μF Mylar 50V
C18	470μF/16V elec
C19	2200μF/16V elec
C20	2200μF/16V elec
C21	.01 disc
C22	.1μF (Rated for ac line service)

MISCELLANEOUS ITEMS*

Part	Value
F1	1A fuse
S1	SPST toggle switch
T1	120V:12V 450 mA RS P/N 273-1365 or equal

RCA phono jacks as reqd.
PC Board
Cabinet (Radio Shack)
Hardware as required
Ac line cord and plug 3-wire with ground

*Miscellaneous items are *not* included in kit. Purchase locally at Radio Shack.

Fig. 5-29A. Carrier current receiver parts list FM. B. Carrier current receiver parts list AM.

PROJECT 12: UNIVERSAL SHORTWAVE CONVERTER

This converter is a rather simple but very efficient shortwave converter that covers any one MHz segment of the frequency range 5 to 30 MHz. It was designed originally as a converter for the reception of shortwave broadcast signals on an auto radio, but should work well with an ordinary AM receiver covering the standard broadcast band (550 kHz to 1600 kHz). It uses three transistors—an rf amplifier, mixer, and oscillator, all 2N3563 types. It operates from a 12V supply and draws less than ten milliamperes. The converter features high sensitivity (under one microvolt) and works well with a standard 31" auto radio AM-FM whip antenna, although a longer whip works slightly better below 10 MHz. It uses coils made from old TV set i-f coils, which can be taken from a junked TV set usually available free for the asking at any TV repair shop. If preferred, standard 1/4" dia. slug tuned forms can be used. The PC layout has generous size pads so many different variations of terminal configurations that are found on printed circuit coil forms can be accommodated. This simplifies construction for the experimenter with a limited parts inventory. Suggested coil dimensions are given in Fig. 5-32 for various inductances specified in the table in Fig. 5-31 of coil inductances for the various circuits.

In Fig. 5-30, input signals from J1 are routed to switch 1 (S1), a 3-pole double throw switch. In the off (BC AM-FM) position the signals simply go to S1B and to P1, and S1-C opens the 12V lead, turning off the converter. This gives standard AM-FM broadcast reception. P1 is a plug on the end of a length of cable (keep this as short as you can) that plugs into the antenna socket of the auto radio. The auto antenna is plugged into J1. Therefore, the converter goes between the auto antenna and the auto radio.

When shortwave reception is desired, S1 is placed in the SW position. Signals from the auto radio antenna are routed to the link (2 turns of wire) wound around L1. This couples the signals into L1, which is resonated at the input frequency by C1 and C2. C1 is primarily for tuning and C2 matches the tuned circuit to Q1. Q1 is a grounded base amplifier. Signal developed across R1 – C2 (R1 is a bias resistor for Q1) is fed to the emitter of Q1. R3 and R2 bias the base of Q1, and C3 is a bypass capacitor to keep the base at ac ground. The common base configuration lends itself to easy matching from a tuned circuit over a wide range of frequencies and is less likely to suffer from rf instability at the frequencies this converter covers. R4 is a parasitic oscillation suppressor that prevents Q1 from oscillating at VHF-UHF spurious frequencies. Q1, Q2, and Q3 are VHF transistors and even at the highest frequencies this converter will be useful to (up to 50 MHz) they are "loafing" and have quite a high gain. C4 and L2 are tuned to the converter input frequency and serve as a load for rf amplifier Q1. C6 couples the amplified signals from the auto radio antenna to Q2, a common base marker which mixes incoming rf signals with the signal from the local oscillator (LO). C7 couples the LO signal to the emitter of Q2. R6 provides bias for Q2 and R7-R8 bias the base of Q2, which is ac grounded for rf signals. Dc power is supplied to the rf amplifier via R5 and C5 which is a bypass capacitor,

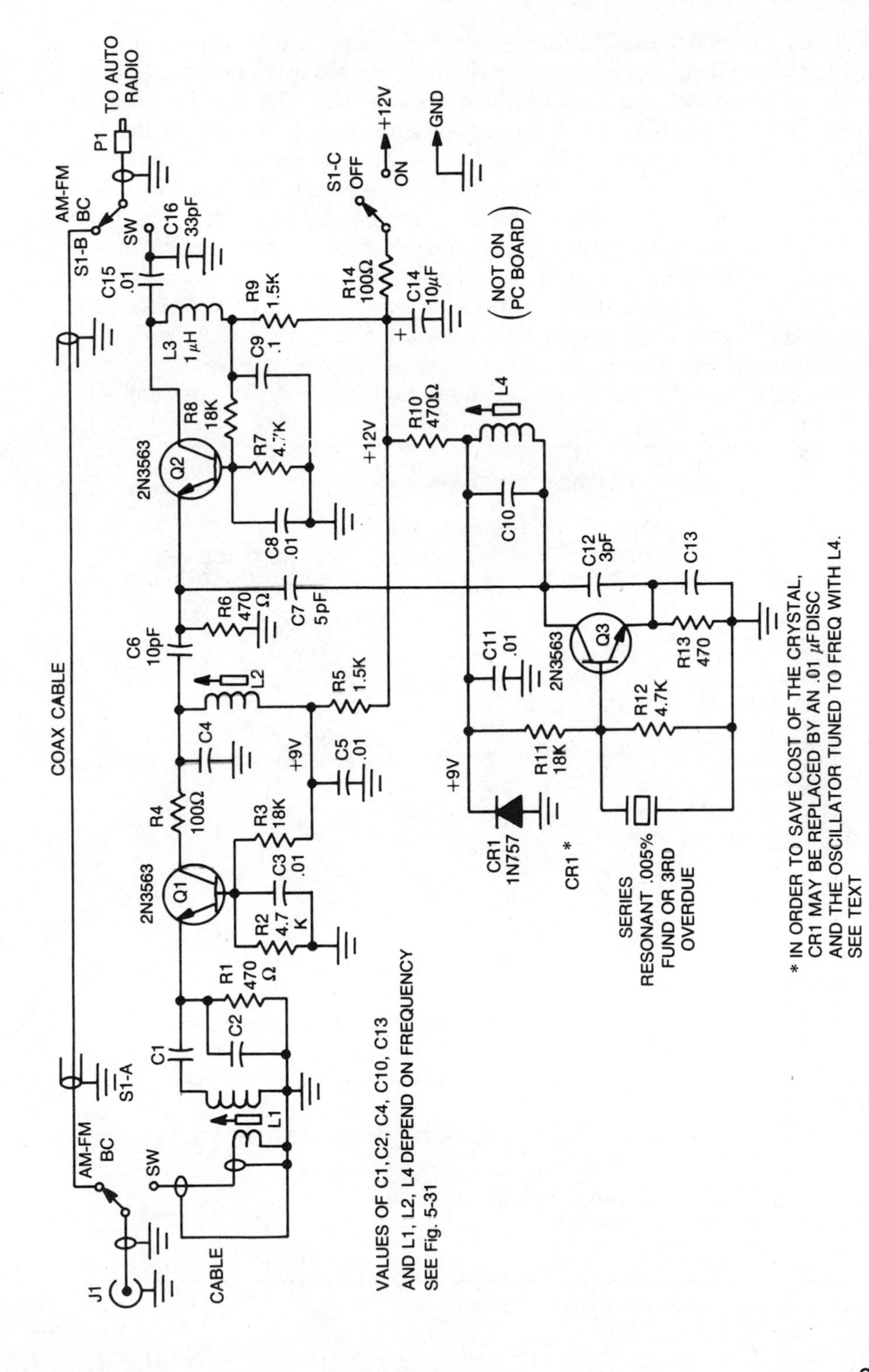

Fig. 5-30. Schematic of universal shortwave converter.

and R9 and C9 to the mixer. The mixer output is developed across L3, a 1 millihenry rf choke. C15 is a dc blocking capacitor and C16 is a matching capacitor. C16 also helps bypass unwanted mixer products to ground. The output across C16, which is equal to the input signal frequency minus the local oscillator frequency, is fed to S1-B and to P1 into the auto radio. As an example, if the LO frequency is 11.0 MHz (11,000 kHz) and the auto radio is tuned to 830 kHz, signals at 11.830 MHz (11,830 kHz) will be received on the auto radio.

The local oscillator is a colpitts type with the frequency control element in the base. Q3 is the LO. R13 provides emitter bias. R11 and R12 provide base bias. CR1, and C11 and R10 provide a regulated +9V to the LO stage. Oscillator stage Q3 is tuned via L4 and C10. C12 and C13 form a feedback network. CR1, a crystal, is a series resonant, fundamental or third overtone type, which ac grounds the base of Q3 only at its series resonant frequency. This prevents

UNIVERSAL SW CONVERTER

FREQ RANGE	CR1 XTAL FREQ (MHz)	TUNING INDUCTANCE μH* L1	L2	L4	CAPACITANCE pF C1	C2	C4	C10	C13
5.5 – 6.5 MHz 49 Meters Band	5.0	5.5	5.5	8.0	150	1000	120	220	100
9.2 – 10.2 MHz 31 Meters Band WWV	8.7	3.0	3.0	3.2	100	820	91	150	100
11.5 – 12.5 MHz 25 Meters Band	11.0	2.2	2.2	2.1	82	680	68	100	100
13.2 – 14.2 MHz 21 Meters Band	12.7	1.8	1.8	1.6	82	680	68	100	100
14.5 – 15.5 MHz 19 Meters Band WWV	14.0	1.6	1.6	1.6	82	470	68	82	82
17.5 – 18.5 MHz 17 Meters Band	17.0	1.3	1.3	1.1	68	470	56	82	82
21.0 – 22.0 MHz 13 Meters BC	20.5	0.9	0.9	1.0	68	470	56	68	68
25.5 – 26.5 MHz 11 Meters Band	25.0	0.8	0.8	0.74	56	330	47	56	56
26.5 – 27.5 MHz 11 Meter CB Band	26.0	0.76	0.76	0.72	56	330	47	56	56

*Should be adjustable −30 to +50% of value shown.
See coil winding table for suggested dimensions.
L1 has 2 turn link around cold end.

Fig. 5-31. Component value for various frequencies.

Q3 from oscillating anywhere except at the crystal frequency. If CR1 is replaced by a .01μF capacitor, Q3 will oscillate wherever L4 is resonant with its tuning capacitance (approximately C10 + C7 + C12 plus strays). This fact can be used to eliminate the crystal CR1 and save a few dollars, but the LO frequency stability will not be as good. Below 10 MHz this may be OK but at 20 MHz or higher excessive frequency drift will most likely be troublesome. Values for L4, C10, and C13 are given in Fig. 5-31 as they are dependent on frequency. Also, see Fig. 5-32.

Microhenries	# Turns (approximate)
0.75	8
1.3	10
1.8	15
2.2	17
3.0	19
5.5	27

Coil form 1/4" dia. with slug inductance range depends on slug, but typical TV coil SW will give −30 to +50%.

Fig. 5-32. Suggested coil dimensions.

The converter is constructed on a PC board 2" × 4 1/4". The PC board is single sided, G-10 is preferred as a base material. First, install resistors, then capacitors. Transistors are next installed, then the coils L1, L2, L3, and L4. Wind a two turn link over the winding of L1 near the grounded (cold) end. This link connects via a short length of cable, to S1-A. R14 is installed off the PC board, between the PC board and S1-C. The PC board can be mounted in a small metal box that can be later mounted under the dash of the automobile, or in applications where this converter is used with an ac powered receiver, near or on the receiver. The box should be big enough to house the board, and also J1, S1 A, B, C, a 3PDT switch. A suitable size might be 3" deep × 5" long × 1" high. Preferably, the dc (12V) power lead should be provided with a 0.5 or 1.0 ampere fuse. Figure 5-33 shows a suitable mounting arrangement.

To test the converter, connect it to a +12Vdc supply. Apply power and check for about +2 volts at the emitter of Q1, Q2, and Q3. Check for +9 volts across CR1. Next, an auto radio or other AM BC receiver is needed. Connect the converter between the antenna and the AM radio. Now, if a frequency counter is available connect it across C13 and adjust L4 until the crystal oscillator operates. Now tune the radio over the AM broadcast band. You should hear some SW signals. Pick a weak signal near the frequency you are most interested in or else tune the radio to one as near 1 MHz as you can find. Adjust L1 and L2 for best reception. There should be a definite point of maximum response. If not, add or subtract a turn from L1 or L2 as required, and try again. In particular, in the shortwave broadcast bands between 6 and 15 MHz, plenty of signals should be heard at most times. The lower frequencies generally are best at night, while the higher (15 – 30 MHz) are best during daylight hours. However, this is not always the rule. If no signals are heard, recheck your wiring and solder connection.

This completes the alignment and testing of the SW converter. If you desire several different frequency bands, a switch can be used to switch in various values of components and crystals, but it is probably easier to build several of the converters and simply switch the power and signal leads. A suitable PC artwork is given in Fig. 5-34 and parts mounting diagram in Fig. 5-33. Figure 5-35 is a photo of a completed PC board. See Fig. 5-36.

A kit of parts consisting of the printed circuit board and all parts that mount on the board is available from:

North Country Radio
P.O. Box 53E, Wykagyl Station
New Rochelle, NY 10804

Price: $32.50 plus 2.50 postage and handling.
New York residents must include sales tax.

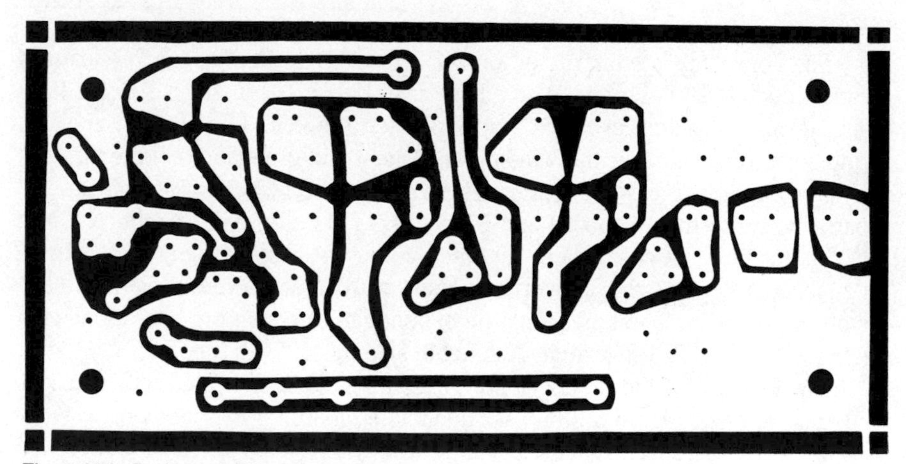

Fig. 5-33A. Project 12 PC layout.

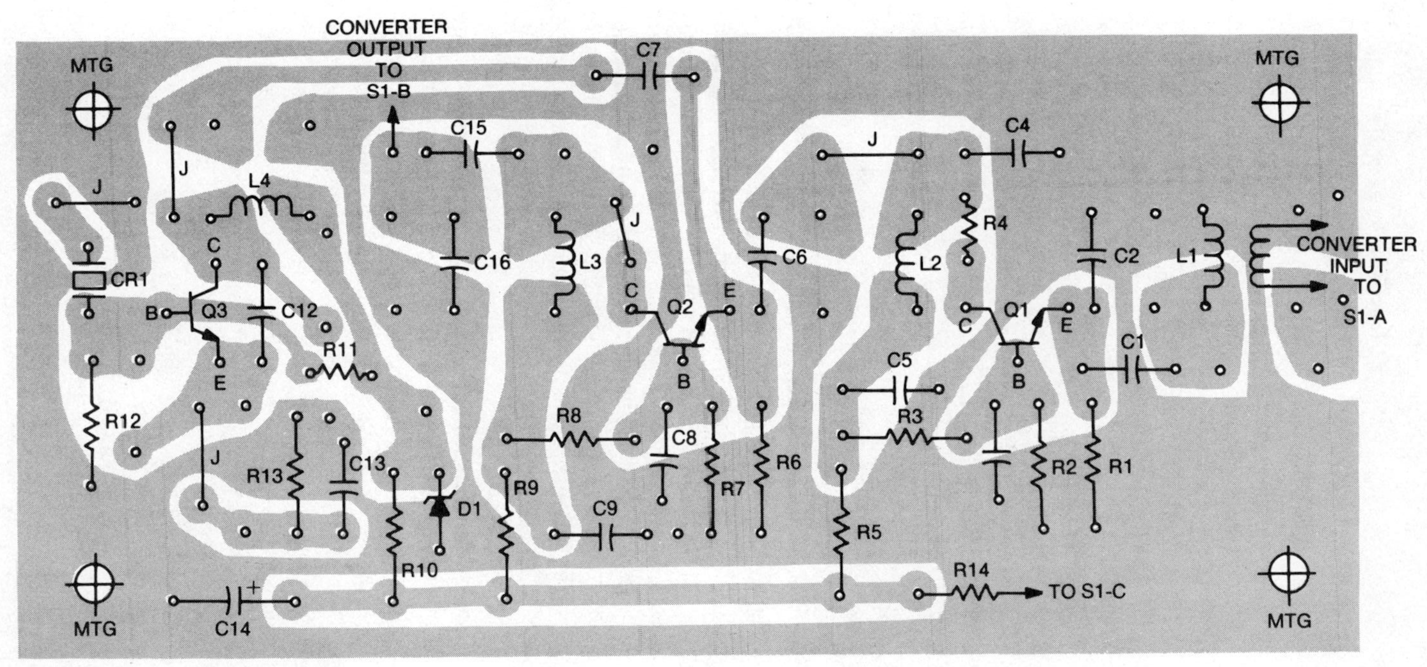

Fig. 5-33B. PC artwork for parts placement.

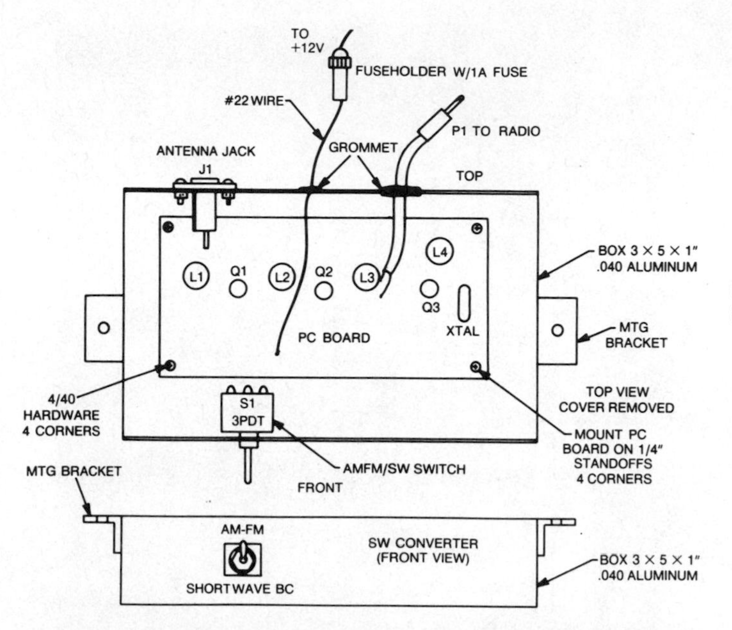

Fig. 5-34. Shortwave converter maintaining arrangement.

PROJECT 13: LOW FREQUENCY ACTIVE ANTENNA/CONVERTER

The frequency range below the standard AM broadcast band, from 10 to 550 kHz, has not generally been included on many communication receivers. For the most part, the extra sets of coils and increased cost and complexity required in the rf circuits has not justified the inclusion of this range. However, a simple converter is not difficult to construct, and will provide reception of signals in this frequency range. Among the signals found below 550 kHz are maritime mobile, distress, radio beacons, aircraft weather, European longwave AM broadcast, and point to point communications. A brief summary of these signals is found in Fig. 5-37.

The low frequencies of these signals make the use of conventional receiver techniques found in shortwave receivers impractical. In the case of the usual four band general coverage receiver, four sets of coils are required for coverage of 550 kHz to up to 25 to 30 MHz, the usual upper limits of short wave reception. To cover the low frequency range completely, at least three and more likely four coil sets are needed, since a 3:1 range is all that can be easily obtained with satisfactory tuned circuit tracking and alignment. In addition, with the usual 400 pF variable capacitors, inductances of around six hundred millihenries are needed to reach the low frequency limit of 10 kHz that is used for radio communications. Obviously, this is somewhat impractical, but it can be done if cost is not a factor.

Many radios produced for the European markets cover the range of 150 kHz to 400 kHz; however, they are usually designed for AM reception and are not required to have a high sensitivity figure, since European longwave (LW) stations generally run 50 or 100 kW, or more to the antennas.

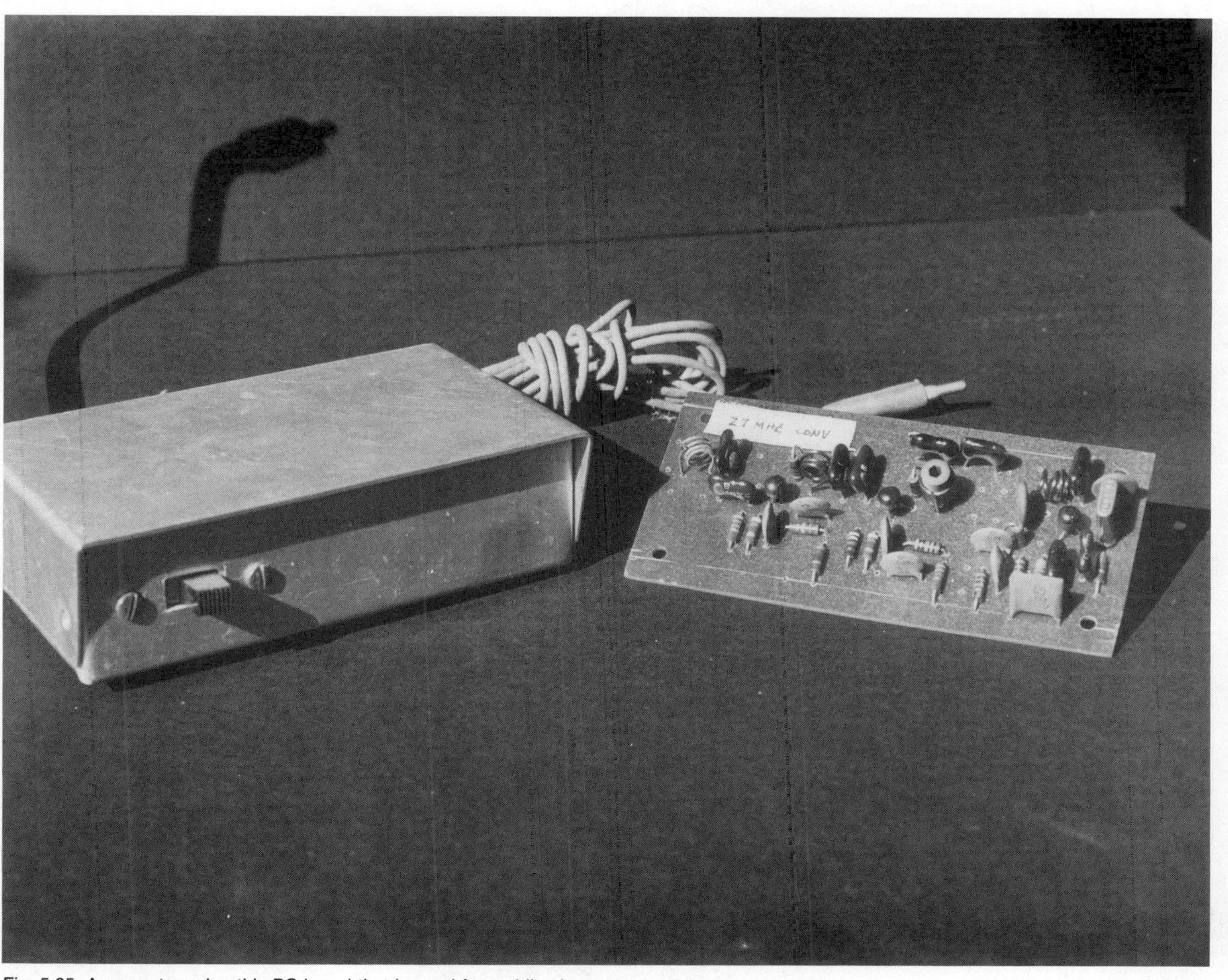

Fig. 5-35. A converter using this PC board that is used for mobile shortwave reception on auto radio.

RESISTORS		COILS & CHOKES	
R1, R6, R10, R13	470Ω	L1, L2, L4	See text
R2, R7, R12	4.7k Ω	L3	1mH RFC
R3, R8, R11	18k Ω		
R4, R14	100Ω	**CAPACITORS**	
R5, R9	1.5k Ω		
		C1, C2, C4 C10, C13	See table
CAPACITORS			
		MISC. MATERIALS	
C3, C5, C8, C11, C15	.01 disc		
C6	10pF NPO	S1	3PDT switch
C7	5pF NPO	J1,	To suit
C9	.1μF Mylar	P1	interface requirements
C12	3pF NPO		
C14	10μF 16V	CR1	Crystal as reqd. See table
C16	33pF NPO		
TRANSISTORS		*Also:* RG58U cable, wire as read, PC board, enclosure or case to suit application.	
Q1, Q2, Q3	2N3563		

Fig. 5-36. Project 12 parts list.

It is much more practical to up-convert the range of 10 to 550 kHz to a higher frequency range. The converter to be described converts the LF range to the range 1 to 1.55 MHz. It can be thought of as adding 1 MHz to the frequency of all received signals. Therefore, by connecting the converter to a communications receiver or AM broadcast receiver covering this range, all the features of the receiver can be used for longwave reception. Calibration is unnecessary. Signals are received at a dial setting corresponding to their frequency plus 1 MHz. A 100 kHz signal is received at 1100 kHz, a 335 kHz signal at 1335 kHz, etc., just drop the first digit and you directly read the longwave signal frequency.

A problem at low frequencies is man-made noise. Many of our everyday devices and appliances are notorious in this regard. Motors, fluorescent lighting, light dimmers, computers, TV receiver sweep radiation, and many small household digital devices generate much ''hash'' in the spectrum below 550 kHz. Fortunately, most of this is chiefly carried on power lines and does not radiate very far from them. It might be thought that tremendous antennas are needed for longwave reception. It is true that at the lower portion of the range a good, high longwire antenna is an advantage. With the use of active antenna techniques, very good reception can be had with a simple vertical antenna a few meters long. The converter to be described even picks up quite a few signals with a clip lead 30 centimeters long.

Low frequency signals picked up on an 8′ whip antenna (a standard CB/HAM 10 meter whip) goes to the gate of Q1, a source follower. Q1 matches the

FREQ RANGE	SIGNALS FOUND	REMARKS
510 – 535 kHz	Misc. radio beacons	________
500 kHz 415 – 490 kHz	Distress (CW) Maritime mobile (CW)	Ship – shore, interesting if you copy CW – about 18 wpm
285 – 400 kHz	Radio beacons, weather aeronautical & marine	Weather info, AM Voice XMSN
190 – 285 kHz	Radio beacons, weather European longwave broadcast	*Also* Carrier current (power line) transmissions
160 – 190 kHz	Fixed public, license free experimental, European longwave broadcast, fixed	Some experimenters run 1 watt transmitters in this band; no license needed
110 – 160 kHz	Maritime mobile, lowest freq. longwave broadcast fixed (point to point)	Tends to be noisy, also some RTTY transmissions
90 – 110 kHz	Loran navigation	________
30 – 90 kHz	Fixed, moblle standard freq. & time signals	RTTY transmissions some CW, noisy
14 – 30 kHz	Submarine communications worldwide high power VLF—military, commercial	RTTY transmissions, some CW heard at times, noisy
10 – 14 kHz	Omega signals freq. standards atmospheric phenomena, whistlers	Lowest part of radio spectrum, frequently used
Below 10 kHz	Atmospheric noise, whistlers, experimental transmissions – military	Experimental

Fig. 5-37. Signals found below the AM broadcast band.

high impedance of the whip (which looks like a small capacitor of about 20 to 30 pF) to the low pass filter formed by C1, C2, C3 and L1, L2, L3 and L4. R1 provides a dc ground for the gate of Q1 and D1, D2, D3, D4 bleed off accumulated static charge on the antenna while having no effect on signals less than about 1 volt on the antenna. The low pass filter rejects signals above 500 kHz, preventing break through or cross modulation from strong broadcast and shortwave signals. R2 provides a return path for the drain current of Q1. Low frequency signals are coupled through C4. C4 is selected to alternate signals below 10 kHz. The signals are fed to doubly balanced mixer IC1, an MC1496L, which has been around for many years and is a proven reliable, easy to use device. R3, R5, control R7, R4, and R6 provide an adjustable bias network for the input pins

1 and 4 of IC1. R9, R8, R10 and R14 provide correct dc operating voltages and bias levels for IC1. C5, C6, and C7 are bypass capacitors. IC1 has balanced (dual polarity) inputs and outputs, but single (unbalanced) ended inputs and outputs as used here may be accommodated simply by using only one (either will do in this case). R11 sets the gain (about three) of IC1. R12 and R13 feed bias to the local oscillator inputs of IC1. LO signal from Q2 is fed to pin 8 of IC1. It is a fixed frequency signal at 1.000 MHz. Actually, if desired, other frequencies may be used. For example, if it is desired to use an 80 meter (ham band) receiver to tune the LF range, use a 3.500 MHz oscillator. In this case, a crystal controlled oscillator might be a better bet if the utmost in stability is wanted. Signals in the 10 to 550 kHz range are converted to 1010 to 1550 kHz. Also, a 990 to 450 kHz output is produced, but it is ignored, because direct readout of frequency is not obtained and most AM receivers only cover down to 530 kHz or thereabouts.

Mixer output appears at pins 6 and 12. Pin 12 is used arbitrarily (easier PC layout). R21 is used as a PC board jumper and is not critical in value. R18 and R20 provide bias to the output stages of IC1. R19 and C12 are for decoupling of the dc power supply. Q3 transforms the high output impedance IC1 to around 100 ohms so as to better match the receiver input. C15 is a dc blocking capacitor, and feeds signals from the emitter of Q3 (1010 – 1550 kHz) to the output jack J3. R22 is a dc return and bias resistor for Q3. L6 is used to feed dc voltage also carried on the signal cable from the receiver. This saves a connecting wire and simplifies use of this device as in remotely mounted applications. C14 and C13 provide dc supply filtering. Q2 is a Hartley type oscillator operating at 1 MHz. R15 and C9 are supply decoupling components. R16 provides source bias to Q2. L5 and C11 are a resonant circuit tuned to 1 MHz by a slug in L5. C10 couples the oscillator transistor Q2 to the top on the tank circuit L5-C11. Signal from the LO is taken off via R17, which sets the level of LO signal at pin 8 of IC1, and dc blocking capacitor C8.

The entire unit, converter and antenna is best remotely mounted, as far away from ac wiring and interfering devices as possible, but if you live in the country and have a quiet location, or are willing to tolerate some line noise, the converter can be used near the receiver. Figure 5-39 shows suggested mounting arrangements. For local use (nonremote) the end of L6 converted to output jack J3 may be disconnected from J3 and fed +12Vdc directly from a convenient supply source. About 11 to 15 milliamperes is required. (See Fig. 5-38.)

For remote applications, the cable from J3 is brought out to a small shielded box containing J4, J5, and dc blocking components C16 and L7, and filter capacitors C17 and C16. Dc is fed in via J6. It should be very well filtered (less than 1% ripple). In this case, an extra volt or two of dc is recommended to make up for losses in L6 and L7 which have about 60 ohms dc resistance each. Figure 5-40 shows the complete hookup.

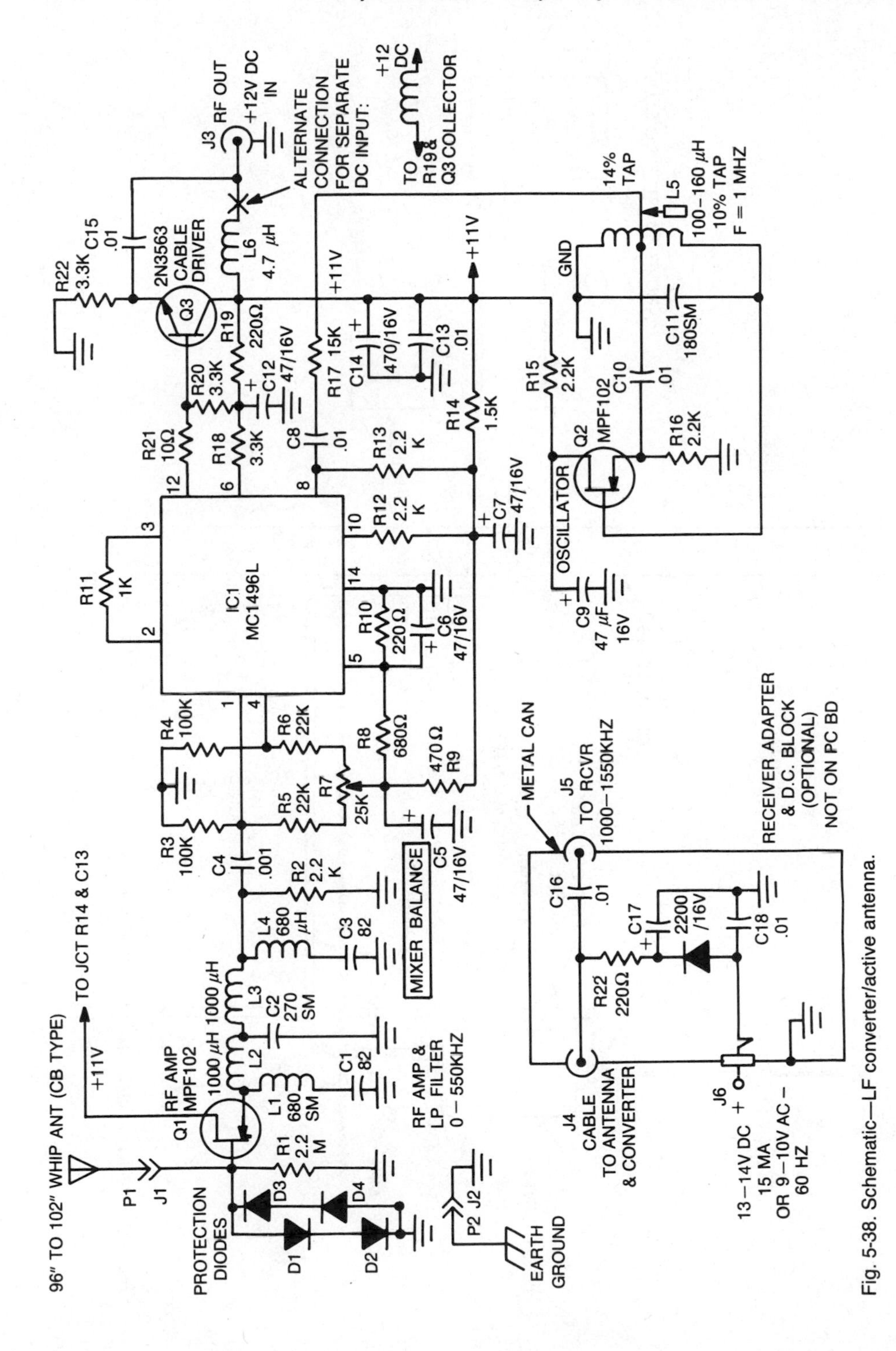

Fig. 5-38. Schematic—LF converter/active antenna.

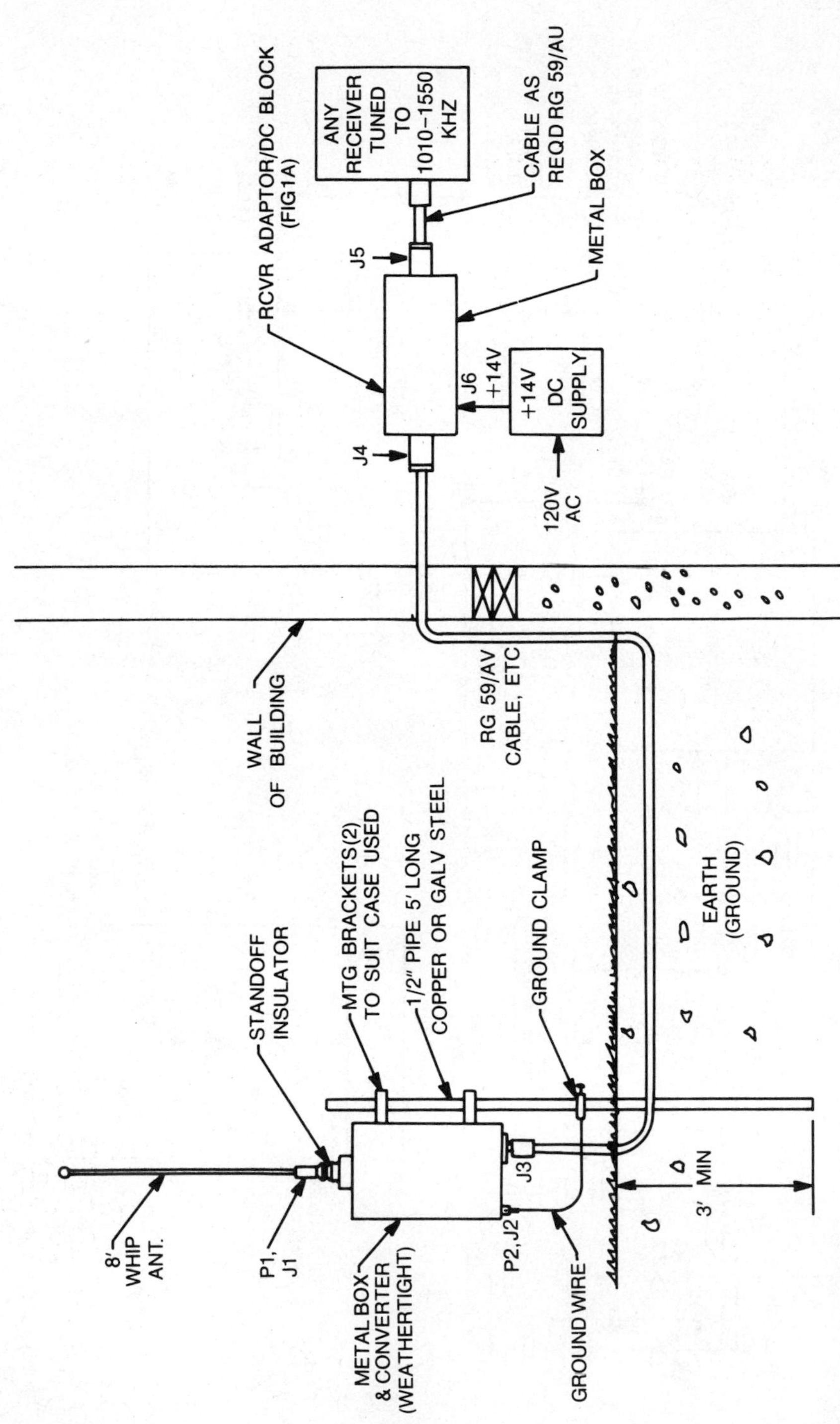

Fig. 5-39. Installation of LF converter/active antenna.

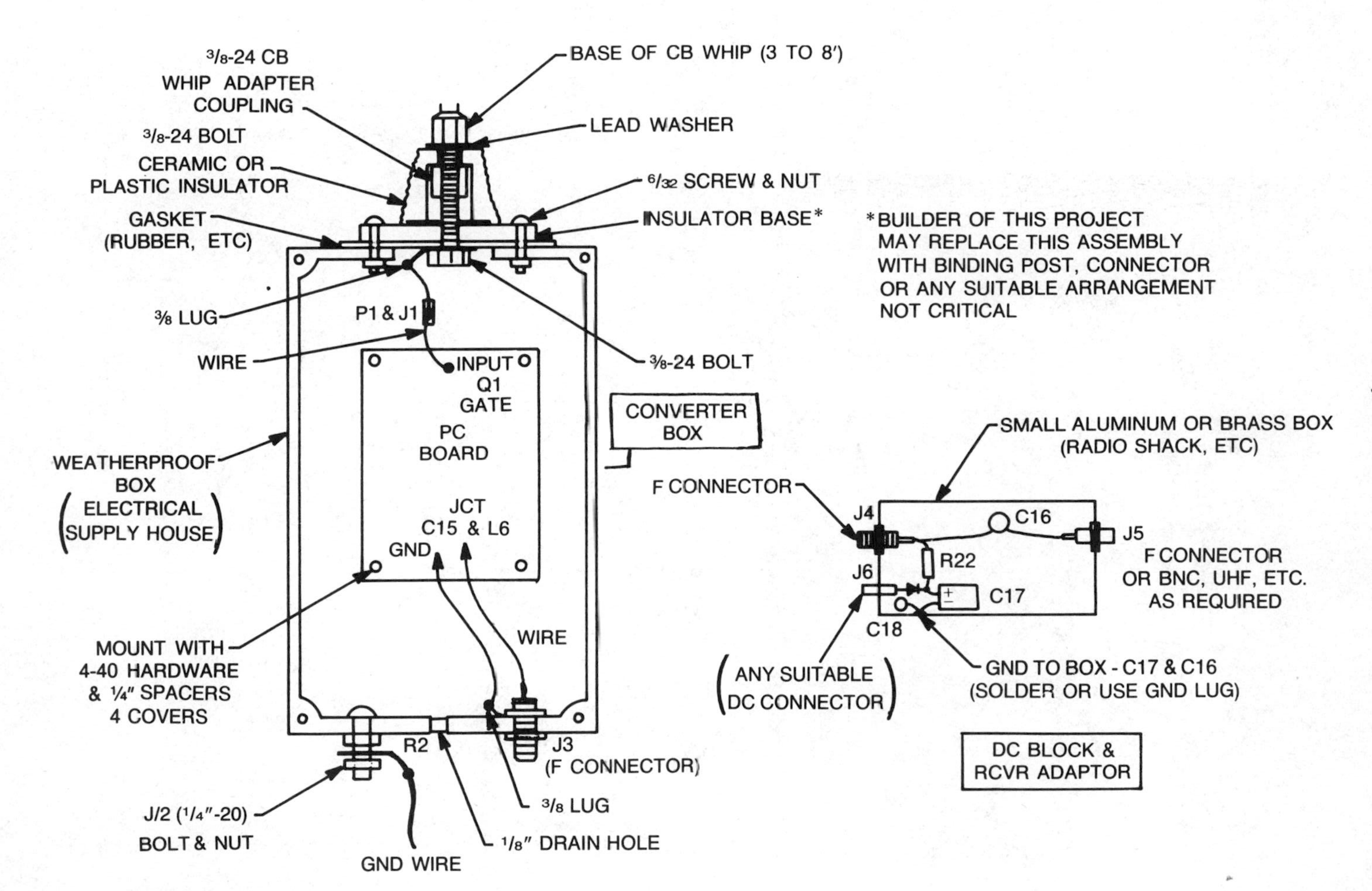

Fig. 5-40. Packaging of LF converter/active antenna.

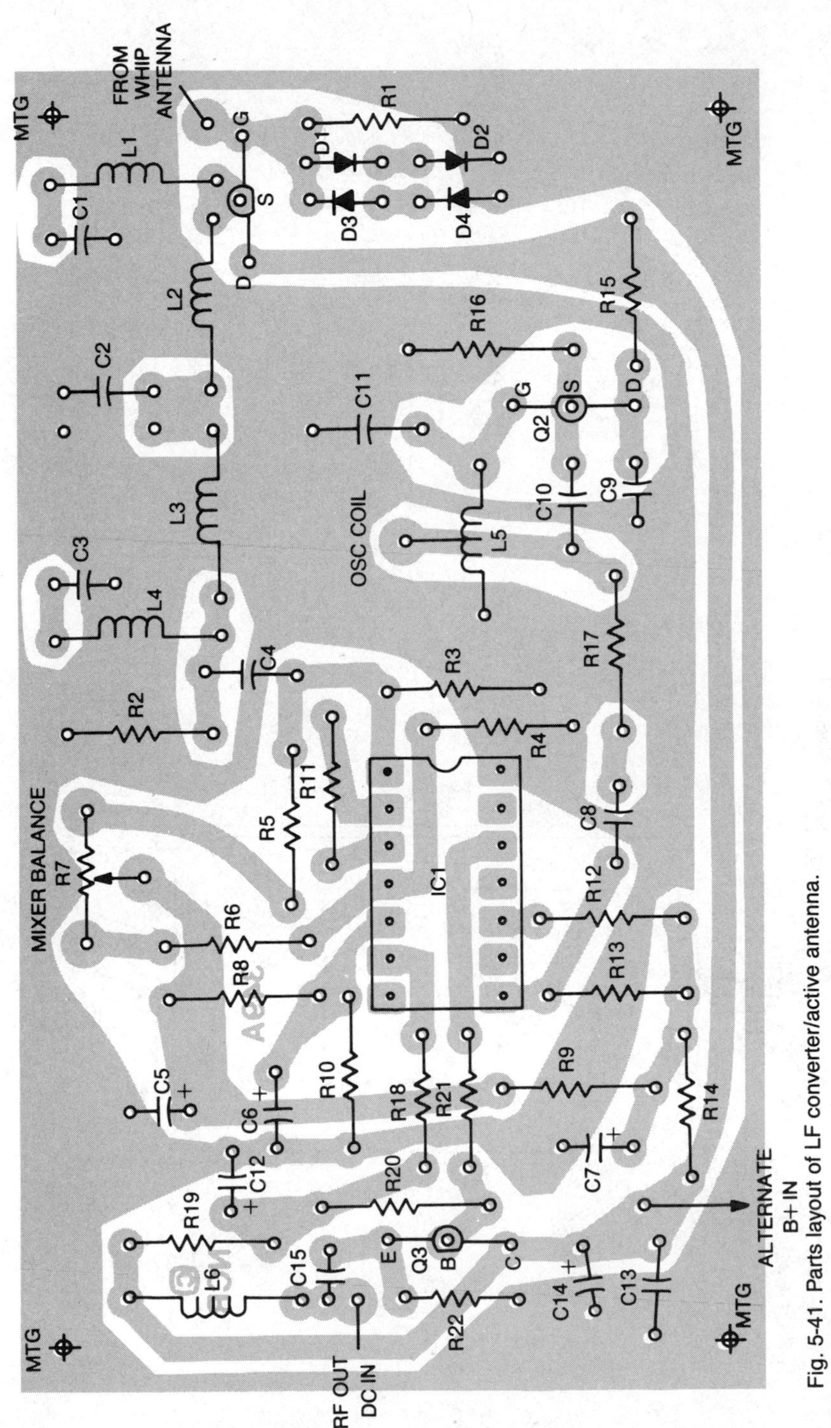

Fig. 5-41. Parts layout of LF converter/active antenna.

To make this converter, use the PC layout in Fig. 5-41 and artwork in Fig. 5-42. As usual, first mount resistors, then capacitors. A socket for IC1 is desirable but unnecessary. Then mount D1 through D4, Q1, Q2, Q3, and last install L1 through L6 and IC1. If you are using a remote installation, next make up the receiver adapter and dc block shown in Fig. 5-38. Layout is not critical, but be sure to use a completely shielded metal box to avoid pickup of strong AM stations in the 1010 – 1550 kHz range.

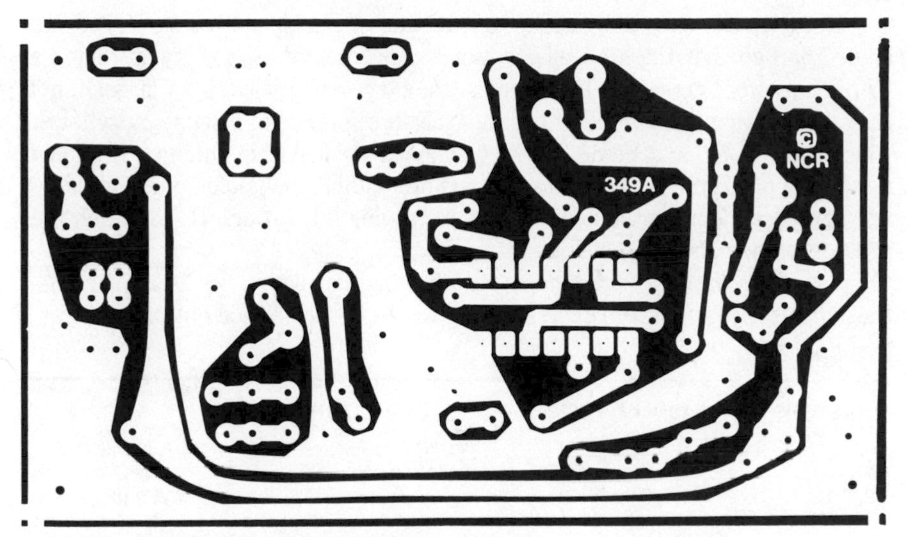

Fig. 5-42. Parts artwork of LF converter/active antenna.

Tune-up is very simple. First, check all wiring and components for correct placement. Connect −12V if everything checks out, to adapter box J6. Next, check for +12V at J3, (or L6 if you are not using the remote setup). Next, verify +11 volts or so across C14. Measure current drawn from the +12 dc supply. More than 15 mA may indicate problems.

If all is OK so far, check for:

1. Jct. L1 and L2 +0.5− +2.0 volts
2. +2−+4 volts at source of Q2
3. +8−+10 volts across R22
4. +5−7 volts pins 8 and 10 IC1
5. +3.5−5 volts across C5
6. +0.8−+1.5 volts pin 5 IC1
7. +8−+10 volts pins 6 and 12 IC1

Connect a receiver covering the AM broadcast band to J3. Tune the receiver to 1.000 MHz. Adjust L5 for the strongest signal. Next, adjust R7 to minimize this signal. R7 should cause a definite null around the middle of its range.

If not, check IC1, R7, and R3 to R6. Next tune the receiver between 1010 and 1550 kHz. You should hear signals in the longwave range. At 1100 kHz a loud, rattling noise will be heard in most areas of the US and Canada. This is the LORAN navigational signal. No AM broadcast stations should be heard. If they are, check your cables and grounding. Make sure shielding is adequate, especially the box used for the dc block. As a last resort, check L1 through L4, C1, C2, C3, and the setting of R7. The circuit is noncritical and should work the first time it is tried.

Occasionally, extremely strong AM broadcast signals may cause "crud" to come through that rides on all longwave signals heard. If so, try installing a 47 pF capacitor across R1. If this helps, then try smaller values (or larger) until the smallest value is found that reduces this interference to a satisfactory level. This problem also may be dealt with by using a smaller whip antenna. It should not be encountered, however, unless you are within a few miles of a high powered broadcast station. You can also try removing D1 through D4 although the protection they afford Q1 will be lost.

As a final warning, again *do not* mount this unit within 50′ of any ac high tension lines (power poles) or near any ac or telephone service entrance cables. See Fig. 5-43.

RESISTORS 1/4W ±10% or ±5%

R1	2.2 Megohm
R2, R12, R13, R15, R16	2.2k Ω
R3, R4	100k Ω
R5, R6	22k Ω
R7	25k trim pot
R8	680 Ω
R9	470 Ω
R10, R19, R22	220 Ω
R11	1k Ω
R14	1.5k
R17	15k
R18, R20, R22	3.3k
R21	10 Ω

MISCELLANEOUS

J1, J2, J5, J6	Suitable converter of your choice
J3, J4	"F" chassis connector

1 - Box, weatherproof (main converter)
1 - Box, small metal (dc block)
1 - CB whip, 102″ & insulator
1 - PC board
Hardware, wire, cable as reqd.

CAPACITORS

C1, C3	82 pF ± 5% NPO ceramic
C2	270 pF ± 5% NPO ceramic
C4	.001μF 50V Mylar
C5, C6, C7, C9, C12	47μF/16V electrolytic
C8, C10, C13, C15, C16	.01 disc ceramic 50V
C11	180 pF ± 5% NPO ceramic
C14	470μF/16V
C17	2200μF/16V

SEMICONDUCTORS

D1, D2, D3, D4	1N914B
Q1, Q2	MPF102
Q3	2N3563
IC1	MC1496L

COILS

L1, L4	680μH ± 5%
L2, L3	1000μH ± 5%
L5	100 - 160μH* tapped 14%
L6	4.7μH rf choke

Fig. 5-43. Parts list for LF converter/active antenna.

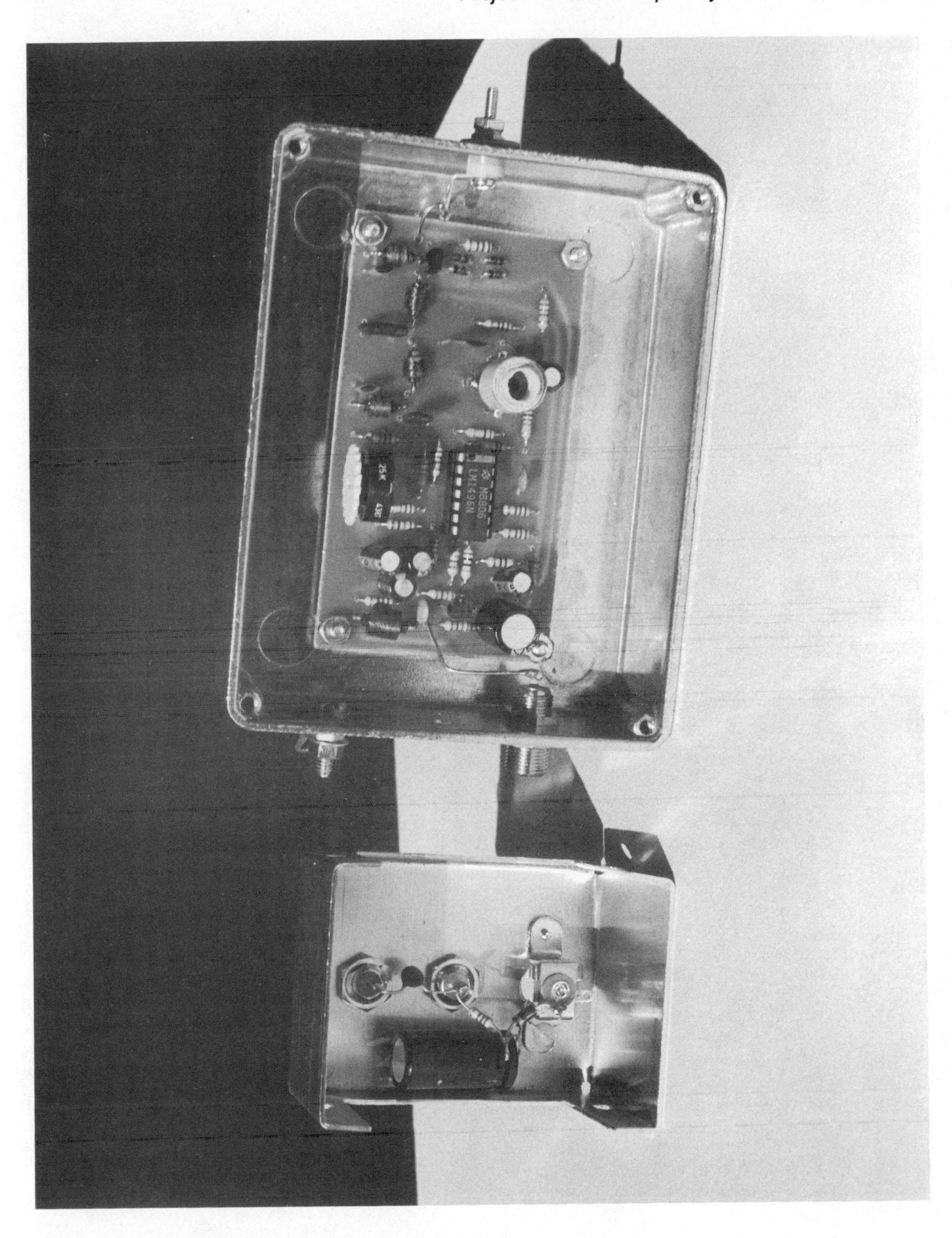

Fig. 5-44. Photo of completed LF converter box and receiver adapter.

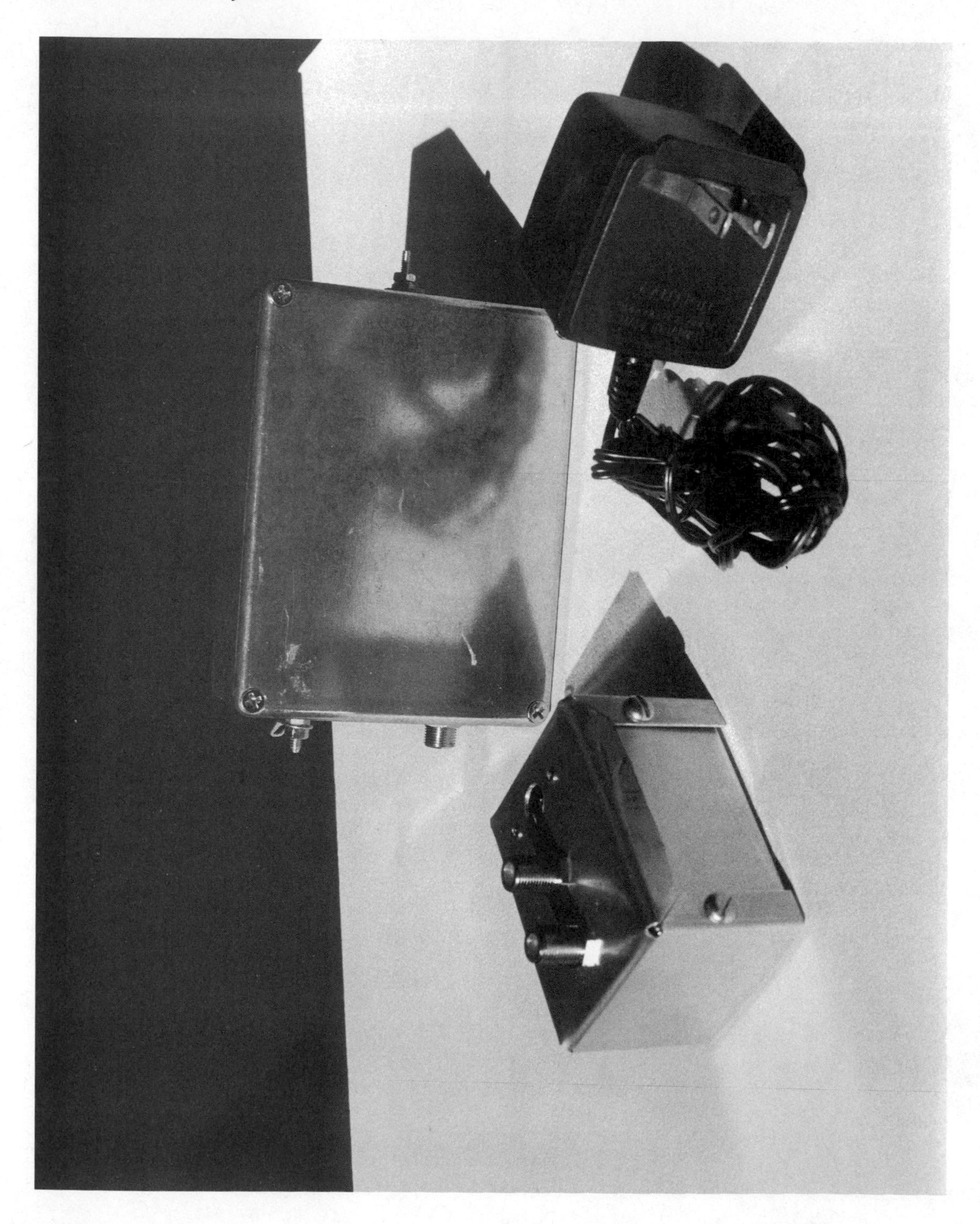

Fig. 5-45. Complete LF receiving system—components.

A kit of parts consisting of the printed circuit board and all parts that mount on the board is available from:

North Country Radio
P.O. Box 53E, Wykagyl Station
New Rochelle, NY 10804

Price: $33.75 plus $2.50 postage and handling.
New York residents must include sales tax.

Part 6
Electro-Optical, Photo, and Measurement Projects

Recently many devices using infrared technology have come into the electronic field. For many years much infrared research has been confined to military and industrial use, and much infrared technology probably still remains classified. Such items as snooperscopes, night vision devices, thermal cameras, heat seeking missile guidance systems, and remote sensing satellites use infrared technology. Most of the new TV set remote controllers use infrared remote controllers, and even some remote controls for hi-fi systems are made in a similar manner. A prominent manufacturer of two-way radio systems has a remotely mounted control panel for the radio functions (digital readout included). A fiber optic cable using infrared radiation connects the radio control panel with the main chassis. Undoubtedly more uses of infrared (abbreviated IR) will come forth in the future. This section discusses IR communications devices. Construction details for an IR transmitter and receiver for audio are presented using available parts kits. In addition, applications of these devices will be discussed.

INFRARED COMMUNICATION

Infrared radiation (IR) is broadly defined as any radiation lying between the limits of visible light at the red end of the spectrum, and the upper limits of the radio spectrum. While there are no sharply defined limits, this would be taken as radiation having a wavelength of from 700 nanometers (0.7 micron wavelength) and radiation having a wavelength of up to 1,000 microns (1 millimeter). This corresponds to the present upper limit of the microwave radio spectrum, around 300 GHz. This would mean that visible light of 0.5 micron (blue-green light) would have a frequency of 600,000 GHz, or 600 million megahertz. Therefore, the infrared frequency range has approximately a 1,000 to 1 spread.

Although there is no theoretical reason that similar circuits and other techniques could be used for visible light, infrared, and microwave, the wide variation of frequency and the extremely small wavelengths made for the necessity of using widely different techniques, as will be obvious from later discussions.

A diagram of the infrared spectrum is shown in Fig. 6-1. There are several distinct regions:

1. Near IR—visible light 0.7μ to 3.0μ
2. Middle IR—3.0μ to 6.0μ
3. Far IR—6.0μ to 15.0μ
4. Extreme IR—15.0μ to 1,000μ (microwave rf)

One micron (1.0μ) is one millionth of a meter or one thousandth of a millimeter. A millimeter is about 0.04 inches (1/25th of an inch). A micron is 40 one-millionths of an inch. A human hair is about 75 – 150 microns thick, to give you an idea of the dimensions we are talking about. We are mainly concerned with the near IR in this article, but will discuss the entire spectrum of IR radiation briefly.

If nature endowed us with the ability to perceive IR radiation, or we had an optical ''up converter'' so that we could convert any desired radiation wavelength to visible radiation, we could ''see'' many, many other ''colors'' to which our eyes are totally blind. Let us start in the visible spectrum. Imagine we are looking at a scene lit by the sun, of a scenic spot somewhere in the country. It is a beautiful day and everything appears ''normal.'' The sky is blue, with clouds and in the distance, hills can be seen. Figure 6-2 is a photograph of this typical scene. Now imagine we would see selectively only one color at a time, as though we were looking through a transparent but strongly colored glass or plastic filter. If the color of the filter is ''removed'' by photography using black and white film, or a monochrome video system, we would see a black and white scene as Fig. 6-2. This scene seems perfectly normal. Now lets ''tune'' down the spectrum towards the red end. If a red filter were to be placed on the camera, the picture would show the clouds more prominently and the sky would be darker. Foliage would appear darker. The scene would seem to have greater contrast. Distant details stand out. Now, let us go beyond the red end of the spectrum. We are nearly or totally blind to radiation at 800 nm (0.8 microms) and are now in the near IR. Figure 6-3 shows the scene taken with infrared sensitized film. The sky is black, and foliage, which is strongly reflective at this wavelength, appears white. It is a strange appearing scene, with distorted tonal values. Infrared rays are scattered less by the atmosphere and penetrate haze better than visible rays.

As we go lower into the IR range, the scene would take on a strange appearance that becomes very different than what our eyes see. At about 3.5 microns, window glass becomes somewhat opaque and houses would appear to

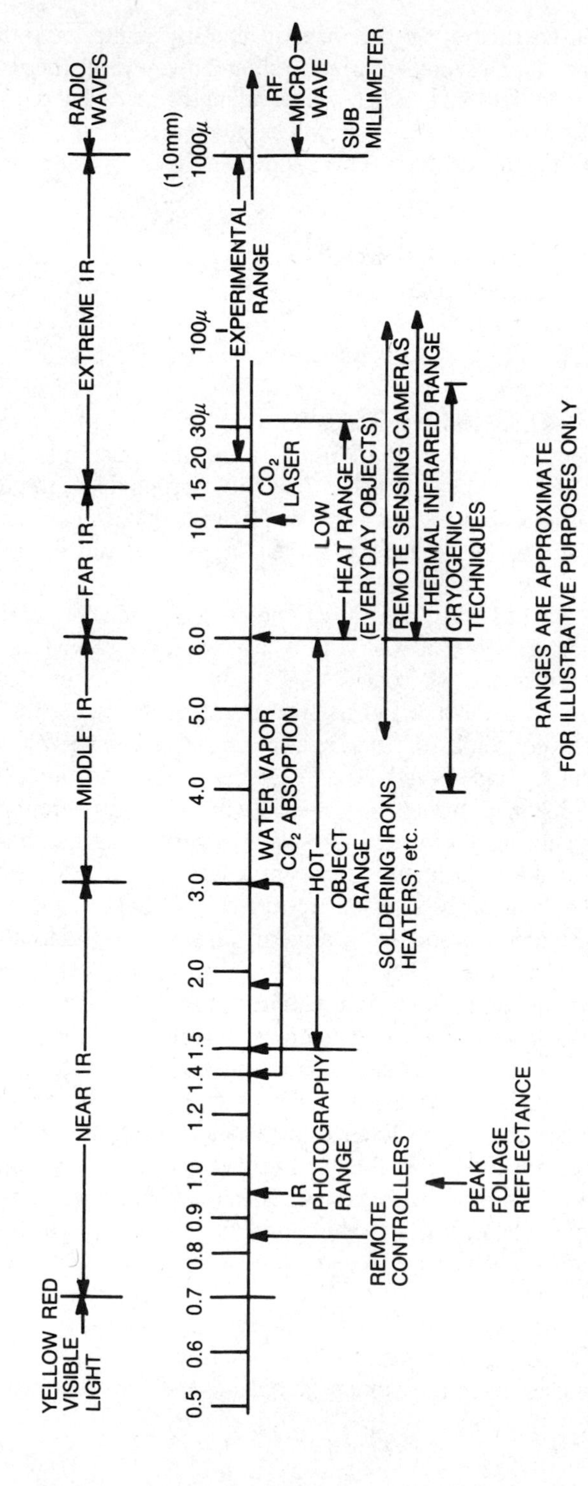

Fig. 6-1. Infrared spectrum.

Fig. 6-2. Normal photograph.

Fig. 6-3. Infrared photograph.

have "black" windows. The sky would be totally black. The scene would gradually darken, and become dark at certain wavelengths corresponding to those wavelengths where high absorption of IR radiation occurs. Water vapor, carbon dioxide, and oxygen molecules cause these "holes." The spaces between are called "windows." One such hole occurs at about 1.85 microns. Another occurs from about 2.5 to 3.0 microns, and from 4.2 to 4.5 microns. There are also others.

As we go lower and lower, some strange things would occur. Objects would appear to "glow." A hot flat-iron would glow very brightly. Warm objects could be seen to glow against the cooler surroundings. We are seeing "heat" radiation. A much confused definition must be clarified at this point. Heat is the result of an object having its molecules and atoms in vibration. The rays or radiation given off by that object is a result of this molecular motion. If you doubt this, try the following experiment. Face the sun making sure your eyes are closed. Now place a sheet of cardboard in front of your face. You will sense loss of warmth on your face. The cardboard is obstructing the IR radiation, which your skin detects as warmth. This experiment will also work with any other source of radiation, such as an electric heater, wood stove, or a hot plate. Most objects at environmental temperatures radiate in the 8 to 20 micron range. This is why most longwave IR detectors must operate with cryogenic (low temperature) detection techniques. In order to differentiate radiation from objects at room temperature from that of the detector itself, the detector must be operated down near liquid air temperatures (100°K or lower). 0° Kelvin is absolute zero (−273°C). To convert °Kelvin to °C, simply add 273.

Beyond 10 or 15 microns is largely experimental territory. This has long been a "gray area." It has been difficult to generate or detect waves in this region, until recently. At these wavelengths, detectors made of semiconductor materials such as germanium doped with zinc, copper or other materials can be used as detectors. Bolometers and thermopiles are used, in which incident radiation is absorbed and converted to heat, producing a change in resistance or a generated voltage.

Finally, we would come to the more familiar region in the submillimeter microwave spectrum. Much work presently is being done in this frequency range, and undoubtedly the extreme IR to microwave range will eventually become more manageable as techniques are refined. At these frequencies, extremely directional antennas would be very small. A high gain parabolic dish might be the size of a thimble. These frequencies may be used in space communication and for very wideband data transmission, at gigahertz/sec., data rates, over fiber optic or other such media, using dielectric wave guides.

The advent of LED devices and integrated circuits makes it possible to construct reasonably priced IR control and communication devices, with operation in the near IR (around 0.9 microns) portion of the spectrum. A block diagram of such a system in shown in Fig. 6-4.

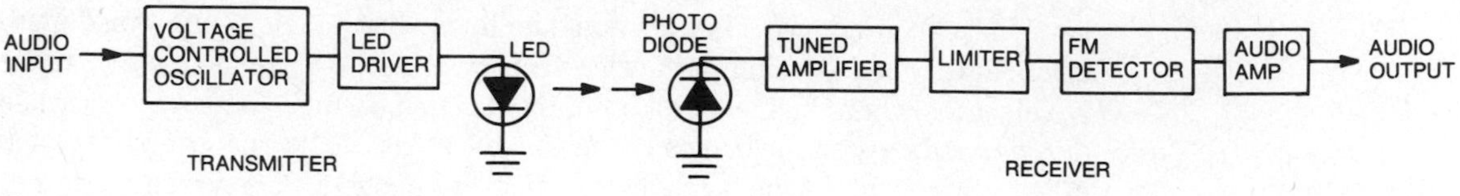

Fig. 6-4. IR receiver block diagram.

Why would you use such a system rather than radio methods? There are several advantages:

1. Low cost
2. No FCC problems or legal problems
3. Relative freedom from interference
4. Can be made very directional
5. Can be made secure with little trouble

Many applications require only short range (10 – 100 ft.) audio transmission. A few applications would be:

1. Wireless links between audio equipment
2. Remote controllers
3. Computer data interfaces
4. Wireless intercoms
5. Aids for the hearing impaired
6. Cordless headphones

For these applications, IR methods are useful and can be superior to ordinary radio methods. A schematic of a transmitter is shown in Fig. 6-5 and a receiver in Fig. 6-6. Depending on the optics and techniques used, a system such as this is useful for ranges from 20 to several hundred feet (250 ft. with a simple convex lens of 1½ in. diameter and about 4-inch focal length). If a parabolic reflector is used at the transmitter in addition, ranges up to half a mile or more are possible. However, the directivity will be pronounced. See Fig. 6-7. This may be of some advantage for certain applications. By using two transmitters and two receivers with a suitable optical arrangement, a directional, tight beam "microwave telephone" can be set up, with excellent range. A suitable "antenna" arrangement might be two parabolic reflectors about 6 inches in diameter (concave mirrors) with the LED and photodiode at the focus, as close together as possible. Suitable mirrors may be obtained from dime store makeup or shaving mirrors, or old flashlight parabolic reflectors. This is an interesting experimental project in itself.

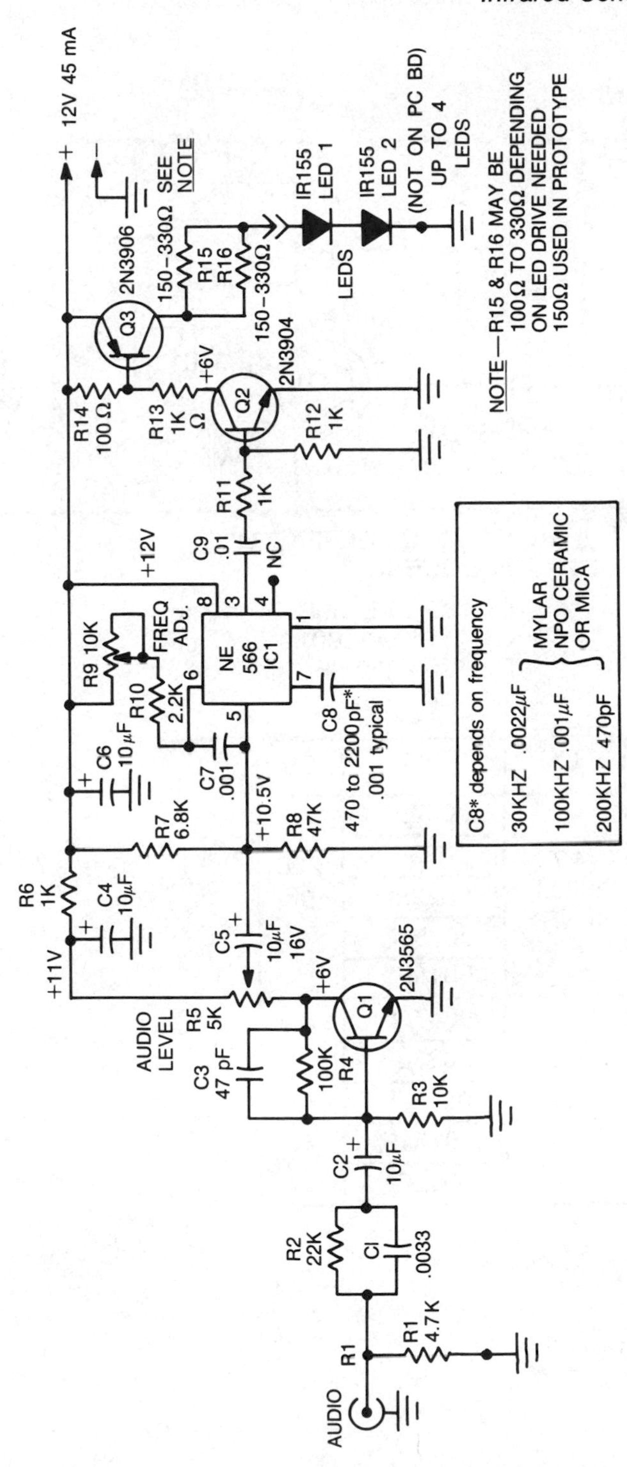

Fig. 6-5. Schematic—IR transmitter.

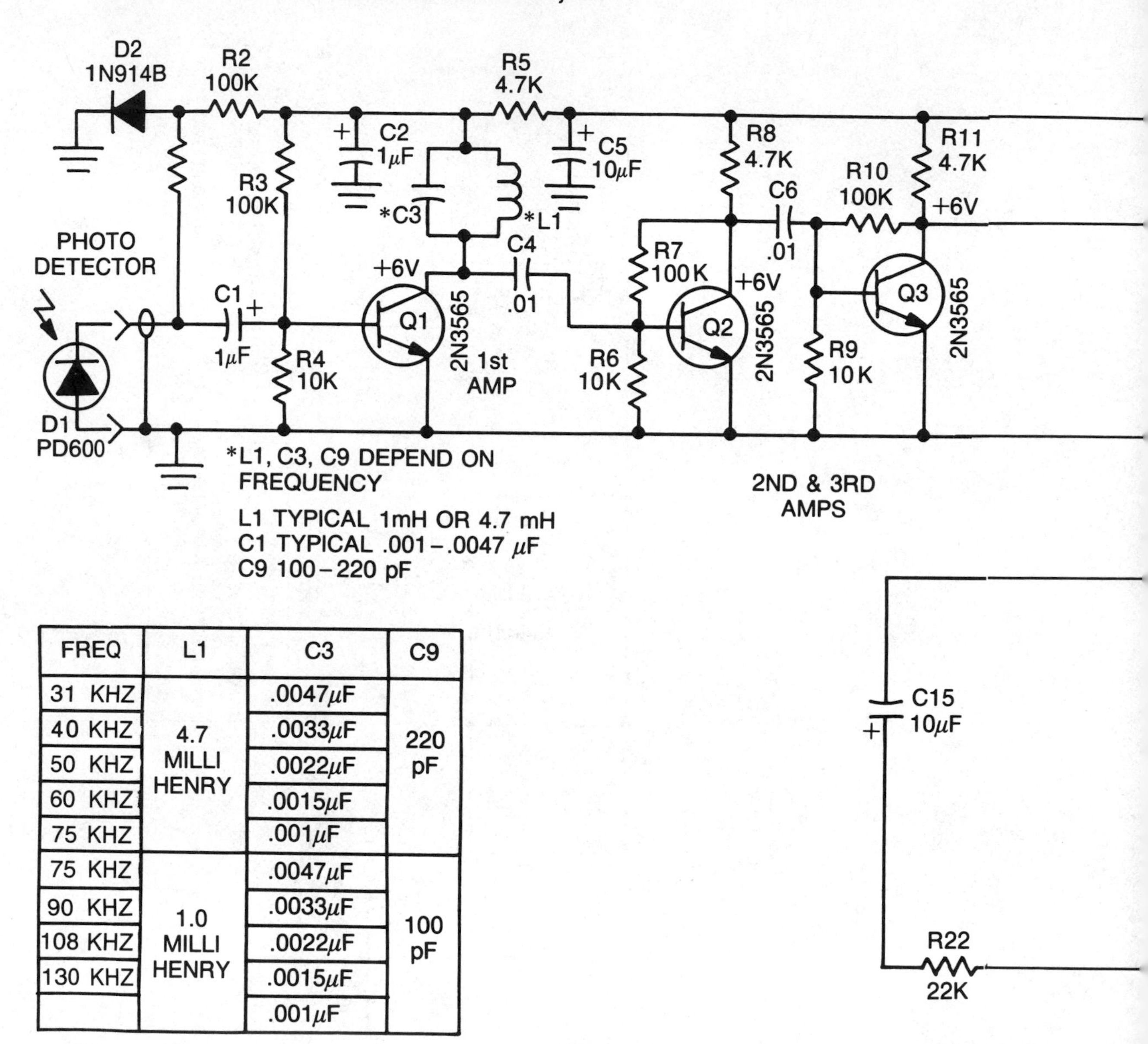

FREQ	L1	C3	C9
31 KHZ	4.7 MILLI HENRY	.0047μF	220 pF
40 KHZ		.0033μF	
50 KHZ		.0022μF	
60 KHZ		.0015μF	
75 KHZ		.001μF	
75 KHZ	1.0 MILLI HENRY	.0047μF	100 pF
90 KHZ		.0033μF	
108 KHZ		.0022μF	
130 KHZ		.0015μF	
		.001μF	

Fig. 6-6. Schematic—IR receiver.

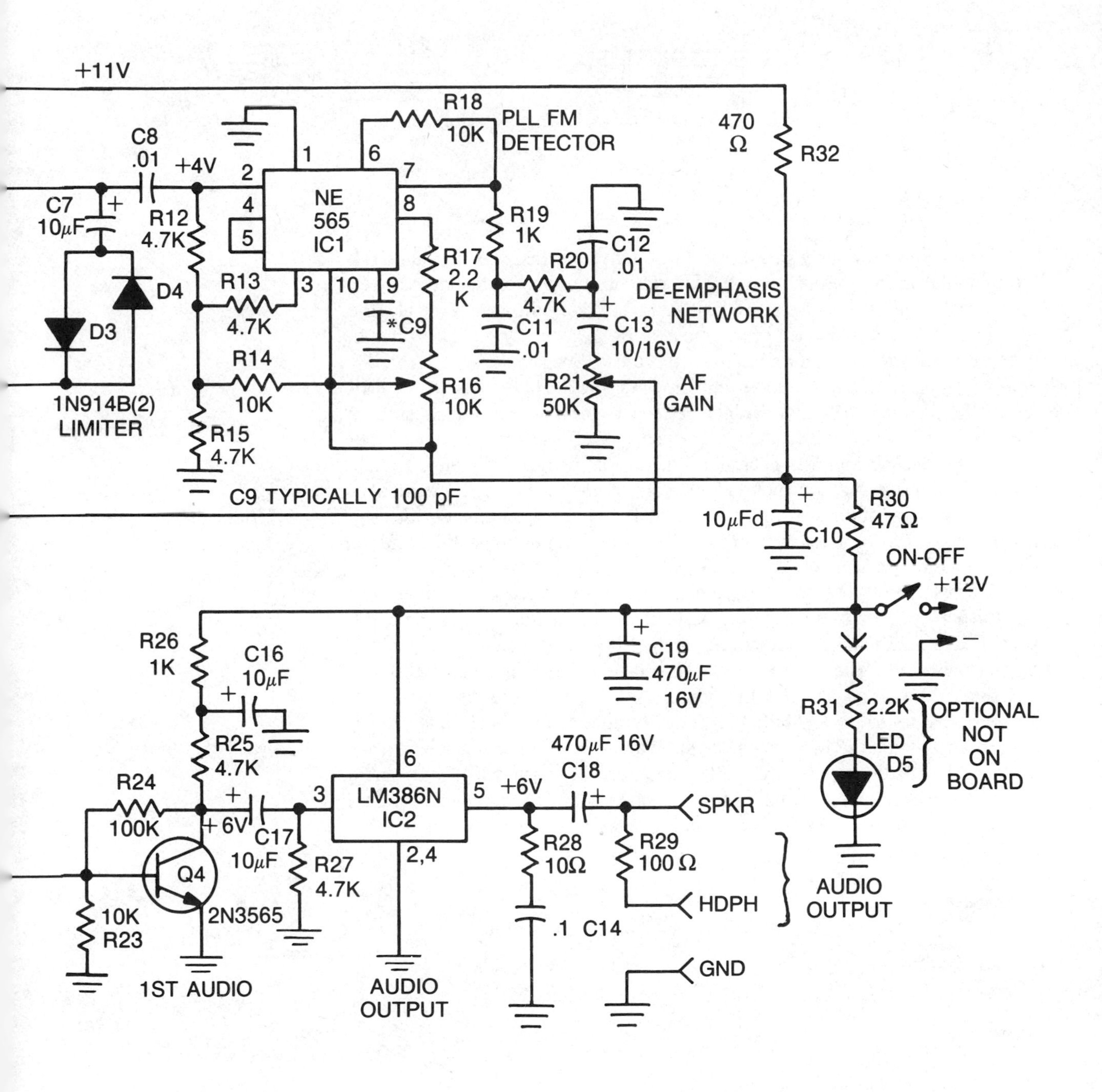

+11V
C8
.01
+4V
C7
10μF
R12
4.7K
D4
D3
1N914B(2)
LIMITER
R13
4.7K
R14
10K
R15
4.7K
NE
565
IC1
1
2
4
5
3
10
9
6
7
8
*C9
R18
10K
PLL FM
DETECTOR
R17
2.2
K
R16
10K
R19
1K
R20
4.7K
C11
.01
C12
.01
C13
10/16V
DE-EMPHASIS
NETWORK
R21
50K
AF
GAIN
470
Ω
R32
C9 TYPICALLY 100 pF
10μFd
C10
R30
47 Ω
ON-OFF
+12V
−
R26
1K
C16
10μF
R25
4.7K
R24
100K
+6V
C17
10μF
R27
4.7K
Q4
2N3565
10K
R23
1ST AUDIO
LM386N
IC2
6
3
5
2,4
AUDIO
OUTPUT
+6V
470μF 16V
C18
C19
470μF
16V
R28
10Ω
R29
100 Ω
.1 C14
SPKR
HDPH
GND
AUDIO
OUTPUT
R31
2.2K
LED
D5
OPTIONAL
NOT
ON
BOARD

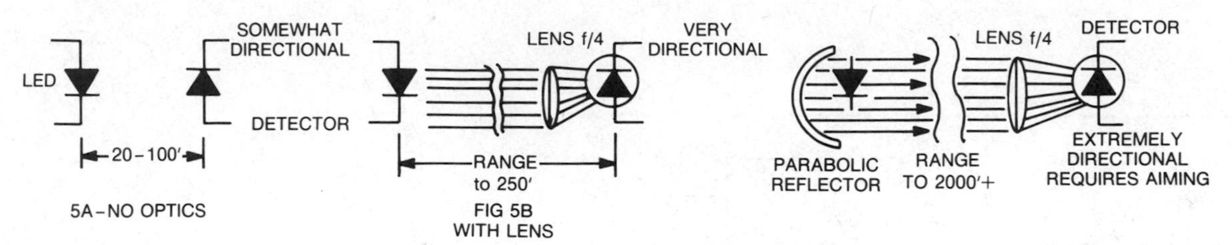

Fig. 6-7. Typical range of operation of IR system.

While it was previously stated that the system to be described would be relatively free from interference, since rf interference does not extend into the IR spectrum frequencies (above 300 GHz), this is not really the case. Actually, the IR region of the spectrum, especially the near IR, is generally very noisy due to low frequency amplitude modulated optical sources. Such sources are incandescent lamps (120 Hz hum), fluorescent lamps (also 120 Hz but "spiky"), light flashes from stray reflections, daylight, reflections from moving objects such as foliage moving in the wind, and other such unlikely sources. If you do not believe this, look at Fig. 6-3. Foliage is "bright" in the near infrared. A simple AM modulated system would not do for simply interference reasons mentioned. Most stray IR radiation is AM modulated, and would easily swamp out the signals from the LEDs. The problem of 120 Hz hum from room lighting sources alone would be insurmountable. See Fig. 6-8.

A simple way to overcome these difficulties is to modulate or "chop" the IR radiation at a high frequency. It is perfectly feasible to do this at around 100 kHz. The audio can then be modulated on the IR beam by modulating the chopping or pulse rate. The receiver would then detect the IR beam as a 100 kHz FM signal. The disadvantage in this approach is that an FM receiver with high gain is necessary to connect to the IR detector rather than a simple audio amplifier. However, with the IC devices currently available, the FM receiver is easy to construct, and contains only a little more circuitry than a high gain audio amplifier system. The transmitter would require some suitable oscillator that can be frequency modulated by the audio information, and also a means to drive the LED emitter.

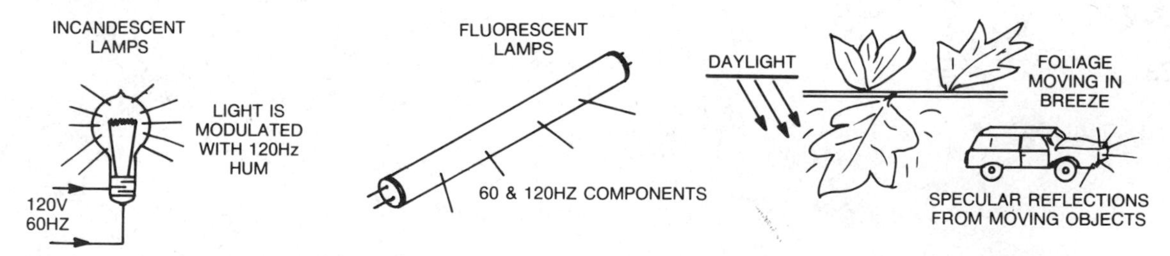

Fig. 6-8. Sources of IR "interference."

Transmitter

The schematic of a suitable transmitter is shown in Fig. 6-5. It operates as a voltage controlled oscillator, driving two LED emitters. Operation is as follows. Waveforms are shown in Fig. 6-9.

Audio signals in the 20 Hz to 15 kHz range, with a level of 50 millivolts to 1 volt is fed to J1. These signals appear across R1.

Audio across R1 is fed to compensation network R2 and C1. This network boosts high frequencies (above about 2,000 Hz) to improve the signal to noise ratio at these frequencies. C2 couples audio to the base of Q1 to a collector voltage of about 6 volts, and a collector current of about 1.0 milliampere. Q1 is a low noise high gain audio transistor, a 2N3565. This amplifier has a gain of about 5× voltage (set by R4 and R2). C3 reduces response above the audio range. C4 and R6 decouple the power supply to Q1. C4 bypasses any stray signals to ground, R5 is the audio gain control. C5 couples audio to pin 5 of IC1, an NE566. Level is about 0.25 to 0.5 volts p-p. This IC is an oscillator whose frequency is set by the value of R10, R9, the frequency adjust pot, and C8. It

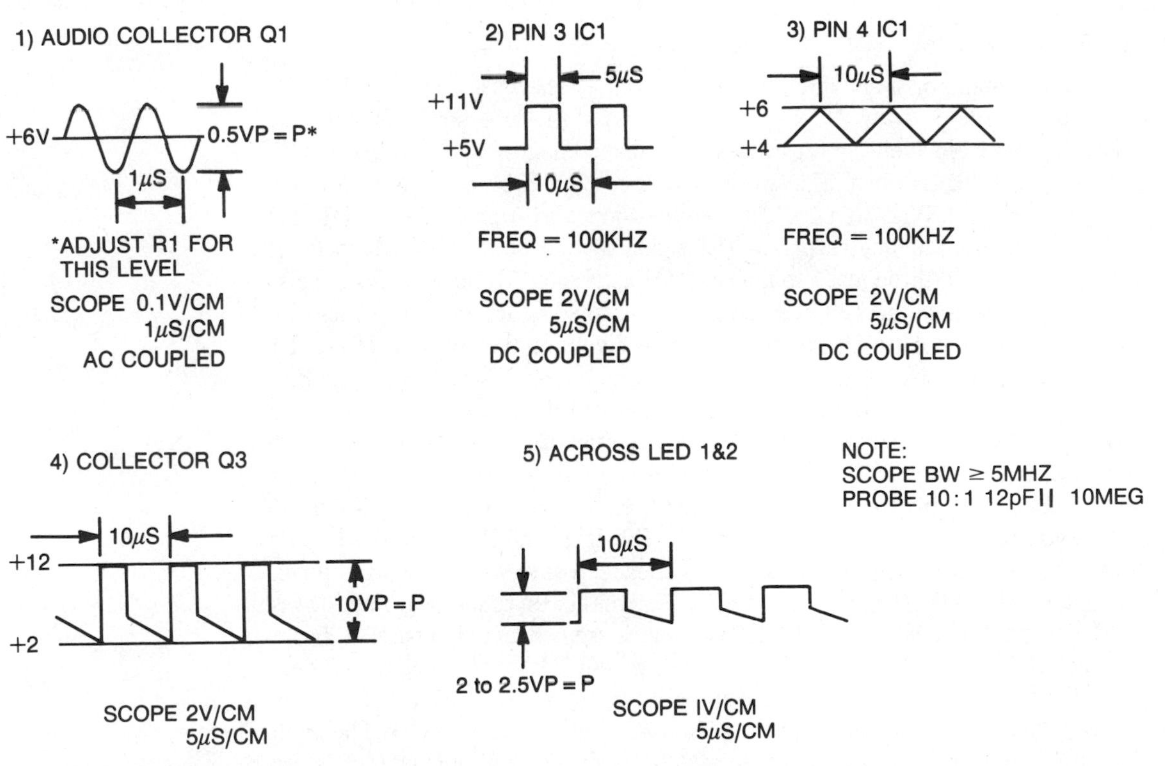

Fig. 6-9. Transmitter waveforms.

operates in the 30 kHz to 250 kHz range, depending on application. C7 connected between pins 5 and 6 is to suppress parasitic switching transients. Capacitor C8 from pin 7 to ground sets the oscillation frequency range that adjustment of R9 can produce. It is about a 4:1 range, depending on C8. A value for C8 is typically .001 microfarad. Resistors R7 and R8 bias the frequency control pin (pin 5) with about 1.5 volts negative with respect to pin 8. Up to ±60% frequency variation can be obtained if pin 5 swings ±1 volt with respect to pin 8 of IC1. Supply ground is pin 1. A square wave of about 5 to 6 volts p-p appears at pin 3 of IC1. A triangle wave of about 2V p-p appears at pin 4 but is not used. C9 couples the square wave to R11 and R12, which drives Q2 (a 2N3904) on and off. Driver Q3, a 2N3906 is switched via pulses from the collector of Q2. Resistors R13 and R14 provide the proper drive for PNP switch Q3. The collector current through R15 and R16 bias LED 1 and LED 2, the IR emitters. Up to 4 LEDs can be used. Peak current is about 100 milliamperes, at a 50% duty cycle. The reason for the Q2-Q3 configuration is that the LED supply can be returned to ground, and only one "hot" lead from R15-R16 is necessary. R15 and R16 set the current level through LED 1 and LED 2. A 12-volt supply is used to power the transmitter. IR energy output from the two LEDs is in excess of 12 milliwatts.

IR Receiver

A schematic or a suitable infrared receiver is shown in Fig. 6-6. D1 is a PD600 photodiode with a built-in infrared filter, to reduce stray visible light. Infrared energy incident on D1 causes it to conduct a small reverse current. R2 and D2, (a 1N914B) form a network to derive +0.5V reverse bias to D1. R1 couples the +0.5Vdc to D1. This reduces junction capacitance of D1 and improves high frequency response.The signal (modulation) on the IR incident energy is coupled through C1 to amplifier Q1 a 2N3565 low noise audio transistor. This signal is in the 100 kHz range. Q1 is biased by R3 and R4. R5 and C2 form a decoupling network and provide about 6 volts to the collector of Q1. L1 and C3 form a tuned circuit broadly resonant to about 30 – 150 kHz (see Fig. 6-9). This reduces low frequency noise (60 and 120 Hz, etc.). The signal at the collector of Q1 is coupled to the base of Q2 through C4. C5 decouples the B+ line. Q2 is biased by R6, R7, and R8 at the same dc voltages and currents as Q2, 6 volts, one milliampere. Q2 amplifies the signal and feeds Q3, biased the same way as Q2. Q3 is also a 2N3565, and R9, R10, and R11 form the biasing network. At the collector of Q3 signals can exceed several volts peak to peak. In order to limit these signals to 1 volt p-p a limiter consisting of D3 and D4 is connected to the collector of Q3. These diodes are connected in parallel, back-to-back. C7 isolates the dc level at the collector of Q3 from affecting the diodes. The limited signal of 1 volt p-p is coupled to IC1 pin 2 through C3. IC1 is an NE565 phase-locked loop. It consists of a phase detector and a VCO. The input to pin 2 causes the internal VCO to lock to the input signal frequency (pin 2). An internally generated dc voltage controls the VCO. This dc voltage comes from an internal phase detector and is available at pin 7. The phase detector output is connected to the internal VCO control input (pins 4 and 5 respectively) via an external jumper wire. Pin 3 is an alternate input and is not used. The internal VCO frequency is out by R16, R17, and C9, connected to pins 8 and 9 respectively. A resistor of 10k is connected to pin 7 and pin 6. This resistor restricts the hold-in range of ±40% of center frequency. If this resistor is left out, a lock range of ±60% is possible. This permits restricting the range of lock to that required of the input signal plus some error in tuning. R16 is the tuning control. The resistor R16 is adjusted so the VCO free running frequency is the same as the signal frequency (normally 30 – 150 kHz). Recovery of the audio signal is obtained at pin 7, since this voltage tracks any frequency variation of the input signal at pin 2.

Therefore, the NE565 acts as an FM detector. R19 and C11 suppress high frequency (100 kHz) components. R20 and C12 provide de-emphasis and improve signal to noise ratio of the recovered audio. C13 couples audio to R21, the AF gain control. About 50 to 100 millivolts of received audio is available. C15 couples this audio or a portion of it to the AF amplifier stage consisting of Q4 and associated components. Q4 is also biased at 6 volts, 1 milliampere by R23, R24, and R25. R26 and C16 decouple the power supply line. R22 limits the gain of the stage Q4 to about five. Up to 0.5 volts of audio is available at the collector of Q4, more than enough to drive the audio output stage. C17 couples audio from the collector of Q4 to input pin 3 of IC2, the audio power amplifier. IC2 is an LM386N and will provide over 0.5 watts of audio to a loudspeaker. Output is taken from pin 5. R28 and C19 form an oscillation suppressor and stabilizing network. C18 couples audio to a loudspeaker or to R29. R29 is a limiting resistor for use with 16 ohm or 32 ohm headphones, and is not needed for loudspeaker only application, C20 is the main power supply bypass capacitor. R31 and D5 form a pilot lamp, not necessary but optional. It should be left out for battery operation applications to serve drains. Either a 12 volt power supply or battery can be used. Drain is about 50 to 75 milliamperes, depending on the audio output level. Eight AA cells can be used as a battery supply if desired. For battery applications, headphone operation to conserve power of loudspeakers is not necessary. The receiver will detect about 1 microvolt of 100 kHz signal so it is a high gain device and should be carefully constructed.

Construction is not too critical. Take the necessary precautions as you would in any high gain device. Output should be separated as far as possible from input. All low level leads to the detector photodiode should be shielded (keep short to minimize capacitance). Use of specified transistors is desirable to keep noise low, and to be sure of reproduceability. The LM565 IC is often sold under several different part numbers such as NE565, etc. Use a physically small coil at L3 to reduce stray inductive coupling. Use only the types of capacitors specified. While the range you will obtain depends on many factors, use of a matched photodiode detector and LED combination is highly advisable for the

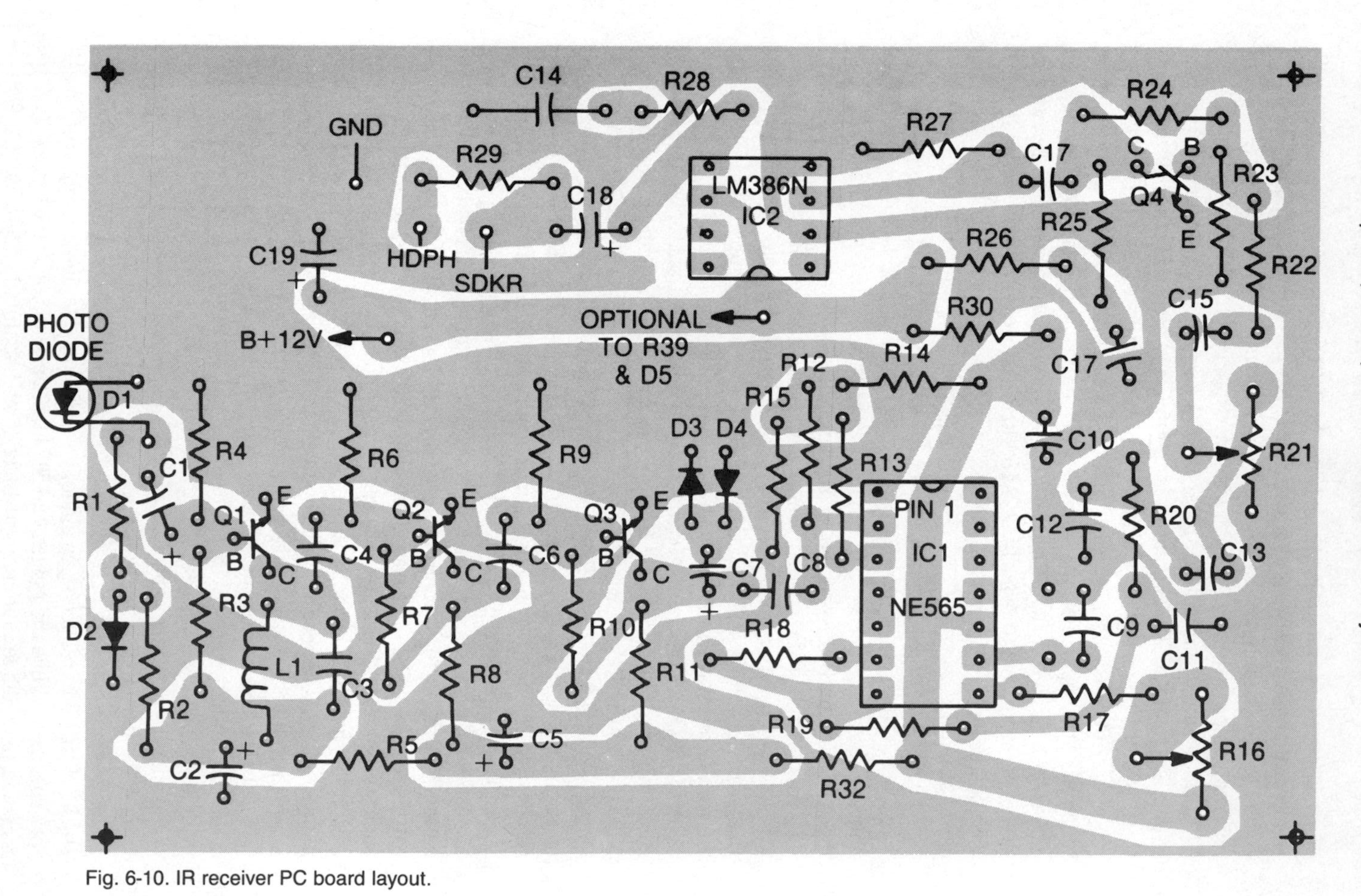

Fig. 6-10. IR receiver PC board layout.

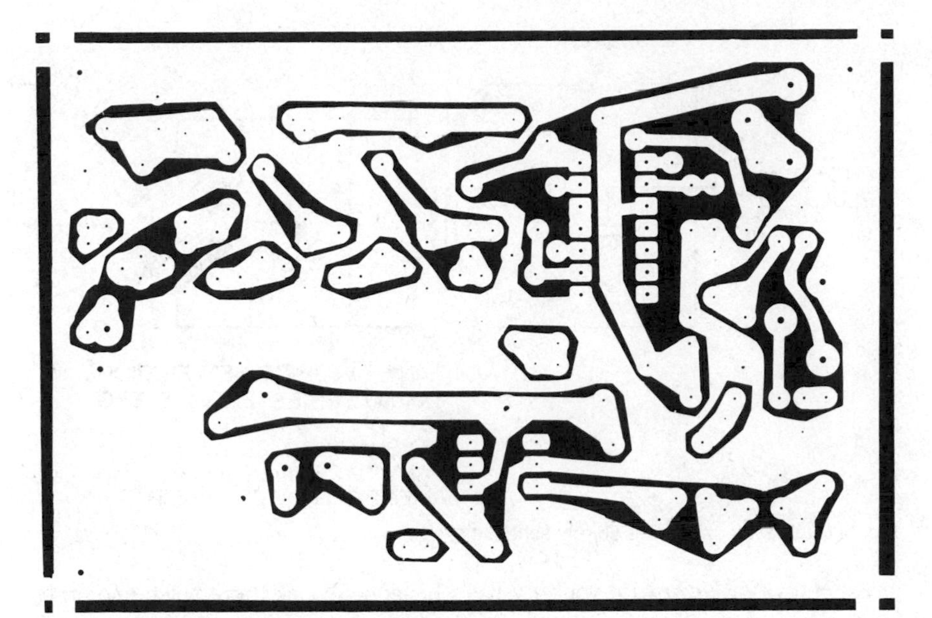

Fig. 6-11. IR receiver PC board foil pattern.

best results. The PC layouts in this article are to be recommended. If you would rather not make your own PC board, a drilled and etched set of PC boards is available from the source at the end of the article. Also available is a complete kit of parts including the photodiode and two LEDs. We recommend the use of two LED devices. However, one will work but range will be slightly less, and coverage area will be limited to that of one LED. However, for applications where auxiliary optics are used, one LED will be adequate.

Applications for the IR transmitter and receiver are many. Several will be discussed. See Fig. 6-12 for a schematic of a suitable supply to run these modules on 120Vac. Parts are commonly available at most Radio Shack or other parts stores, or through surplus or mail order houses, and should be no problem to obtain. All that is needed is 12 – 14Vdc at about 75 milliamperes. The schematic in Fig. 6-12 shows two separate outputs with separate filtering. The value of the 47 ohm resistor can be raised or lowered to adjust output voltage.

PROJECT 14: WIRELESS HEADPHONES

Figures 6-15A, B, and C include a wireless headphone setup. This should prove invaluable to anyone wishing to view a TV program without disturbing others, or to the hard of hearing. The transmitter audio is obtained from the TV speaker or headphone jack. (*Be careful* if you have to open up the TV set to make connections—some sets are constructed with "live" chassis where the ground may be directly returned to one side of the ac power line. These sets

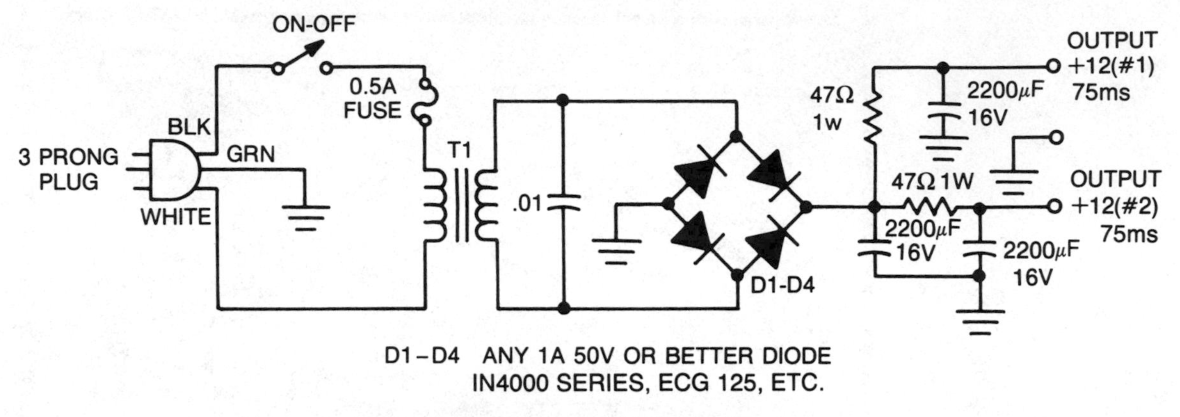

Fig. 6-12. IR system—power supply schematic.

present a *safety hazard*.) If your TV has a headphone jack there will be no problems if it was manufacturer installed. The receiver module with a battery pack (8-AA cells) can be mounted in a small plastic case (Radio Shack has several sizes in stock), and fitted with a headphone jack. Ordinary Walkman-type headphones are excellent. You will be able to hear the TV audio in the receiver anywhere in the room, up to 30 to 50 feet away. If your walls and ceilings are reflective to IR, this will increase range somewhat. The detector and LED should ideally face each other. Since the IR receiver has its own volume control, the listener can suit himself. The range of the unit will exceed comfortable viewing distance, even for a 45-inch projection TV set.

Figure 6-16 shows a fiber optic setup. A length of fiber-optic tubing (with suitable terminations) connects the transmitter and receiver. This setup is useful for experiments with fiber optics. Any audio source, such as a radio, tape deck, CD player, or microphone can be used.

IR Perimeter Security System

Figure 6-17 illustrates an IR perimeter security system. If no audio is used to modulate the transmitter, the receiver audio output will be zero. When the IR beam is broken, the receiver will produce a loud, rushing noise. This noise can be used to drive a relay circuit or other alarm. The noise can be rectified with a diode and RC filter so a dc level is produced. This dc level can control the alarm system. The IR beam is completely invisible. Path lengths of several hundred feet can be obtained, but probably will require optical components. This is desirable anyway in order to keep the IR beam as narrow as possible. This is left to the constructor or experimenter.

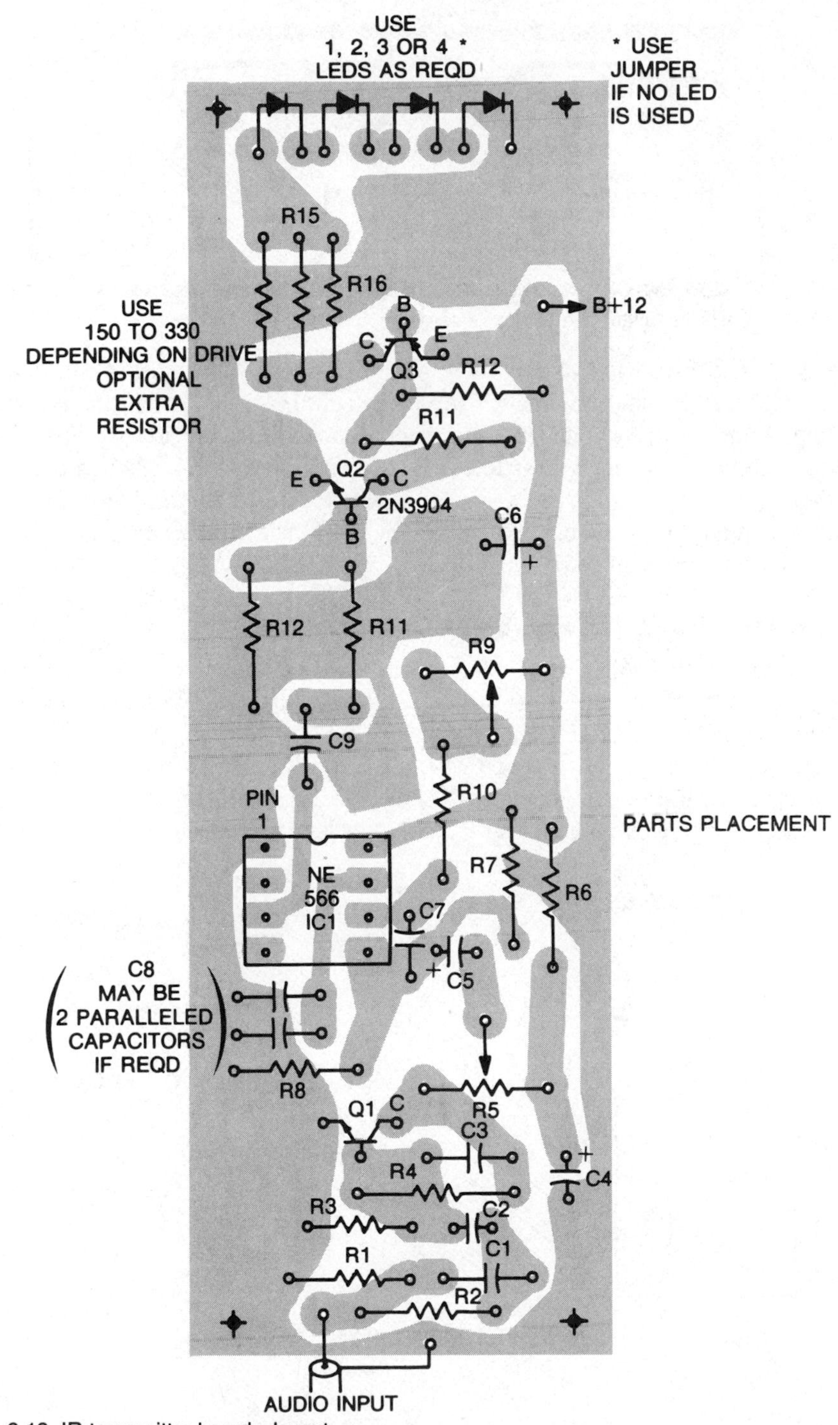

Fig. 6-13. IR transmitter board—layout.

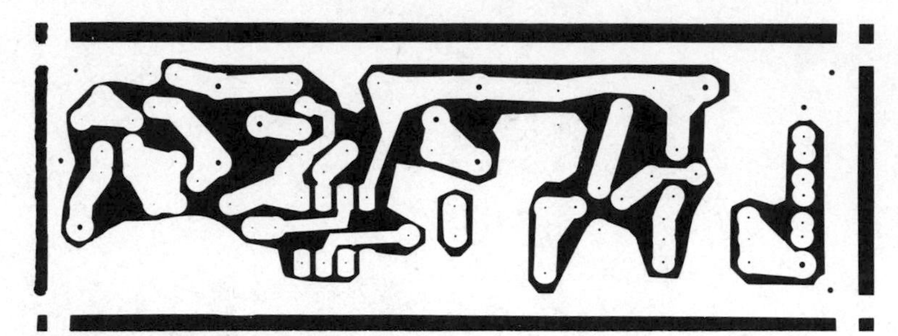

Fig. 6-14. IR transmitter board—foil pattern.

Figure 6-18 shows a secure IR communications system using optical components. A transmitter and receiver at each end, together with suitable switching of power supply voltages, and a microphone and loudspeaker or headphone are all that is needed, bypass R2, Fig. 6-7. This should not really be necessary, however. The microphone may also be used to switch the transmitter and receiver if it has a suitable internal switch, as some CB mikes do (a Radio Shack hand mike, or Astatic D104 will do).

PROJECT15: WIRELESS SPEAKERS

Figures 6-19A, B, and C show a wireless speaker setup. It is useful in conjuntion with Walkman pocket stereo sets. Audio from the Walkman headphone jack feeds two separate transmitters, one for each channel, right and left. Two inexpensive speakers, in wood cabinets, have a receiver module and battery pack in each one. The speakers used for small hi-fi sets will do nicely, with plenty of room to mount an IR receiver module, battery pack, and on-off switch.

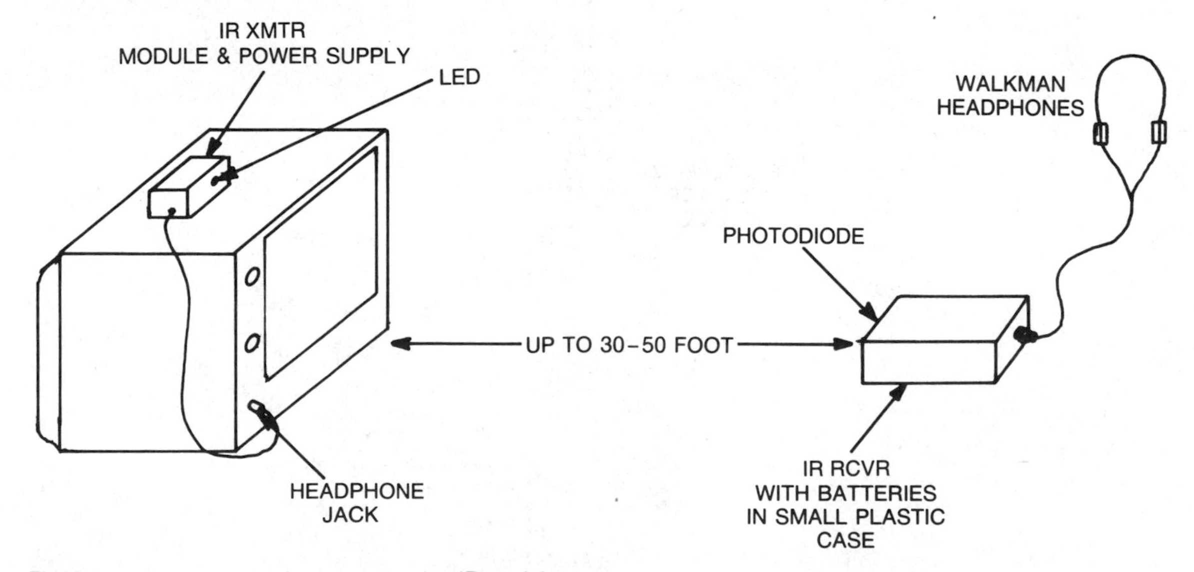

Fig. 6-15A. Wireless headphone setup using IR modules.

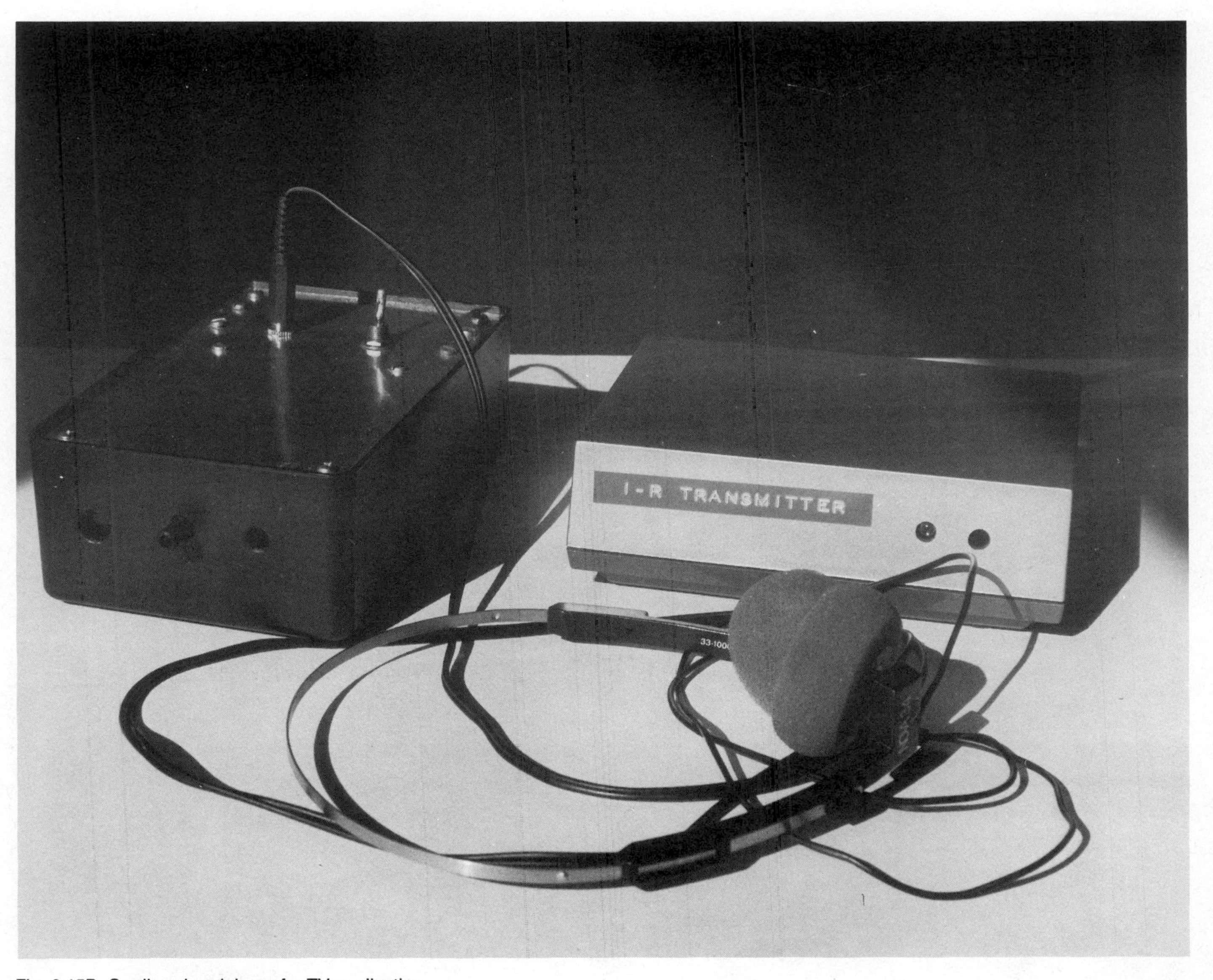

Fig. 6-15B. Cordless headphone for TV application.

Fig. 6-15C. Internal views of wireless headphone system.

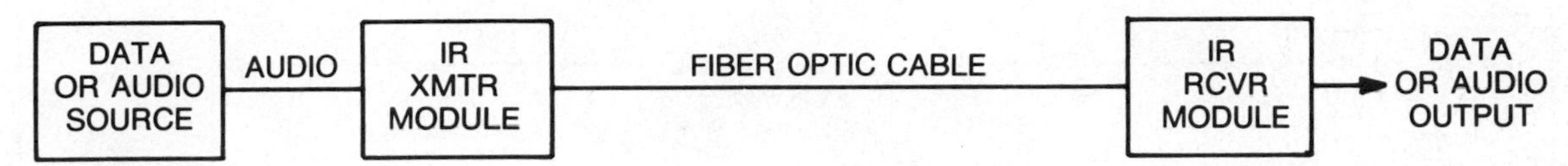

Fig. 6-16. Fiber optic line with IR modules.

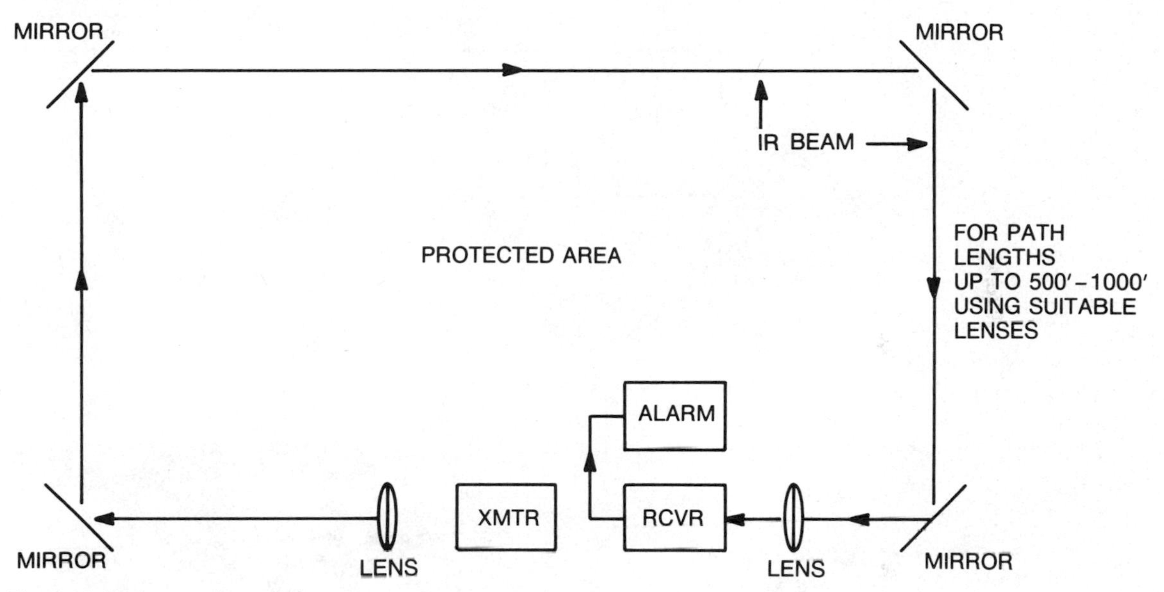

Fig. 6-17. Perimeter alarm system.

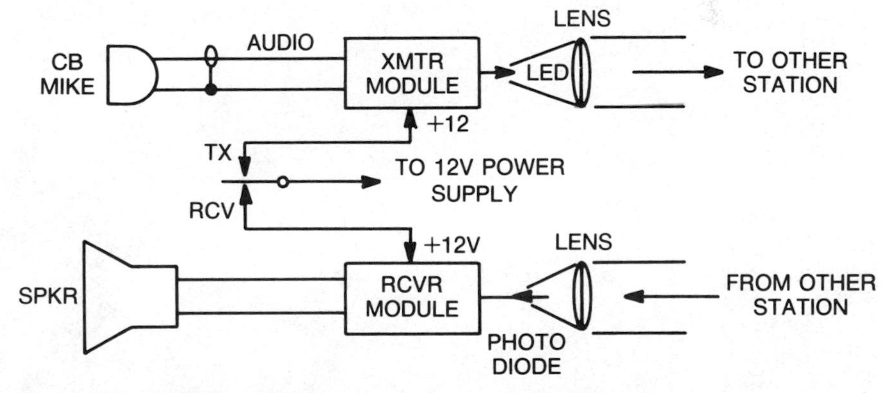

Fig. 6-18. IR XMTR/Receiver setup.

The speakers can be placed anywhere in direct view of the transmitter box. In this case, the transmitters should be operated on two widely spaced frequencies, and it is best to aim the LEDs in opposite directions. Receiver selectivity is somewhat limited, and these measures help to avoid mutual interference. This is a handy way to add speakers to a Walkman.

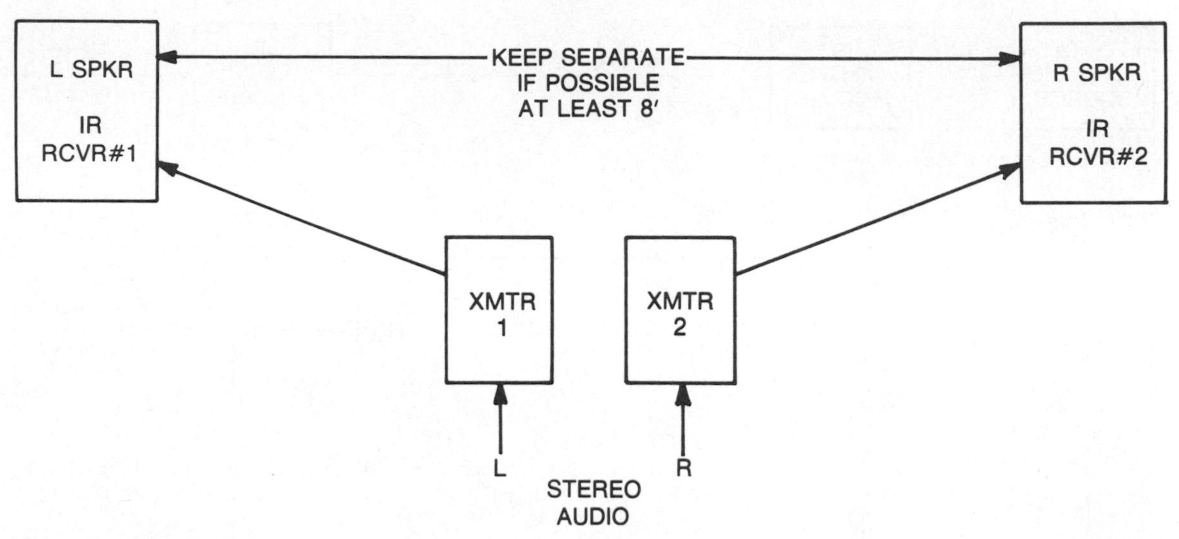

Fig. 6-19A. Wireless speakers.

Fig. 6-19B. IR transmitter.

Fig. 6-19C. IR receiver.

In addition, a receiver may be made up consisting of two additional modules to drive a pair of headphones. A hi-fi system can be used as the source of audio as in Fig. 6-20. We used frequencies of 80 kHz and 140 kHz in our setup. Do not operate at *exactly* 100 kHz since there is a powerful loran navigation signal on this frequency and the receiver might pick it up. 100 kHz may not be a problem in some areas away from the East Coast (US), but the loran pickup can cause objectionable interference. Frequencies from 30 kHz to 250 kHz have been tried with good results. A system that transmits stereo in a manner analogous to an FM stereo station, using subcarriers, has been built by the authors and will be described in another article. In this method, only one IR transmitter is required, but the circuit complexity is greater, since a multiplexed signal must be generated.

Tune-up of the completed modules is relatively straightforward and simple. A frequency counter helps but is not necessary. The transmitter (it is best to build the transmitter first, since an IR source is needed to test the receiver) is hooked up to a power supply of 12V, or 8-AA cells in series. If batteries are used, use alkaline cells. Use of a VOM will be necessary. First, check for about (refer to Fig. 6-7) 12Vdc on pin 8 of IC1. Next test for the following dc voltages:

1. +6V (5 to 8V OK) on collector of Q1
2. +9V on pin 3 IC1 (8 – 10V OK)
3. +6V on collector Q3 (LEDS CONNECTED TO CKT ONLY!)

Place an AM radio tuned to a low frequency station (around 600 kHz) close to the transmitter. Rotate R9. You should hear some squeals from the radio if the NE566 is oscillating. Set the pot somewhere in midrange. By listening for

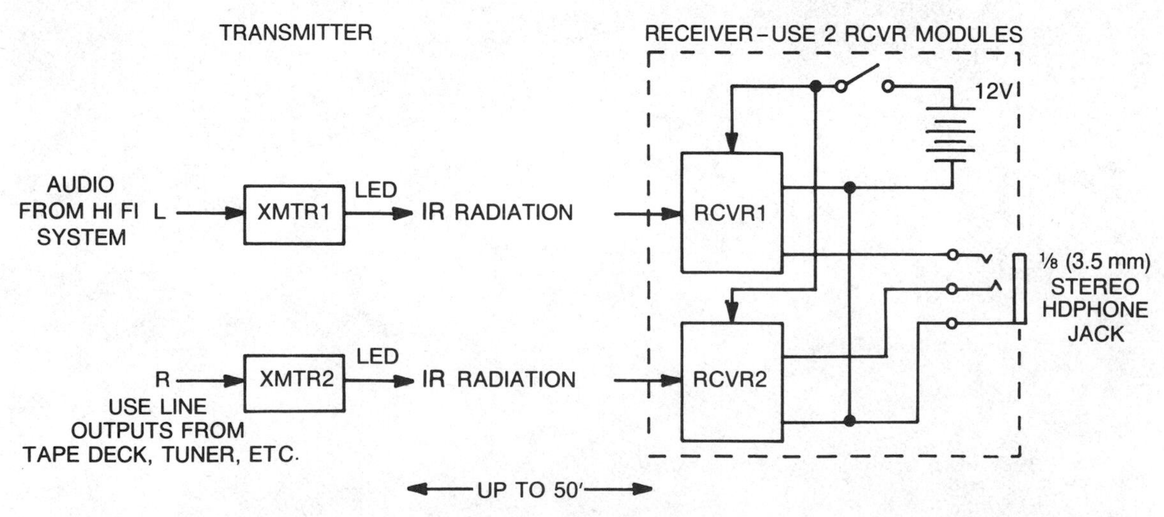

Fig. 6-20. IR stereo headphones.

harmonics on an AM radio, you can tell the frequency of the transmitter. For example, if you hear one on 610 kHz, another on 720 kHz, another on 830 kHz, you would be close to 110 kHz, since the harmonic spacing is 110 kHz, which is equal to the fundamental frequency. If you have a scope, see Fig. 6-9 for the waveform. By the way, the LEDs must be either mounted on the circuit board or separately. We have provided space for them at the end of the PC board, but you can mount them as you see fit. We used old 35mm film cases to mount the LEDs for the transmitter in our photo (actually two transmitters and power supply in a wooden box) and also mounted the receiver photodiodes in these cans on top of the speaker cabinets (see photo, Fig. 6-19C). These are but one way to mount them. There are many others (see Fig. 6-21).

The receiver is checked out as follows: Connect an 8 ohm (no lower) speaker to the speaker audio output terminal. Set R16 about halfway. Set R21 to midpoint. Apply +12V to the receiver B+ line (across C20). You should hear a rushing noise if everything is OK. Check for +6Vdc at the collectors of Q1, A2, Q3, Q4, and pin 5 of IC2. You should read +11V at pin 10 of IC1 and about 44V at pins 2 and 3 of IC1. If you have access to a signal generator with a calibrated attenuator you can check the receiver sensitivity. About 1 microvolt into the base of Q1 (use a 0.1 microfarad coupling capacitor so as not to upset the bias) should cause the receiver to quiet down by about 10dB. If the generator can be FM modulated, set it for 25 kHz deviation and adjust R16 for maximum audio. Generator frequency should be tuned to the resonant frequency of L1 and C3 (generally 50 to 150 kHz). Next, set up the transmitter a few feet from the receiver. Turn on the transmitter. The receiver should instantly quiet. Adjust R9 in the transmitter and/or R16 in the receiver to set them on the same frequency. This should be equal to the resonant point of L1 and C3.

See the table in Fig. 6-6 for the values of L1 and C3. Now, apply about half a volt of 1 kHz signal to the transmitter audio input. A 1 kHz tone should be heard in the receiver. The signal should cease when the IR beam is interrupted. Cover the LED or photodiode to do this. IR will sometimes penetrate visually opaque materials, as you will discover.

At this point, all are operating. Now you can mount the modules in whatever cases you plan to use together with suitable power sources. If you can obtain a TV remote controller, as a further check, you should hear the pulsing of the controller in the receiver. The remote controller must be of the IR type. Do not expect to hear it more than a few feet away. The receiver tends to reject the signal from the controller.

You may wish to check the range. At least 20 feet should be obtained, with good signals. You will find the system somewhat directional. If you have white ceilings 8 ft. high aiming the LEDs at the ceiling should enable complete room coverage of a fairly large room (15×20 ft.). Signals may sometimes be picked up in adjacent rooms by reflections from other objects. Line of sight range should be about 30 to 50 feet. If you position a lens in front of the detector (try a magnifying glass) you should get over 100 feet. In one test we got 300 feet in an

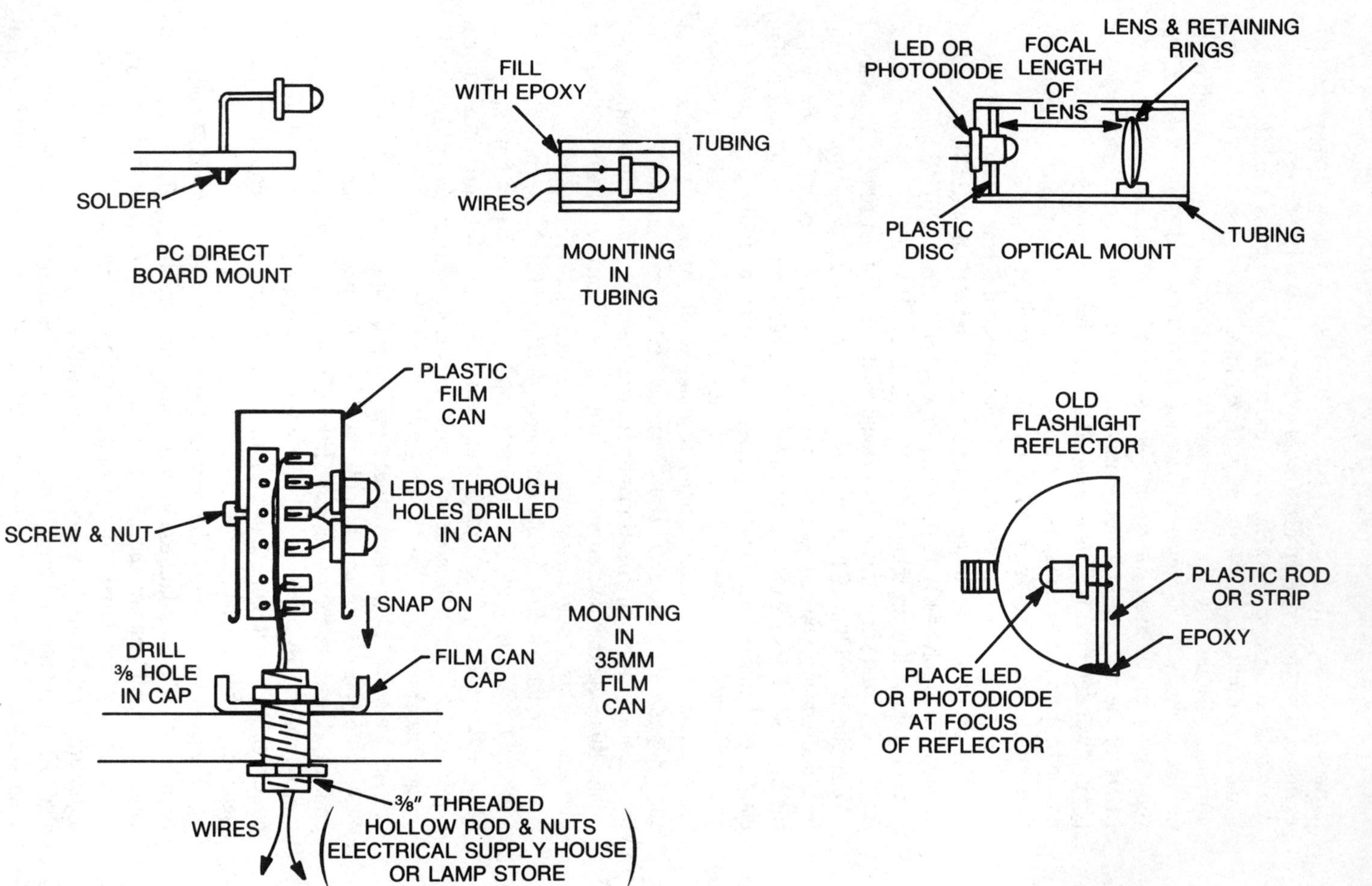

Fig. 6-21. Ways of mounting LEDs and photodiodes.

open field, line of sight. In this test, the transmitter was aimed out a window and a magnifying lens used over the photodiode on the receiver. However, this was very directional. With optics at both ends, up to a 2,000+ foot range should be obtainable. The optics would have to be fairly large, 4 inches or more aperture. Highly polished flashlight reflectors would be suitable. Remember that a telescope can offer even greater ranges than this. If you can see the transmitter with your eye, the detector should respond. However, atmospheric conditions must also permit this.

If you wish to install these modules in an opaque plastic housing, a visibly opaque filter that transmits IR can be made up from stock red and green sandwiched plexiglass. See Figs. 6-22, 6-23, and 6-24.

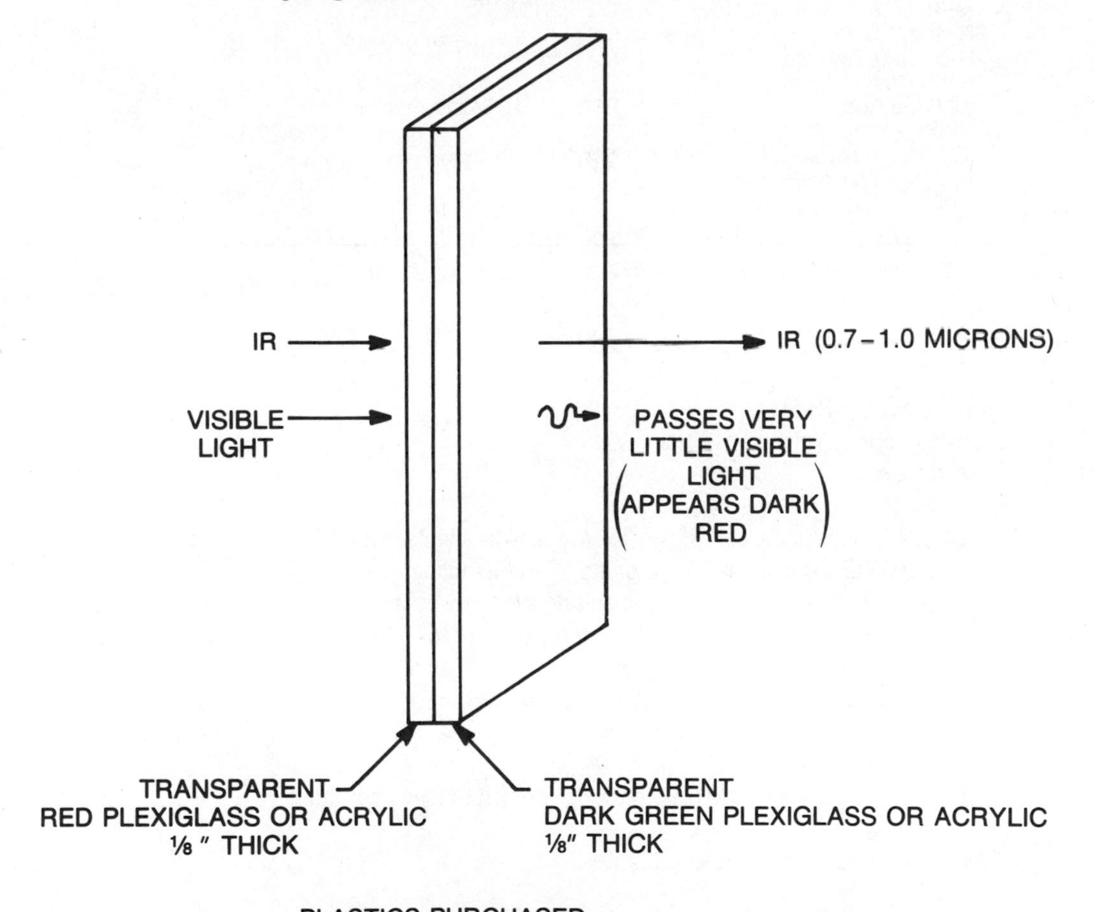

Fig. 6-22. Homemade IR filter.

R1	4.7k	C1	.0033μF ±10% Mylar or ceramic
R2	22k	C2	10μF 16V Elec.
R3	10k	C3	47 pF ceramic or Mica ±10%
R4	100k	C4	10μF 16V Elec.
R5	5k variable trimmer	C5	10μF 16V Elec.
R6	1k	C6	10μF 16V Elec.
R7	6.8k	C7	.001μF ±20% Mylar or ceramic
R8	47k	C8*	470-2200 pF Mylar, Mica, or NPO Ceramic
R9	10k variable trimmer	C9	.01 disc +80 -20%, 25V
R10	2.2k		
R11	1k	Q1	2N3565
R12	1k	Q2	2N3904
R13	1k	Q3	2N3906
R14	100 ohm		
R15	150-330 ohm	IC1	NE566
R16	150-330 ohm		
All Fixed Resistors 1/4W ±10% Tol.		LED1, 2	IR155 LED matched to photodetector
J1	Input jack (RGA type) not on PC board	Misc.	— 1 PC board
			*See Table Fig. 6-5

Fig. 6-23. IR transmitter parts list.

A kit of parts and PC boards are available from:

North Country Radio
P.O. Box 53E, Wykagyl Station
New Rochelle, NY 10804

1. 2×2×1/8″ plastics per Fig. 6-22 are also available with PC boards or kits for $12.50 plus $1.75 for postage and handling
2. Kit of parts for one system consisting of receiver and transmitter PC boards, all components that mount on the boards plus LED1, LED2, and D1.....$39.95 plus $2.50 for postage and handling
3. Two systems (for stereo).....$74.95 plus $2.50 for postage and handling
4. One each receiver and transmitter PC boards.....$12.50 plus $2.50 for postage and handling
5. LED1, LED2, D1.....$16.95 plus $2.50 for postage and handling

New York residents must include sales tax.

R1	1Meg	C1	1μF Elec. See Table Fig. 6-6
R2	100k	C2	1μF Elec.
R3	100k	C3	See Table Fig. 6-9
R4	10k	C4	.01μF +80 −20%, 25V disc
R5	4.7k	C5	10μF 16V Elec.
R6	10k	C6	.01μF +80 −20%, 25V disc
R7	100k	C7	10μF 16V Elec.
R8	4.7k	C8	.01μF +80 −20%, 25V disc
R9	10k	C9	See Table Fig. 6-5
R10	100k	C10	10μF 16V Elec.
R11	4.7k	C11	.01 Mylar ±20%
R12	4.7k	C12	.01 Mylar ±20%
R13	4.7k	C13	10μF 16V Elec.
R14	10k	C14	.1 Mylar ±20%
R15	4.7k	C15	10μF 16V Elec.
R16	10k variable trimmer	C16	10μF 16V Elec.
R17	2.2k	C17	10μF 16V Elec.
R18	10k	C18	470μF 16V Elec.
R19	1k	C19	470μF 16V Elec.
R20	4.7k		
R21	50k variable w/shaft	Q1	2N3565
R22	22k	Q2	2N3565
R23	10k	Q3	2N3565
R24	100k	Q4	2N3565
R25	4.7k		
R26	1k	IC1	LM565
R27	47k	IC2	LM386N
R28	10 ohm		
R29	100 ohm	D1	Photodiode PD600
R30	47 ohm	D2	1N914
R31	2.2k (not on PC board)	D3	1N914
R32	470 ohm	D4	1N914
		D5	LED − Not on PC board
		L1	See Table Fig. 6-6
All Fixed Resistors 1/4W ±10% Tol.		Misc. — 1 PC board	

Fig. 6-24. IR receiver parts list.

PROJECT 16: ULTRA-SENSITIVE PICOAMMETER/ELECTROMETER

(Measure Currents as Low as 10^{-12}A and Resistance as High as 10^{12} Ohm)

One basic instrument for use in electrical or electronic work is the VOM or volt-ohm-milliammeter. However, even electronic VOMs, such as the vacuum tube voltmeter (VTVM) or field effect transistor VOM (FET VOM), have one limitation. They place a load on the circuit being measured, and therefore interfere with the actual circuit or source being measured in some way. Fortunately, for most electrical or electronic work, this is not a problem. However, when one is faced with the problem of measuring minute charges or sources having a very high impedance (geiger muller tubes, photo multiplier circuits, MOSFET gates, electrostatic devices, as well as humidity sensors, ionization gauges, and such) the 10 megohm VTVM or FET VOM simply will not do.

It is possible to use electrostatic instruments that are voltage operated and do not draw any current from the source to be measured. Unfortunately these devices are suitable only for relatively high voltages, over about 50 to 100 volts. Figure 6-25 illustrates one such device, called an electroscope. Invented over a century ago, it uses the principle of electrostatic repulsion of like charges. Two very light metallic leaves, made of gold foil, were suspended from a wire. A charge placed on the wire caused the leaves to repel each other. It is theoretically possible to calibrate such an instrument, but hardly practical.

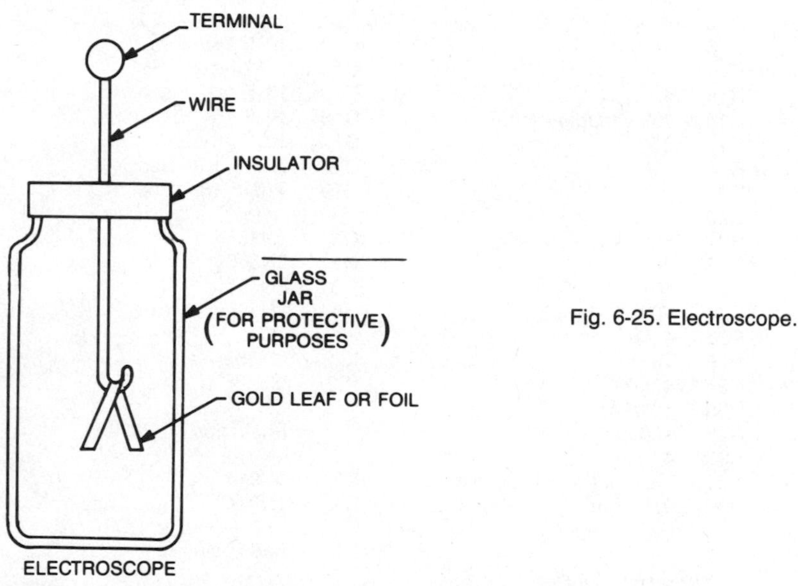

Fig. 6-25. Electroscope.

The invention of the vacuum tube made it possible to use the tube as an amplifier, the grid theoretically drawing no current. However, tubes for this application must be operated at very low voltages so as not to cause the grid of the tube to collect unwanted electrons and interfere with the quantity to be measured. A device such as this is called an electrometer. Very small currents, down to a micro-micro ampere (10^{-12}amp), called a pico ampere, can be detected and measured. However, commercial electrometers are expensive and out of the question for the hobbyist to experiment with. Figure 6-26 shows a basic electrometer circuit using a tube.

In the 1960s, a new device called the MOS field-effect transistor appeared on the scene. Conceived back in the 1920s by a man named Lilienthal, it used the principle of surface conductivity of a semiconductor modulated by an electric field, set up by an electrode called a gate. Figure 6-27 shows a simple MOS-FET (Metal-Oxide-Silicon field-effect transistor). Theoretically (and pretty

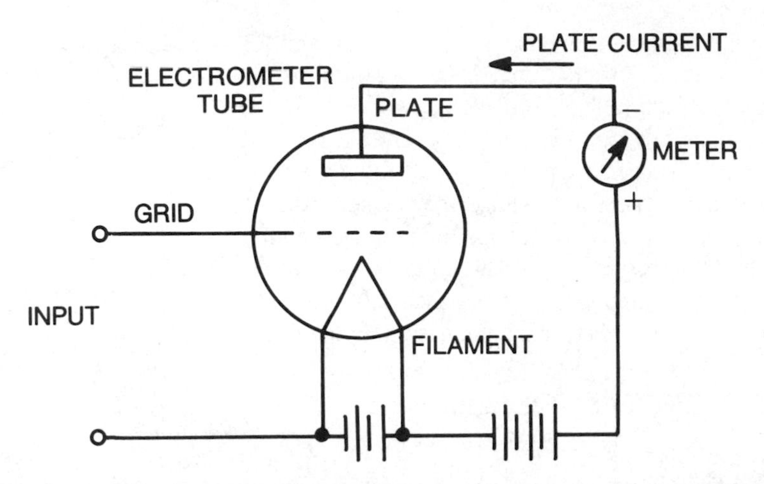

AN INPUT VOLTAGE ON THE GRID CAUSES THE PLATE CURRENT TO CHANGE. A KNOWN CHANGE IN PLATE CURRENT CAN BE CORRELATED TO A CHANGE IN GRID VOLTAGE

Fig. 6-26. Basic electrometer circuit.

closely in practice) the gate draws no current, since it is insulated from the semiconductor structure. However, until the early 1960s, this device could not be made, due to the fact that it was not known how to make a good enough semiconductor surface free of defects. (It is interesting that this concept of the MOSFET predates the junction transistor.) The MOSFET today is, of course, a staple in the semiconductor industry, being the heart of VLSI IC devices.

The MOSFET is an ideal device for electrometer applications, in theory. Practically, early MOSFET devices were unstable and noisy. However, today, operational amplifier IC devices using MOSFETS are common, reliable, and inexpensive. They make it possible to measure very small electrical charges and currents and to use these signals to control very large currents. Devices are available (power MOSFETS) that can switch 10 to 30 amperes at 100 volts or more, and be controlled by megohm-level impedance sources. This was not possible with single transistors using bipolar (NPN-PNP) construction.

What follows describes a simple, inexpensive electrometer type voltage and current meter. Depending on parts on hand, it could probably be built for less than $30. It can be used for low-level measurements, insulation testing at low voltage levels, and many interesting experiments in electronics and electricity. A kit of parts for the PC board is available from the source listed at the end of this article, for those who may want to build this device. Any meter movement of one milliampere sensitivity or so may be used. Sources of these are old or defunct VOMs, surplus houses, junked stereo or CB radios, cheap electrical testers, etc. You may also use your VOM or DVM as an indicator instead of a meter, and even use an old tuning meter from a junked radio as a relative level indicator, if desired. Figure 6-28 shows a basic electrometer amplifier having an input impedance of about 100,000 Megohms (100k Meg). A MOSFET input operational amplifier IC1 is used in a variable gain dc amplifier. R1, R2, and R3 provide dc returns for the input circuit, C1 and R4 provide static protection for IC1. A variable divider consisting of R5 through R8 provide gains of 1, 3.3, 10, and 33.3 times to provide ±1 volt output swing with 1 volt, 0.3 volt, 0.1 volt and 0.03 (30 millivolts) dc or ac input. R21 is a balance control for adjusting the zero point. The negative input of the op amp is kept at a voltage very close to the dc or ac input signal, and the PC trace for this level surrounds the positive (signal) input and R4. Since the potential difference is very small, equal to the output voltage of the amplifier divided by the open loop gain (more than 1,000 times), leakage across the PC board surface due to dirt, humidity, etc., is minimized.

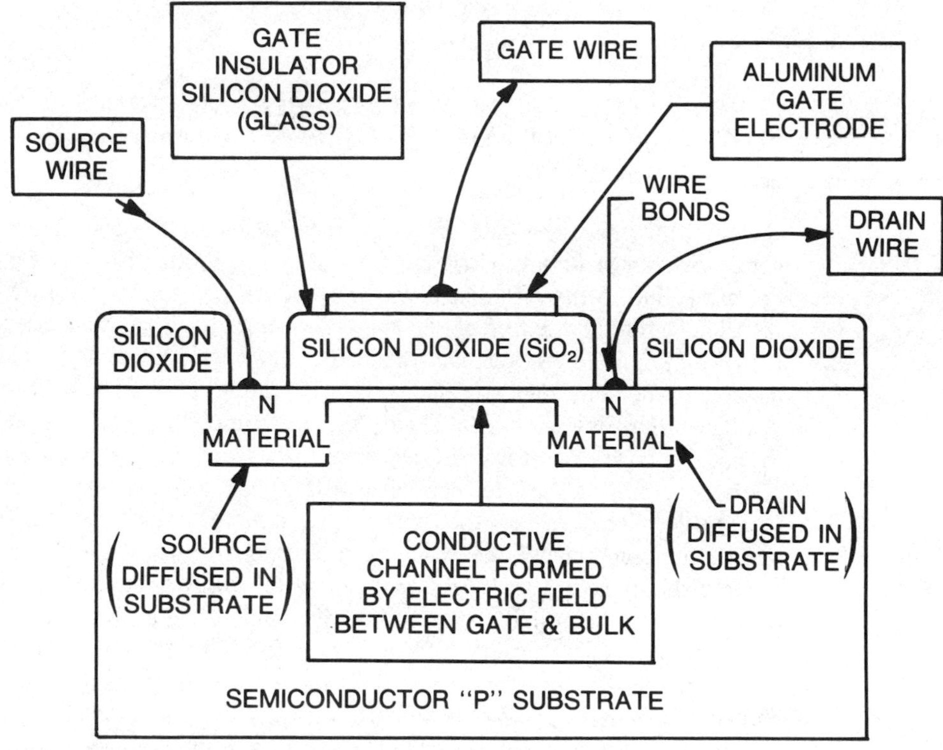

NOTE—IF THE DEVICE IS SYMMETRICALLY CONSTRUCTED SOURCE & DRAIN MAY BE INTERCHANGED

ELECTRIC FIELD BETWEEN P SUBSTRATE AND ALUMINUM GATE CAUSES THE LAYER IMMEDIATELY UNDER THE GATE TO ASSUME CHARGE & EFFECTIVELY "CHANGE" TO "N" MATERIAL, FORMING A CONDUCTIVE CHANNEL BETWEEN SOURCE & DRAIN. THIS DEVICE IS AN *ENHANCEMENT* MODE DEVICE, (NORMALLY "OFF")

A *DEPLETION MODE TYPE* HAS A CONDUCTING CHANNEL THAT IS *CUT OFF* BY THE ELECTRIC FIELD. THIS DEVICE IS NORMALLY "ON"

Fig. 6-27. MOSFET structure.

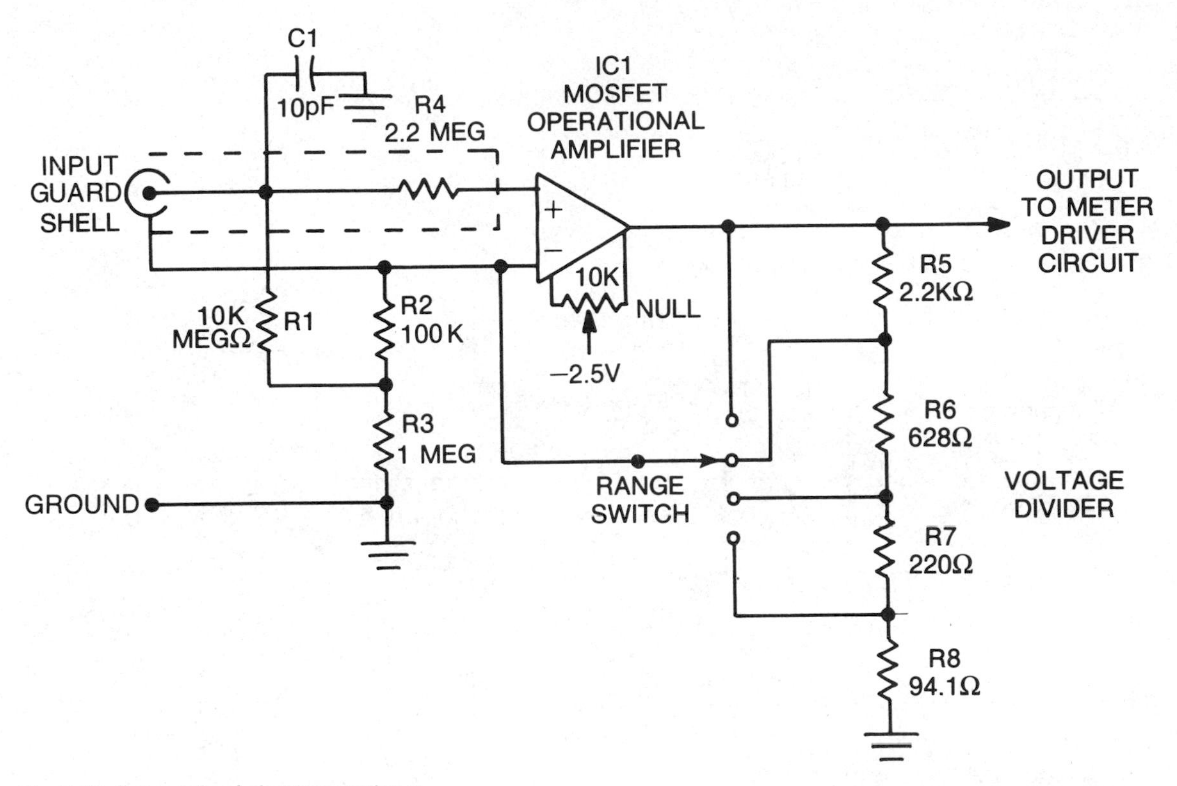

Fig. 6-28. Schematic of electrometer input.

This principle is called "guarding." Note, that the guard terminal is for shielding purposes and must *not* be grounded. Guarding is a technique used for measurement of high resistances, very small currents, very small capacitances, or in measurements where undesired or stray effects may mask or cover quantities to be measured. Guarding generally makes use of a voltage or current, or other signal so as to cancel the unwanted effects.

Figure 6-29 shows several examples. The output of IC1 is fed to a metering circuit to give a readout of the quantity measured, voltages of up to 1 volt peak or either positive or negative polarity. By using an ac metering ac voltages, ac voltages can also be measured.

Note that the input resistor R1 is a special unit having a resistance of 10k Meg (10,000 Megohms). It is returned to the junction of R2 and R3. About 90% of the input voltage appears at the junction. This has the effect of raising the apparent resistance by a factor of ten times, to 100k Meg. Some dc stability is sacrificed. If R2 is reduced to 10k ohms, then 1,000k Meg (1 million megohms) can be obtained although the drift of the operating point may be troublesome. 100k Megohm is sufficient for most purposes. By using the most sensitive scale (.03V) on the meter circuit, input currents of 0.1 pico ampere, or 0.1 micro-micro amperes are detectable with acceptable stability.

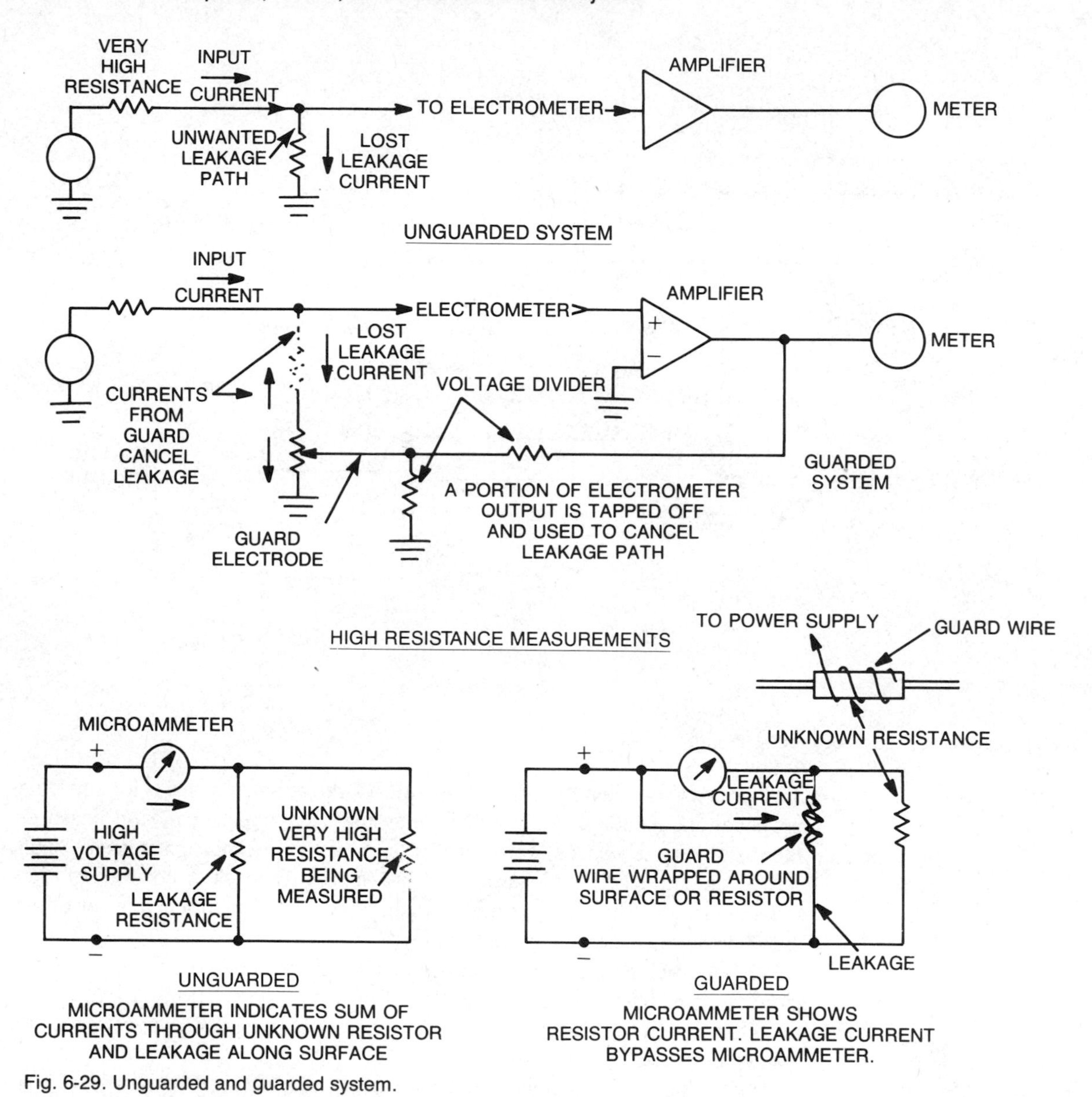

Fig. 6-29. Unguarded and guarded system.

The entire circuit is shown in Fig. 6-30, operation is as follows: IC1 is a MOSFET input operational amplifier with the input applied to the noninverting input across R1. This amplifier has gains of 1, 3.3, 10, and 33 as determined by switch S1 and R5 through R8. R5 and R7 are 2.2k and 200 ohm, but R6 and R8 are matched pairs of resistors selected for 628 ohm and 94.1 ohm respectively.

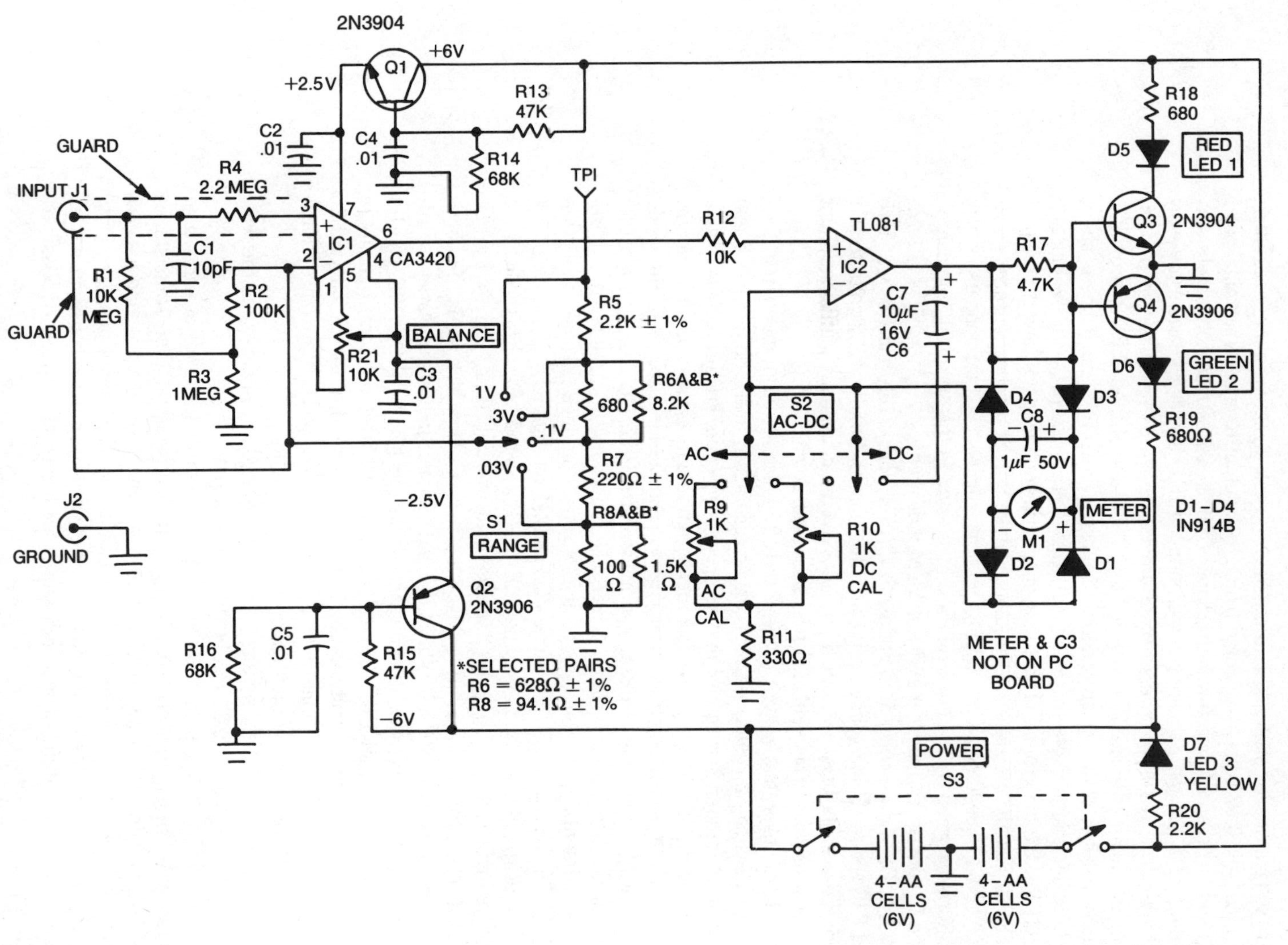

Fig. 6-30. Schematic.

This voltage divider allows IC1 to produce ±1 volt signal for input levels of 1, 0.3, 0.1, and 0.03 volts as selected by S1. The output (common) terminal of S1 connects to the negative (inverting) input of IC1. A PC trace completely surrounding R4 and pin 3 (noninverting) input of IC1 provides guarding of stray leakage paths. A 10k Meg resistor R1 provides a dc bias return for pin 3 (noninverting) input of IC1. R1 is returned to the junction of R2 and R3 to provide a tenfold increase in the apparent input impedance. Effective impedance is about 100,000 megohms. Note that the input voltage is applied between the junction of R1, R4, C1 and ground. The voltage appearing on the guard (inverting input) is almost exactly the same as the input voltage (for all practical purposes). At pin 2 of IC1 depending on the setting of S1, an amplified voltage appears, which is always between 0 to ±1 volt. TP1 is connected to this terminal, and can be used to connect an external DVM or VOM if desired.

Q1, Q2 and associated components C2, C4, R13, R14, and C5, C9, R15 and R16 act as simple voltage regulators to supply a "stiff" ±2.5 volts to IC1 from the ±6V battery, without the need for zener diodes and inefficient shunt regulator circuits. This helps keep battery drain down.

R12 couples this ±1 volt dc level to IC2, which acts as a meter rectifier/driver. A switch S2 selects either ac or dc calibration. In the dc mode, C6 and C7 restrict frequency response and reduce noise pickup. D1 through D4 form a bridge rectifier for M1, or 0–1 mA meter movement. C8 acts as an ac shunt, bypassing the meter. Since D1 through D4 are in the feedback path, the diode nonlinearity is compensated for, and M1 will read the absolute average (dc) value by the signal at pin 3, with a linear scale. R9 and R10 with R11 are calibration adjustments, set so one volt dc at TP1 (or ac) produce full scale deflection of M1. The output of IC2 also drives Q3 and Q4 through R17. If the output of IC1 is positive, Q3 conducts, lighting LED 1 (D5), a red LED. This indicates a positive input voltage. A negative voltage input will light LED 2 (D6) a green LED. Null control R21 is used to adjust the zero of M1, which reads the absolute value of voltage at TP1. This LED arrangement indicates polarity, red if positive, green if negative. Ac input signals will light both LEDs. R21 can be adjusted so that neither LED 1 or LED 2 light, which will only occur at meter zero (balance). A dual 6-volt battery supply is connected to the circuit via S3. Drain is about 10 to 15 milliamperes. R20 and D7 (LED 3) are an optional power-on indicator and not needed for circuit operation.

Figures 6-31, 6-32 show suggested PC board layout, art work, and the wiring diagram. A ready-made PC board, drilled and screened, can be obtained from the source listed at the end of this project, as well as a kit of parts that mount on it. Begin by inserting resistors, then capacitors, then transistors, lastly, the IC devices. Use only rosin core solder. It would be a good idea to clean the PC board with isopropyl alcohol after assembly, and if possible, with acetone (do not get this on the topside of the board—component side) as a final cleaning. Be very careful—these substances are highly inflammable. If possible, do this step outdoors.

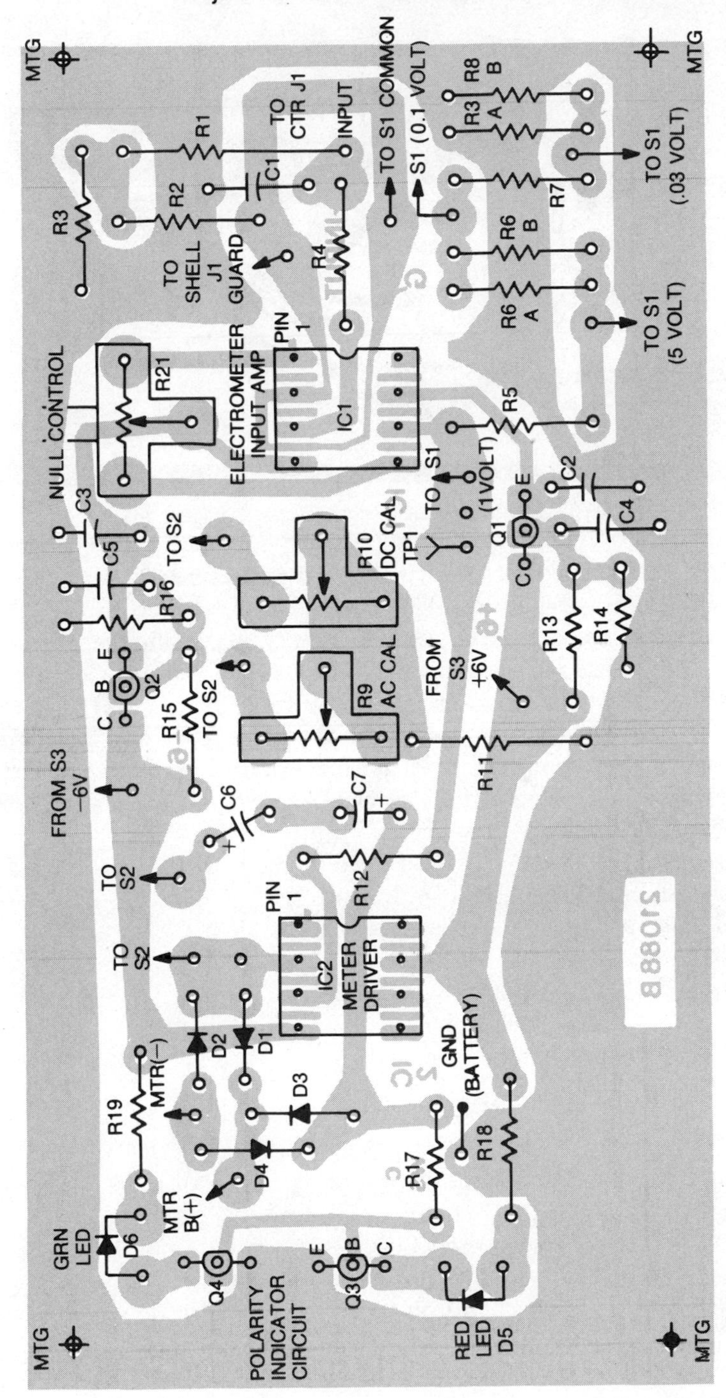

Fig. 6-31. Electrometer PC component placement.

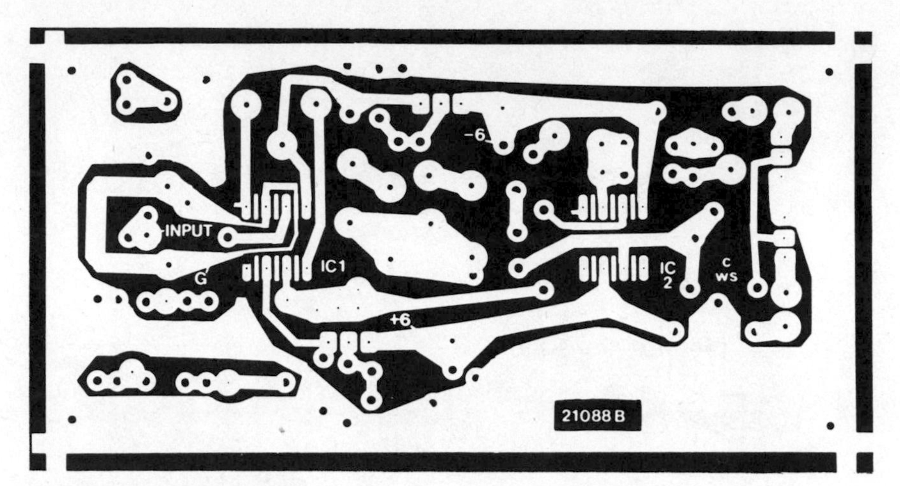

Fig. 6-32. PC artwork—foil side.

Check all connections, solder joints, and make sure all transistors, IC1, IC2, and diodes D1–D4 are correctly inserted. Check to see that C6, C7, and C8 are correctly inserted with regards to polarity. Make sure there are no solder bridges or shorts and all wiring to switches, meter, battery holders, and PC board is correct. Do not insert batteries as of yet.

Use an ohmmeter to read the resistance between pin 7 of IC2 and ground, then pin 4 of IC2 and ground. At least 300 ohms or better should be read. If not, try reversing ohmmeter probes. The higher reading obtained should exceed 300 ohms. If not, check your wiring.

Insert batteries (8-AA cells) and close S3 the power switch. Immediately check to see if LED 3 lights (yellow). Either LED 1 (red) or LED 2 (green) may be lit. Measure and verify the following voltages:

IC1 pin 7 +2 to +3 volts
IC1 pin 4 −2 to −3 volts
IC1 pin 6 −1 to −1 volts, may vary
IC2 pin 7 +6 volts
IC2 pin 4 −6 volts

Connect a wire between the junction of R1, R4, and C1 and ground foil on the PC board. Set S1 (range) to the 1 volt range. Adjust the balance control R21 for a null in the meter reading. If the meter will not zero, turn off S1 and check your wiring. If OK, then check IC1 or IC2 (or both). Set S2 to the dc position, check to see that R21 will null the meter at each setting of S1. Note that on one side of null setting the red LED D1 will light, and on the other side the green LED D2 will light. If not, check IC2, Q3, Q4, and the LEDs.

Set S1 to the 1V range and connect exactly +1 volt dc to the center connector of J1 and ground to J2. The shell of J1 is not connected to anything for this test. The shell of J1 is *not* ground. It is the guard electrode and a dc voltage (or ac voltage) equal to the input will appear here. It is used to connect to a shield for the input lead, and nothing else (see Fig. 6-33). Adjust R10 for full scale meter reading. Next, apply 1 volt RMS ac (60 Hz is OK) to the center of J1, with J2 as a ground. Adjust R9 for full scale deflection. This completes checkout and calibration.

You will find this instrument to be extremely sensitive and it must be either mounted in a metal cabinet or a plastic cabinet lined with foil. *Make sure not to ground the shell of J1, it is the guard electrode (shield of cable, etc.).* Keep wiring to J1 short and direct. Also, keep this wiring to J1 away from other leads.

In order to measure voltages higher than 1 volt, an external divider must be used. This can be either high value resistors or a capacitive divider. The latter method is preferable. Do not attempt to measure higher voltages than about 1 volt without the divider—you may damage IC1. Also, do not touch objects charged to a high voltage to J1. It may be advisable to earth ground (cold water pipe) J2 in some cases. Figure 6-34 shows simple dividers you can use. The capacitive divider has the advantage of drawing no steady dc current from the source. However, since the input resistance at J1 is finite, the dc voltage read across C2 (see Fig. 6-34) will gradually drop even if the input voltage is kept constant to C1. Therefore, you must make sure C1 and C2 are discharged before reading the voltage and immediately take the reading after connecting C1 to the unknown voltage. The electrometer has about 12 pF input capacitance—this should be accounted for in ac measurements. Ac frequency response is about 10 Hz to 10 kHz so low level audio can also be measured. However, realize that the capacitive input impedance is a limiting factor at very high impedances. Remember that, for instance 1 pF is 159 megohms reactance at 1 kHz, for example. At very high impedance levels, this is not negligible.

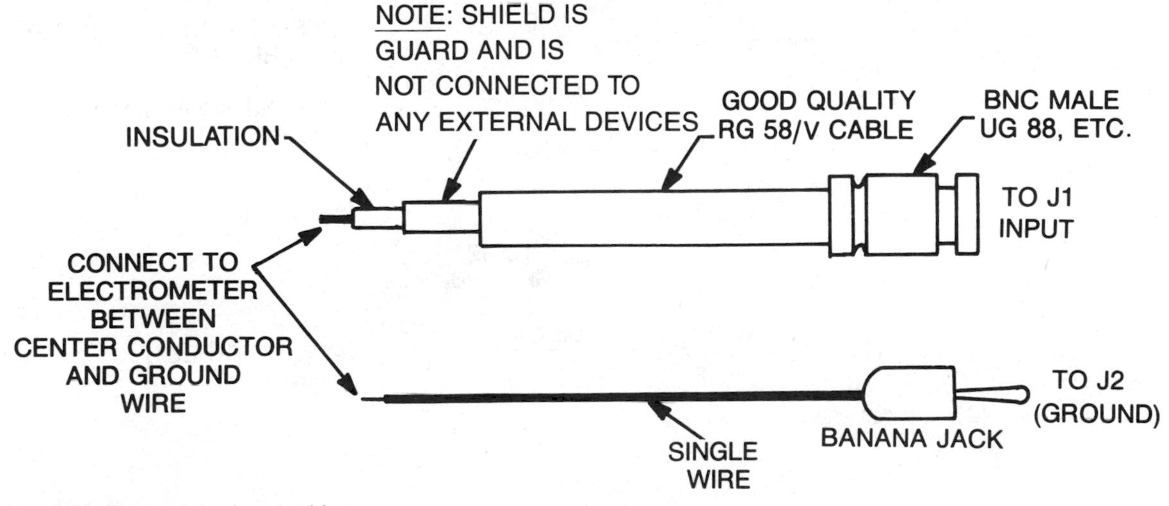

Fig. 6-33. Electrometer input cables.

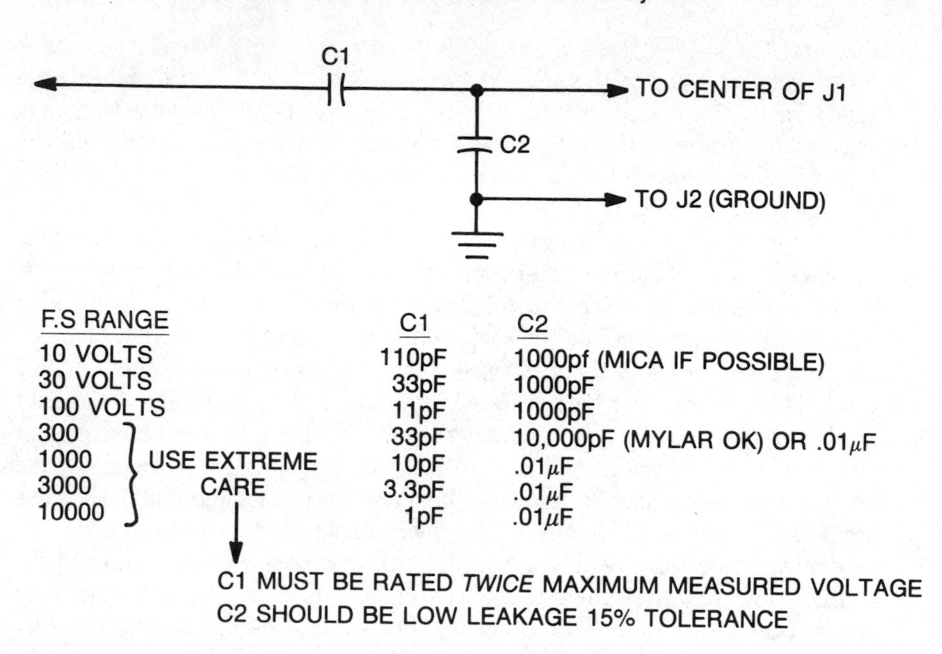

CAPACITIVE DIVIDER

IV
1000 MEGS
3V
TO J1 CENTER
286 MEGS
10V
100 MEGS
30V
42.9 MEGS
TO J2 GROUND

NOTE:
ANY VALUES CAN BE USED IF CORRECT RATIOS CAN BE OBTAINED

VALUES SHOWN ARE AS AN EXAMPLE ONLY

RESISTIVE DIVIDER

(FOR LOW VOLTAGES ONLY)

Fig. 6-34. Electrometer external voltage dividers.

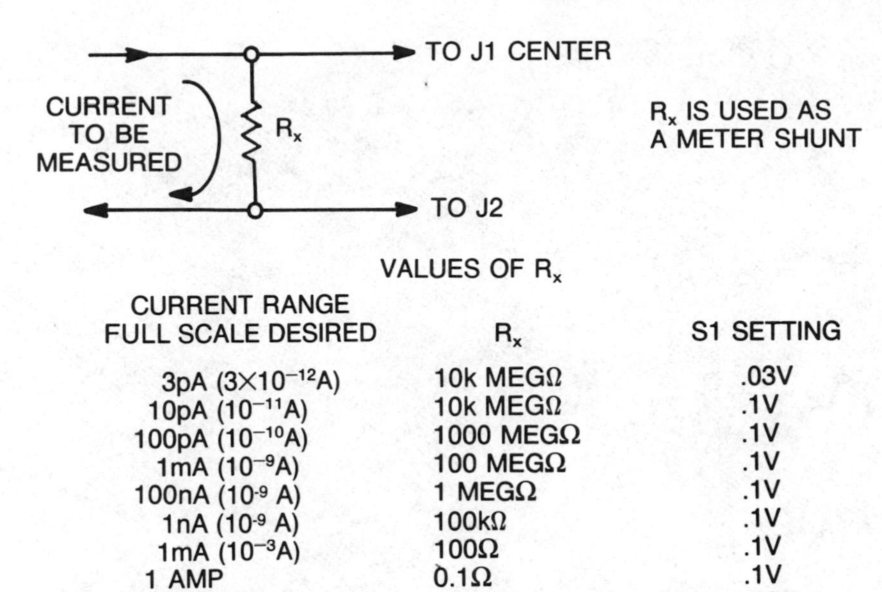

VALUES OF R_x

CURRENT RANGE FULL SCALE DESIRED	R_x	S1 SETTING
3pA (3×10^{-12}A)	10k MEGΩ	.03V
10pA (10^{-11}A)	10k MEGΩ	.1V
100pA (10^{-10}A)	1000 MEGΩ	.1V
1mA (10^{-9}A)	100 MEGΩ	.1V
100nA (10^{-9} A)	1 MEGΩ	.1V
1nA (10^{-9} A)	100kΩ	.1V
1mA (10^{-3}A)	100Ω	.1V
1 AMP	0.1Ω	.1V
30 AMPS	0.001Ω	.03V

Fig. 6-35. Method of current measurement.

This instrument will prove useful and you can perform many interesting experiments in electrostatics and high impedance phenomena that cannot be done with ordinary instruments. A shielded mounting case must be used. See Fig. 6-36 (photo) for suggested layout, using a stock Radio Shack plastic case. The inside was covered with aluminum foil, and this served as an adequate shield. See Fig. 6-37.

For those wishing to construct this instrument, a drilled, etched, and screened PC board as well as a kit of parts is available from:

North Country Radio
P.O. Box 53E, Wykagyl Station
New Rochelle, NY 10804

PC board and all resistors, capacitors and semiconductors (Misc. parts are *not* included)....$37.50 plus $2.50 shipping and handling.
PC board only....$10.00 plus $2.50 shipping and handling.

New York residents must include sales tax.

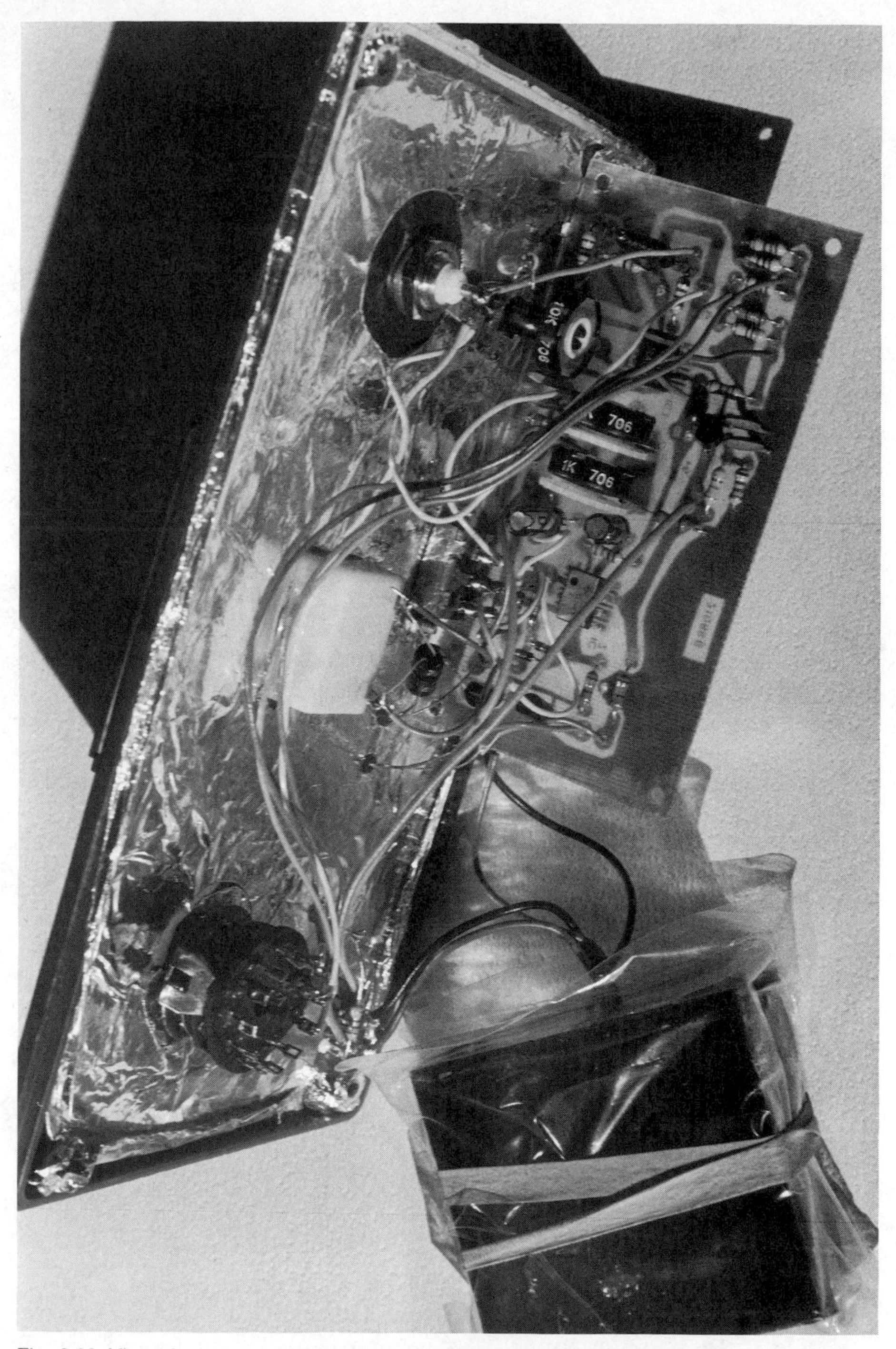

Fig. 6-36. View of completed electrometer. Note aluminum foil used for shielding.

RESISTORS 1/4W 10% (unless noted)

R1	10,000 Megohm ±10%
R2	100k ohm 5%
R3	1 Megohm 5%
R4	2.2 Megohm
R5	2.2k ohm 1% selected
R6A,B	628 ohm 1% selected pair (680 ohm + 8.2k ohm)
R7	220 ohm 1% selected
R8A,B	94.1 ohm 1% selected pair (100 ohm & 1.5k ohm)
R9	1k ohm trimmer pot
R10	1k ohm trimmer pot
R11	330 ohm
R12	10k ohm
R13	47k ohm
R14	68k ohm
R15	47k ohm
R16	68k ohm
R17	4.7k ohm
R18	680 ohm
R19	680 ohm
R20	2.2k ohm (used only if optional LED D7 is used)
R21	10k ohm trimmer w/shaft

CAPACITORS

C1	10 pF Silver Mica ±10%
C2	.01/50V disc
C3	.01/50V disc
C4	.01/50V disc
C5	.01/50V disc
C6	10μF/16V Electrolytic
C7	10μF/16V Electrolytic
C8	1μF/50V Electrolytic

SEMICONDUCTORS

IC1	RCA CA3420 OP AMP
IC2	MOT or TI TL081 OP AMP
Q1	2N3904 or ECG 123A
Q2	2N3906 or ECG 159
Q3	2N3904 or ECG 123A
Q4	2N3906 or ECG 159
D1	1N914B
D2	1N914B
D3	1N914B
D4	1N914B
D5	(LED1) Red LED (Any type)
D6	(LED2) Grn LED (Any type)
D7	(LED3) Yel LED (Optional — Pwr indicator)
1	PC board

∗ MISCELLANEOUS (Radio Shack, etc.)

S1	4 posn. 1 pole switch
S2	DPDT switch
S3	DPDT switch
M1	0– 1 mA meter
J1	BNC female
J2	Binding post
1	Case as read
2	4-AA cell holders
8	AA cells

∗ Not included in kit

Fig. 6-37. Project 16 parts list.

PROJECT 17: ENLARGING LIGHT METER

The use of a simple enlarging light meter (Fig. 6-38) in the darkroom makes the production of prints, both monochrome and color, an easier task and saves time, trouble, and money in otherwise wasted materials. While there are a number of enlarging light meters on the market, the electronics enthusiast who is also a photo hobbyist can produce such a meter at home, with little expense. The meter to be described here can easily be duplicated for $25 or

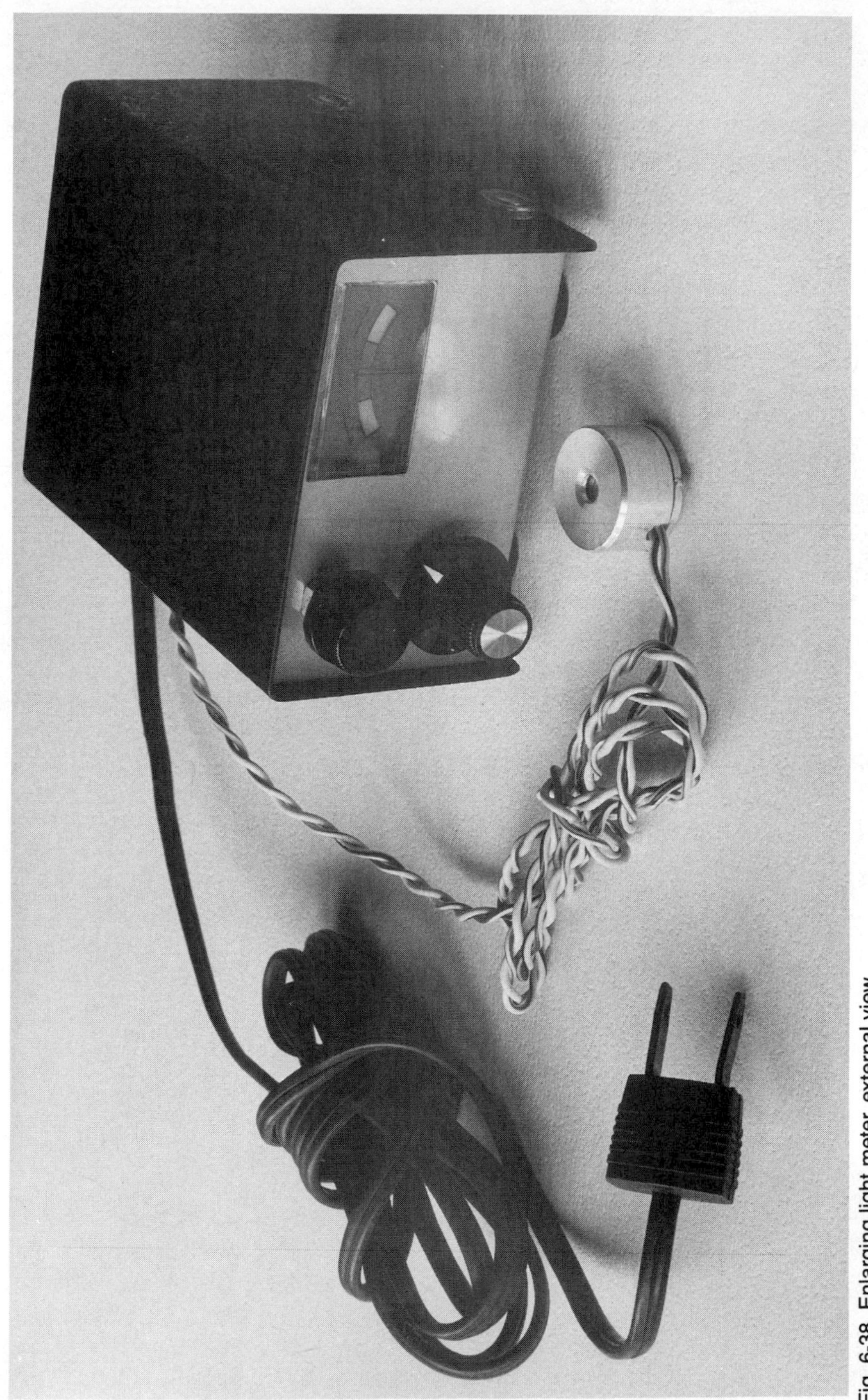

Fig. 6-38. Enlarging light meter, external view.

less, depending on how many materials are at hand. All the parts can be purchased at your local Radio Shack dealer, and many can be salvaged from a junked stereo or CB set.

The light meter is basically a comparator. It is not calibrated against any special standard. This makes it easy to build, set up, and use. Basically, it compares the incident light striking a photocell with a level that has formerly been established as correct based on previous experience. In our case, this level is established with the setting of a potentiometer and a rotary switch.

First, a good print (monochrome or color) is made using your conventional technique. The enlarger settings are recorded (distance between lens and paper, f/stop, and exposure time). Next, the negative used for this print is placed in the enlarger, and the recorded settings are duplicated. The light sensor (CDS cell), which is mounted in a separate housing, is placed on a part of the image that corresponds to a medium gray tone. The meter range switch is used to coarse null the meter (pointer at zero center). Then, the null control is used to exactly null the meter. The settings on the meter are noted and left alone. The meter is now calibrated for your paper and print developing technique.

When another negative is to be printed, simply insert the negative in the enlarger. Compose and sharply focus the picture, using the enlarger lens at maximum aperture usually f/3.5 or larger. Now place the light sensor in a part of the image, that is to be a medium gray in the final print. Preferably this area should lay somewhere in the central portion of the picture, but this is not absolutely necessary. Try to avoid the corners or extreme edges if you can. Now, without changing the enlarger meter settings from those obtained from the first negative, adjust the enlarger lens aperture for a null on the meter. This will be the correct exposure setting. Use the same exposure time and developing method as you used on the first negative.

Insert a sheet of photographic paper into the enlarger baseboard easel, and make the exposure. Develop the print. It should be correctly exposed and look good.

You may also use a lighter gray tone or even total black as a reference point. Just be consistent. We have found the medium gray reference point easiest to judge. Others prefer the darkest area (lightest gray or white on final print), and this works equally well. It is simply a matter of personal preference.

While this meter is primarily intended for B/W printing, it may be also used for color. However, be warned that it will give only correct exposure. The color balance (filter pack in the enlarger) is not measured. However, if you have predetermined the filter pack, this meter will work well for color. Use an area of the negative that is neutral or near neutral in color. Flesh tones work well for this reference point. For landscapes, you can use the sky areas as a reference.

Referring to Fig. 6-39, the operations of the light meter will be discussed. Meter movement M1 (in our case a zero center 50μA meter taken from a junked FM stereo tuner) is driven, through resistors R3 and R4, by the output

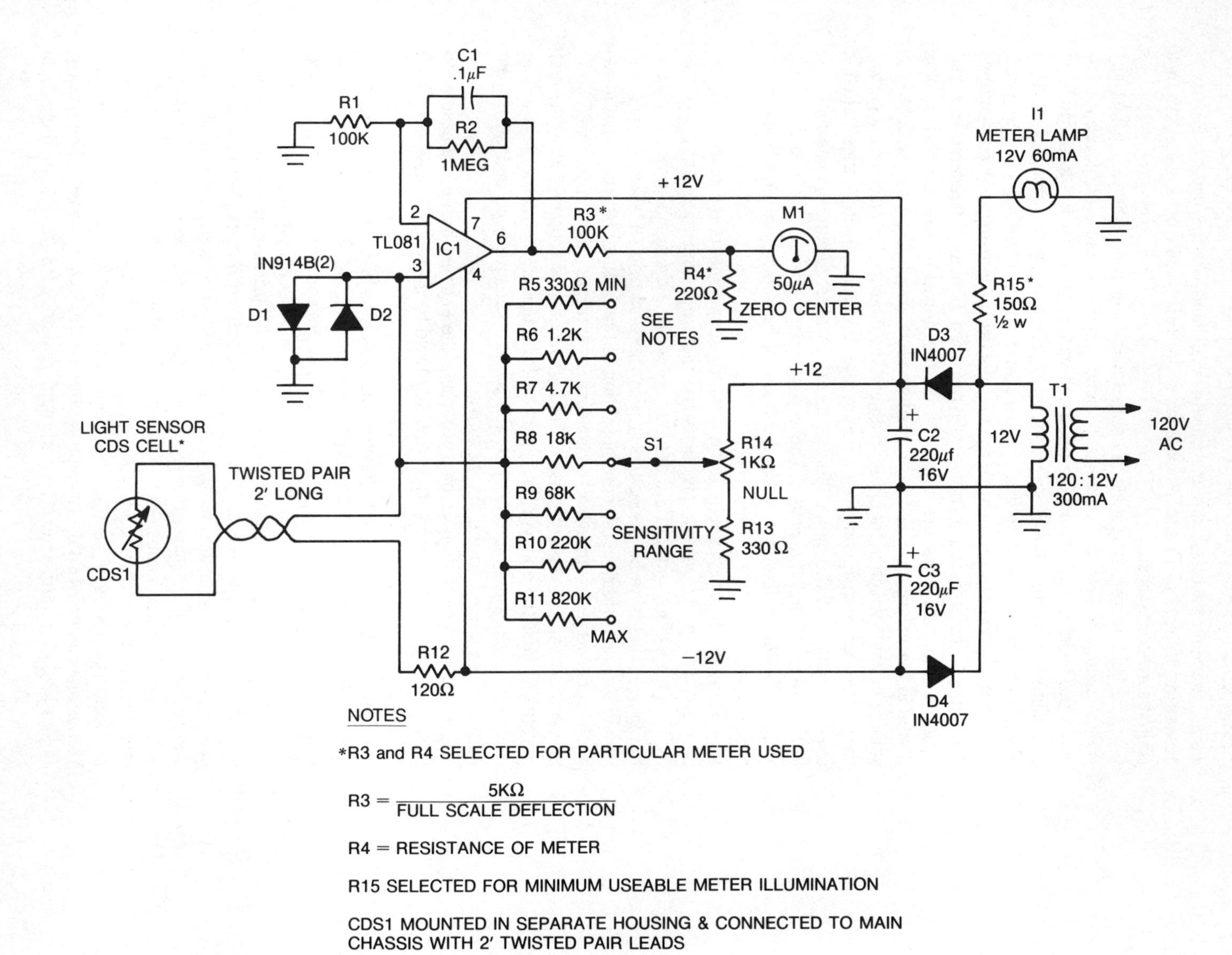

Fig. 6-39. Schematic of enlarger light meter.

of IC1, a FET op amp. R1 and R2 set the gain of IC1 to nominally 11 times. C1 restricts the bandwidth of the amplifier circuit to 1.6 Hz, eliminating possible 60 Hz noise pickup. T1, D3, D4, C2, and C3 form a simple power supply supplying ± 12 volts to IC1 and the rest of the circuitry.

In order to use this circuit as a light level comparator, use is made of the fact that CDS1 exhibits a characteristic such that its resistance decreases with increasing illumination. CDS1 is a Radio Shack item (P/N 276-116) and has adequate sensitivity in low light levels typically used for photographic printing. CDS1 is returned to the -12V supply through current limiting resistor R12. The output current of CDS1 is fed to pin 3 of IC1, D1 and D2 are limiting diodes. At the same time, a variable current determined by the setting of R14 and S1 is fed to pin 3 of IC1. This current can be varied over a 4:1 range. R13 sets the lower limit, and resistors R5 through R11 determine the maximum value of this current. When the current from the output of S1 exactly matches the current from CDS1, zero volts occurs at pin 3 of IC1. This results in meter M1, a zero center meter indicating a null. If the light striking CDS1 increases, the voltage at pin 3 will go negative. This makes the voltage at pin 6 go negative. The pointer on M1 then swings to the right. This tells the user to reduce the light intensity by using a smaller lens opening on the enlarger (smaller f/stop). The opposite occurs if the light is too dim. The pointer swings left.

The circuit is built in hardwired form, since it is not worth using a PC board with such a simple circuit. Layout is not critical—just use good ''common sense'' and all that is necessary is to wire the circuit correctly.

S1 may be a standard 12-position rotary switch with five positions left blank, if a seven position switch is not available. R14 is a linear taper 1k pot, if the users prefer. A lamp I1, a 12V 60 mA ''grain of wheat'' type is used to illuminate the meter scale. It is dimmed by R15 to a faint glow, just enough to be useable in a photo darkroom so that the face of M1 can be plainly seen. If desired, a red LED can be used. If possible, I1 should be painted orange or red to reduce blue light components that photo paper is sensitive to. T1 is a 120V to 12V transformer, not critical. The suggested part for this is Radio Shack #273-1385 if you do not have a suitable transformer on hand.

Packaging is left to the constructor, largely due to the fact that the meter used for M1 will probably determine this. See Figs. 6-38 and 6-40 for our prototype.

CDS1 should be mounted in a separate housing. A plastic bottlecap with a hole drilled in it is satisfactory. See Fig. 6-41. Do not expose CDS1 to very bright light before using the meter, as CDS1 exhibits some ''Memory'' effect which may persist for several minutes. This could throw off your readings for the first few minutes.

A parts list for the meter is given in Fig. 6-42. After the meter is constructed, check for shorts, poor connections, and wiring mistakes. If all looks OK, apply 120Vac to the primary of T1. Check for $+12$ volts at the junction of C2 ($+10$ to $+15$V OK) and for -12 volts (-10 to -15V OK) at the junction of

Fig. 6-40. Enlarging light meter with case opened to show internal construction.

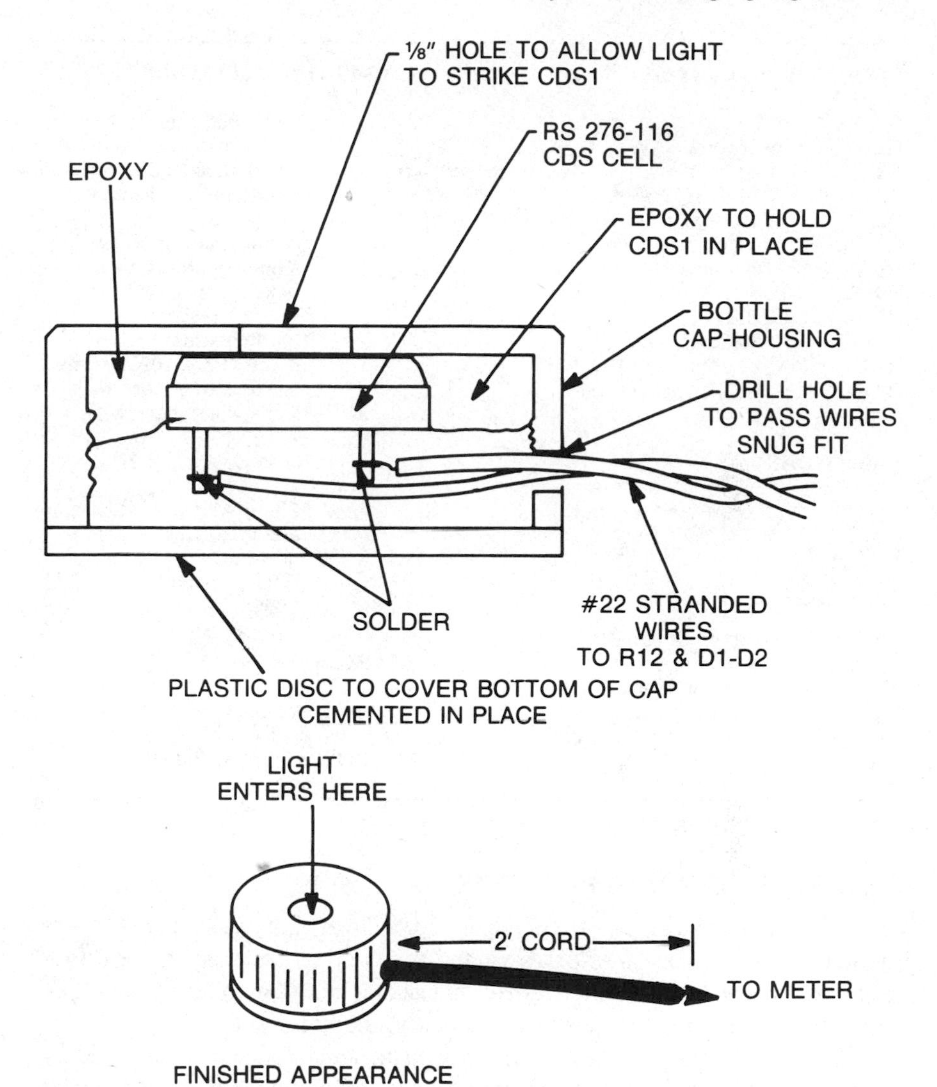

Fig. 6-41. Light sensor construction.

C3 and D4. Next, with CDS1 shielded from light, place S1 in the position corresponding to a maximum sensitivity. It should be possible to null or at least move the pointer of M1 by varying R14. Now, expose CDS1 to a dim light (a 7 watt night light about 3′ from the cell). Vary S1 and R14. You should find a setting of S1 and R14 such that the meter nulls. If not, check your wiring.

Grounding pin 3 of IC1 should null M1. This is a guide check of IC1, the power supply, and the meter. If this test works OK but you cannot get the meter to null otherwise, check CDS1, its wiring, and the wiring of R14, R13, S1, and R5 through R12. However, the circuit is very reliable and no trouble should be encountered.

RESISTORS 1/4W + 10%

R1, R3	100k
R2	1 Megohm
R4	220 ohm
R5, R13	330 ohm
R6	1.2k ohm
R7	4.7k ohm
R8	18k ohm
R9	68k ohm
R10	220k ohm
R11	820k ohm
R12	120 ohm
R14	1k linear pot w/shaft
R15	150 ohm

CAPACITORS

C1	.1μF Mylar or ceramic
C2, C3	220μF/16V Electrolytic

ICs AND SEMIs

D1, D2	1N914B or equiv.
D3, D4	1N4007 or equiv.
IC1	TL081 or equiv.

METERS AND MISCELLANEOUS

M1	50μA zero center (preferred) or *any* meter on hand that can be adjusted to center the power with zero current, of 1 mA or less full scale current (junked stereo or CB set good source of these meters)
I1	12V 60 mA lamp (RS P/N 272-1099) painted red or orange (for illuminating meter M1)
CDS1	Radio Shack P/N 276-116
T1	120V to 12V 100 mA or more (RS P/N 273-1385, etc.)
S1	7 to 12 pole single section Rotary switch (RS P/N 275-1385)

MISCELLANEOUS

1	Cabinet of your choice
1	Ac line cord

Wire and hardware as required

Fig. 6-42. Project 17 parts list.

PROJECT 18: DIGITAL PHOTO TIMER

Many photographic processes require some form of timing. One of these is the photographic process of enlarging and printing. In this case, the enlarger lamp must be turned on for a predetermined time in order to make the exposure, since enlarger lenses generally do not have a timed shutter.

Traditionally, the timing function has been implemented with a mechanically controlled switch, driven by a mechanism employing a clock motor or similar device, which is synchronous with the 50 or 60 Hz power line. Early electronic timers employed a relay driven by a vacuum tube or transistor amplifier, which in turn derives its timing signal from a circuit usually employing a resistance—capacitance time constant. While both of these methods are usually satisfactory, they both lack definite repeatability, since the setting of the time interval depends on setting an analog dial scale or pointer. It is difficult to accurately repeat a given setting. This may not be a problem if long time intervals (20 seconds or so) are involved. However, many of today's papers, both color and monochrome, are quite "fast" requiring short exposures. Also, the use of certain other photographic materials, such as print films for use in slide copy work and production of slides from negatives, require short exposures, usually 1 to 3 seconds. In this case, the typical zero to 60 second mechanical timer leaves something to be desired. It is readily possible, using commonly available IC devices, to construct a simple programmable timer. The timer has an LED readout and data entry (time desired) by thumbwheel switches or, if desired, a keyboard encoder or other source of binary coded decimal (BCD) data. The 60 Hz ac line is used as a clock for timing. Time intervals are set in either 1 second intervals (999 seconds maximum) or 0.1 second intervals (99 second maximum), selected by a switch. Since most enlarger lamps take 100 to 200 milliseconds to turn fully on or fully off, the use of higher resolution than 0.1 second increments is of dubious value.

Ac power control is achieved by a triac and an optical isolator is used so that complete isolation between the timer circuitry and the ac line is achieved. Up to 720 watts can be handled by the triac used (120V at 6 amps). This is adequate for enlargers the average photo hobbyist or photographer is likely to use.

The timer can be also used for film developing as a visual timer, using the LED display. The time must be entered in seconds so that, for example, 10 minutes has to be entered as 600 seconds (60 sec × 10 min.).

Referring to Fig. 6-43, a control latch Q7 is set by a ground signal from the start button. The latch enables counters IC4, IC5, and IC6, which are programmable BCD 74192 TTL counters. Data from these thumbwheel switches preloads the counter to the desired time interval during times that the counter is not running or idle. LED displays DS1 to DS3, driven by seven segment decoders IC1, IC2, and IC3, display the counter status during idle intervals the input data (time setting) is displayed. During a timing cycle, the displays count down to zero, showing the time left to completion of the cycle. This is useful in certain printing operations, such as dodging and burning in various areas of the print, or in additive color printing. At the completion of the cycle, when zero count is reached, a zero count signal resets the control latch.

Clock generator Q4, IC7, IC9, 10, and 11 produce a selectable 1 Hz (seconds) or 10 Hz (tenths of seconds) time base for 0 – 999 or 0 – 99.9 second ranges respectively. IC8 and Q4 provide control of the ac power to the enlarger, and are driven by control latch Q7. Dc power is supplied by transformer T1 diodes D1 to D4, and regulator IC12. IC12 supplies +5 volts to all other ICs as needed. Q1 supplies a variable 0 to +12 volts to the LED displays, for brightness control important in darkroom work.

The timer is built on three PC boards for flexibility in packaging. One board consists of the counters, clock, control latch, and isolator/triac driver circuitry. Another contains the displays and display drivers. A third board, which is

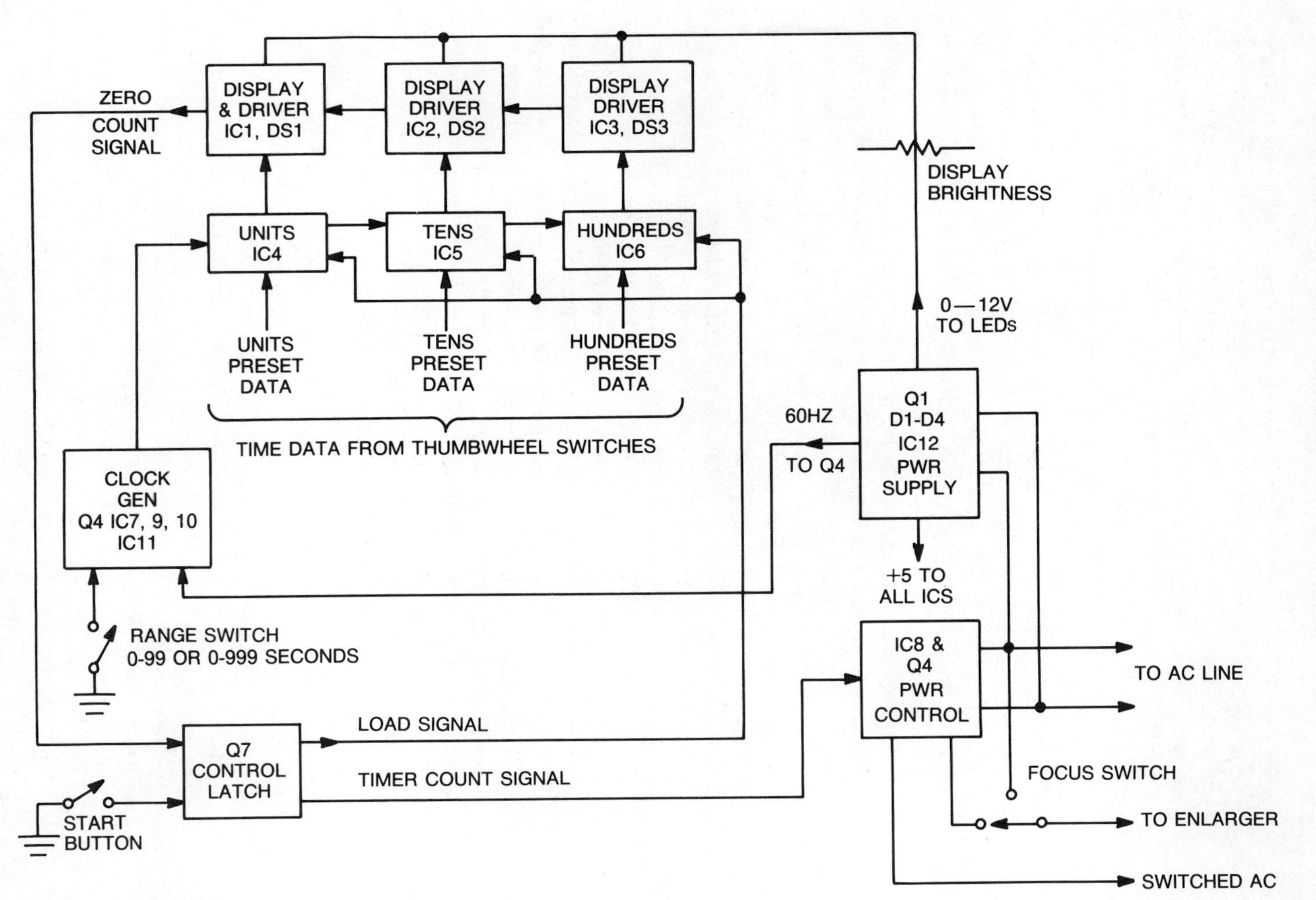

Fig. 6-43. Block diagram of photo timer.

optional, contains the thumbwheel switch and its pull-down resistors. This allows maximum packaging flexibility. All switches and power supply components are also separately mounted, for maximum flexibility. Referring to Fig. 6-44, 60 Hz ac is fed to filter R38 and C18, which removes line noise spikes that could cause false or erratic counts. C17 couples the ac signal to amplifier Q4, a 2N3904 NPN transistor. R37 is a base bias return, and R36 forms a collector load. C15 and C16 reduce stray noise. A square wave of 60 Hz appears at the collector of Q4 and is fed to two gates (cascaded) in IC7. This squares up the 60 Hz wave and level shifts it so it is TTL compatible. C14 is a noise suppressor. The output of the second gate of IC7 feeds IC10, a 7492 counter. IC10 is configured to divide by 6. C13 is a bypass capacitor. Each IC has a bypass capacitor across its +5 volt pin and ground to minimize noise. A 10 Hz square wave appears at pin 8 of IC10. It feeds both speed selector circuit IC11 and counter IC9, a 7490 that divides the 10 Hz signal by ten. A 1 Hz square wave appears at pin 11 of IC9. C11 limits rise time of the 10 Hz square wave. C12 is a bypass for IC9 +5V pin. The output of IC9 (pin 11) feeds selector circuit IC11. IC11 is used to select either the 10 Hz or 1 Hz signal for the counter. If pins 4, 5, and 2 are at a low state (range switch S4 closed) pin 3 of IC11 is a steady high. Pins 6 and 10 are high, so signal at pin 8 of IC10 connected to pin 9 of IC11 also appears at pin 11 of IC11, feeding 10 Hz signal to the counter and also lighting the decimal point of DS1, the tenths display LED. This shows the fact that the 0 – 99.9 second range is selected.

If S4 is open, pins 4J and 2 of IC11 are high. This means that signal from pin 11 of IC9 appears at pin 11 of IC11, feeding 1 Hz signal to the counter, and extinguishing the decimal point of DS1, indicating the 0 – 999 second range. Since pins 6 and 10 are low, pin 8 is a steady high, and the 10 Hz signal is not passed.

IC4, IC5, and IC6 are 74192 BCD counters. Before the start button S1 is depressed, pins 11 of these IC devices is held low. This enters the data appearing at pins 15, 1, 10, and 9 of IC4, 5, and 6. The data is BCD (zero to 9) representing the time entered into the counter via thumbwheel switches or other means. For example, if a time interval 43.5 seconds is necessary, and S4 is closed (0 – 99.9 range) the data at IC4 will be 0101, representing 5. IC5 will have 0011 representing 3 and IC6 will be 0100, representing 4. C4 is used to hold the load line low during power up, which sets the counters to the data programmed in immediately.

When the button S4 is depressed, pin 13 of IC7 is pulled low, since capacitor C5 has discharged through R24. Pin 13 is momentarily pulled to ground as C5 changes to +5V through R23. This causes pin 11 to go high, pin 10 goes high, and since a nonzero count has been programmed into IC4 then IC6, pin 9 of IC6, pin 8 of IC7 goes low. This voltage level is fed to Q4 through R27 and R28, cutting off Q2. This allows bias from R30 to flow into R30 and R31, turning on Q3, activating the optical coupler IC8 via current through R32.This turns on triac Q5 and ac voltage appears at J1, the enlarger power socket.

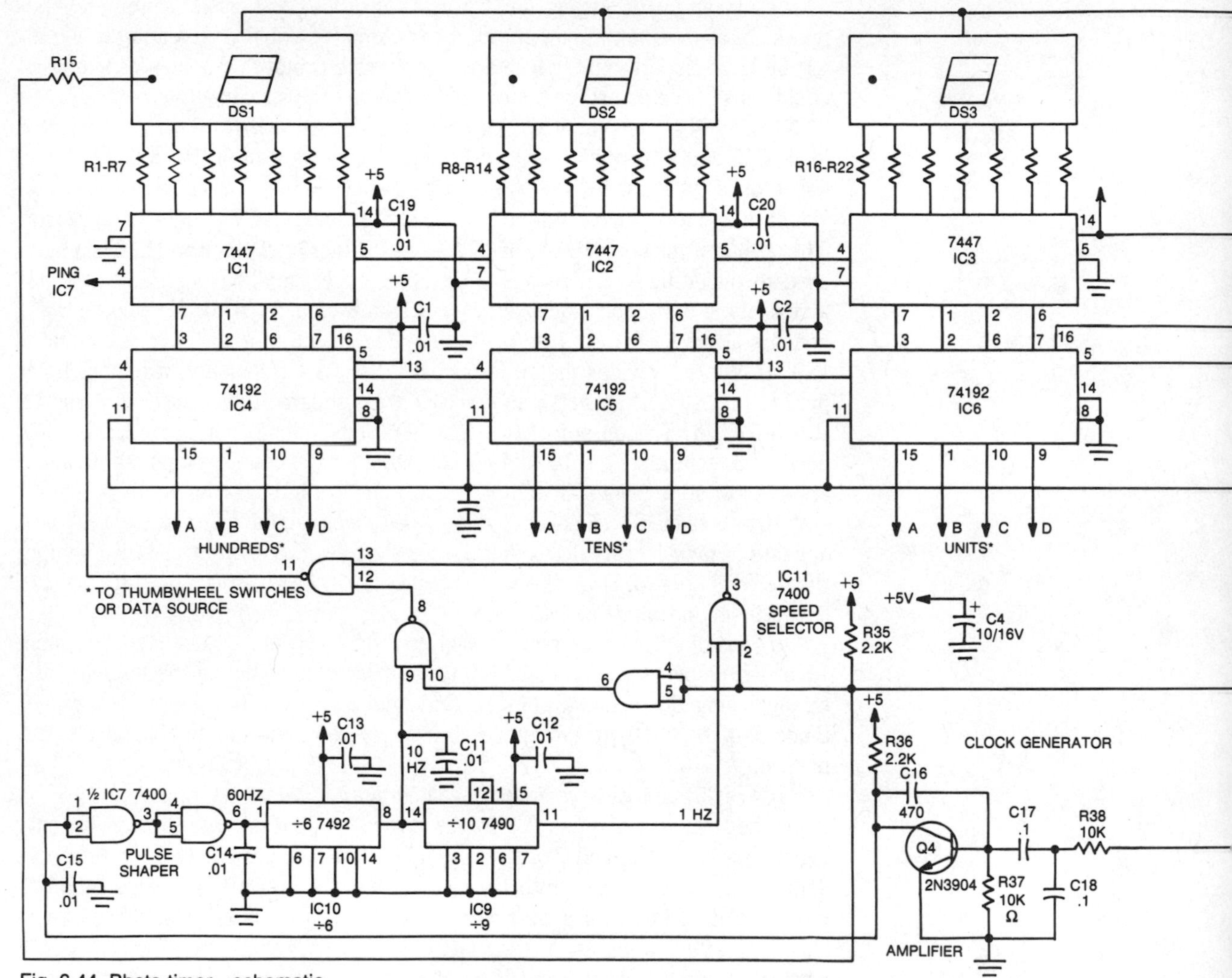

Fig. 6-44. Photo timer—schematic.

Pin 11 of all the counter IC devices IC4 – IC6 goes high. This causes the counter to ignore the data on pins 15, 1 10, and 9 (from thumbwheel switches and start counting down towards zero. IC1, IC2, and IC3 translate the state of the counters into seven segment data, lighting displays DS1 through DS3 to show the value of the data in the counters at a given instant. R1 through R22 are current limiting resistors for the display and are chosen to pass about 15 mA per segment at maximum brightness, enough for darkroom work. C1 through C3 and C19 through 21 are bypass capacitors.

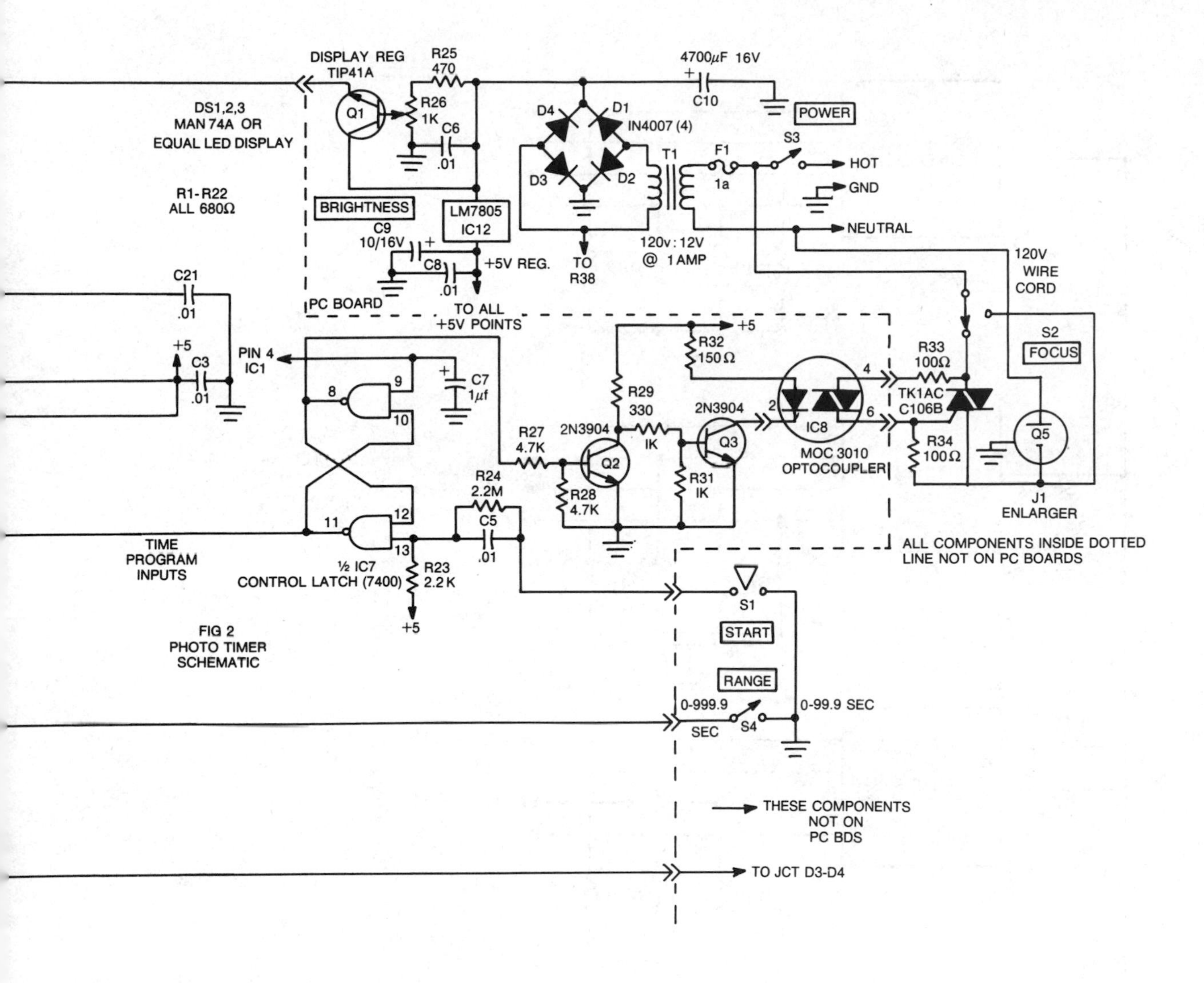

As the count approaches zero, the display DS2 and DS3 blank when they are zero. At the count of zero, all displays blank. This forces pin 4 of IC1 to go low. This causes pin 9 of IC7 to go low, which forces pin 8 of IC7 high. Since pin 13 of IC7 is high once C5 has charged or S4 is open, pin 11 of IC7 is forced low. This resets the counter to the programmed input time. Q2 is turned on, cutting off Q3 and IC8. This results in triac Q5 turning off, cutting off power to the enlarger lamp and ending the timing cycle.

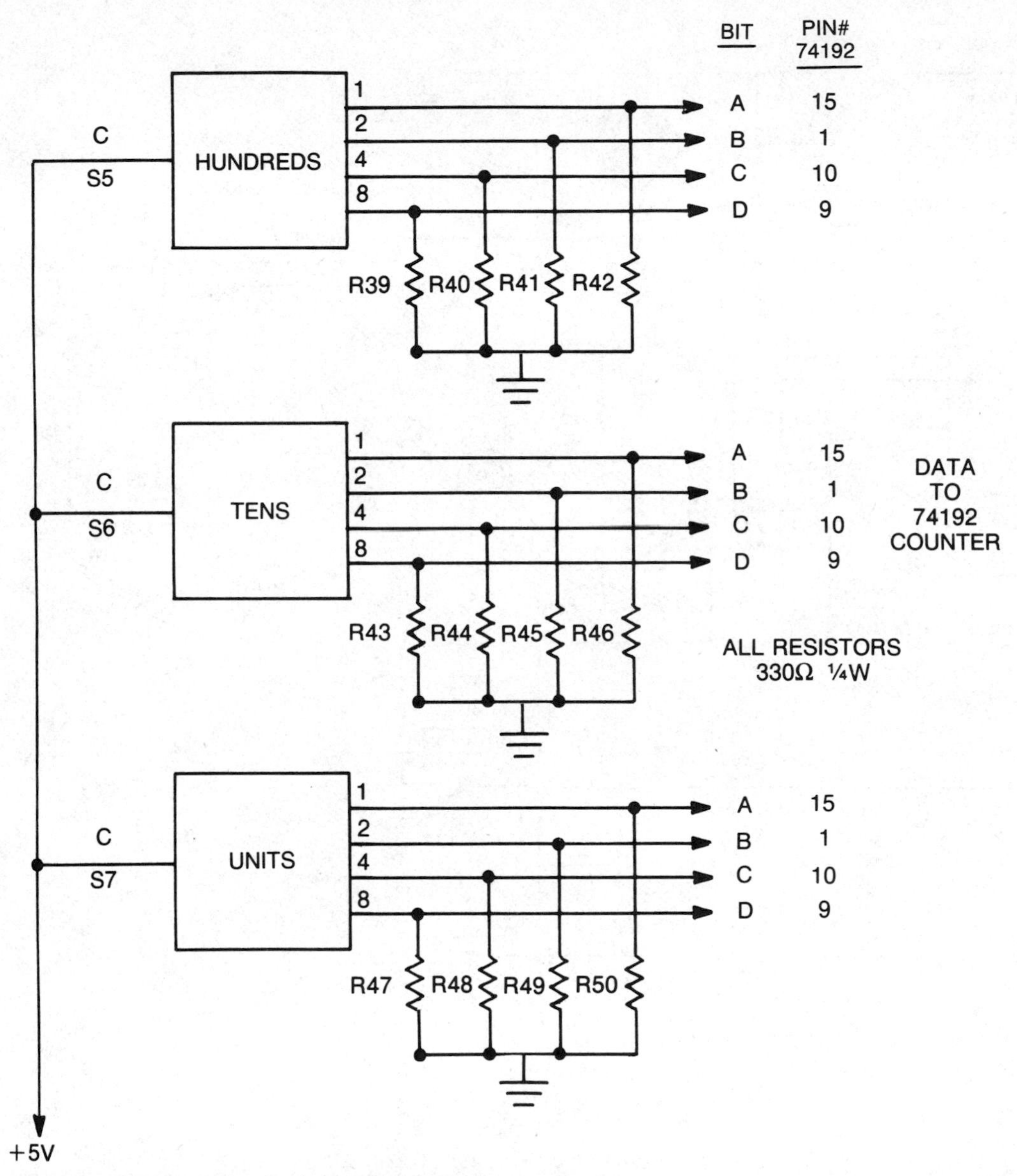

Fig. 6-45. BCD thumbwheel switch connection.

Power is provided by T1, D1–D4, filter capacitor C16, and IC12, to all +5V loads Q1 supplies variable 0–12 volts (depending on the setting of the brightness control R26) to the displays R25 limits maximum voltage available. C6, C8, and C9 stabilize regulator IC12.

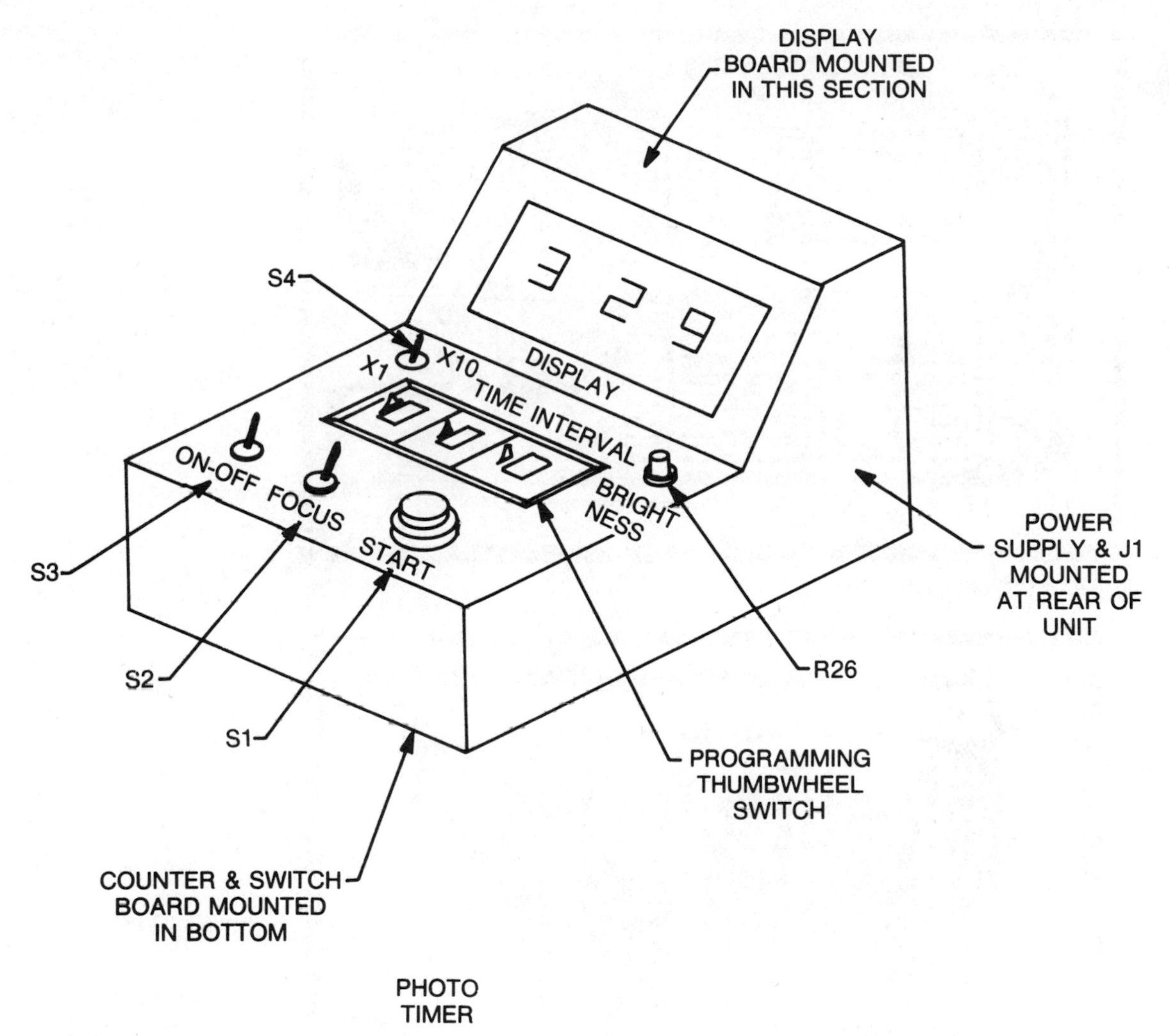

Fig. 6-46. Suggested package configuration.

Construction of the timer is noncritical. No high frequencies are present. But do not get sloppy—the counter can run at speeds of over 10 MHz due to the TTL devices. Stray noise can cause false counts. Keep 120Vac wiring well away from the TTL circuitry. Do not omit any bypass capacitors. The use of the PC layout shown in Figs. 6-47 and 6-48 is highly recommended. The smaller board used with the thumbwheel switches is not critical and if your thumbwheel switches are mechanically different, simply omit the PC board. The pull-down

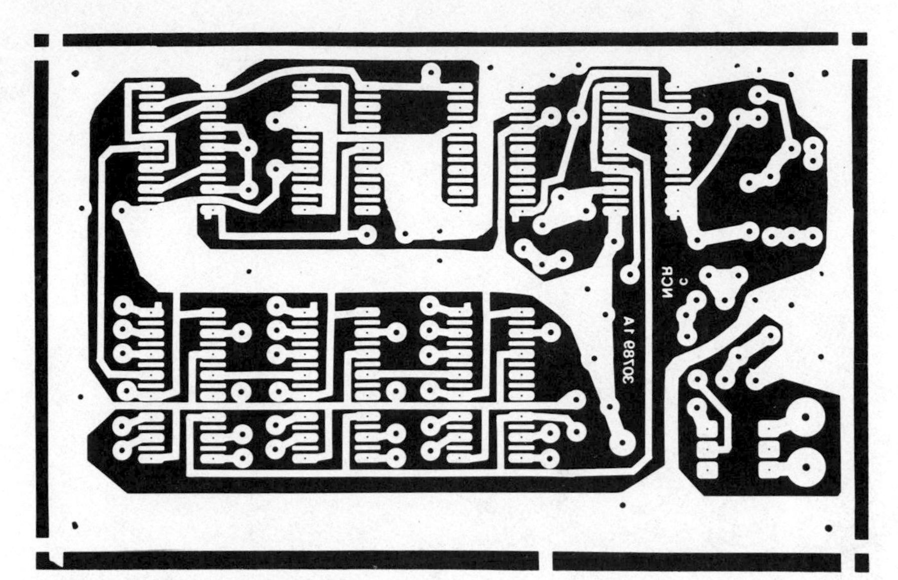

Fig. 6-47. PC board layout for timer.

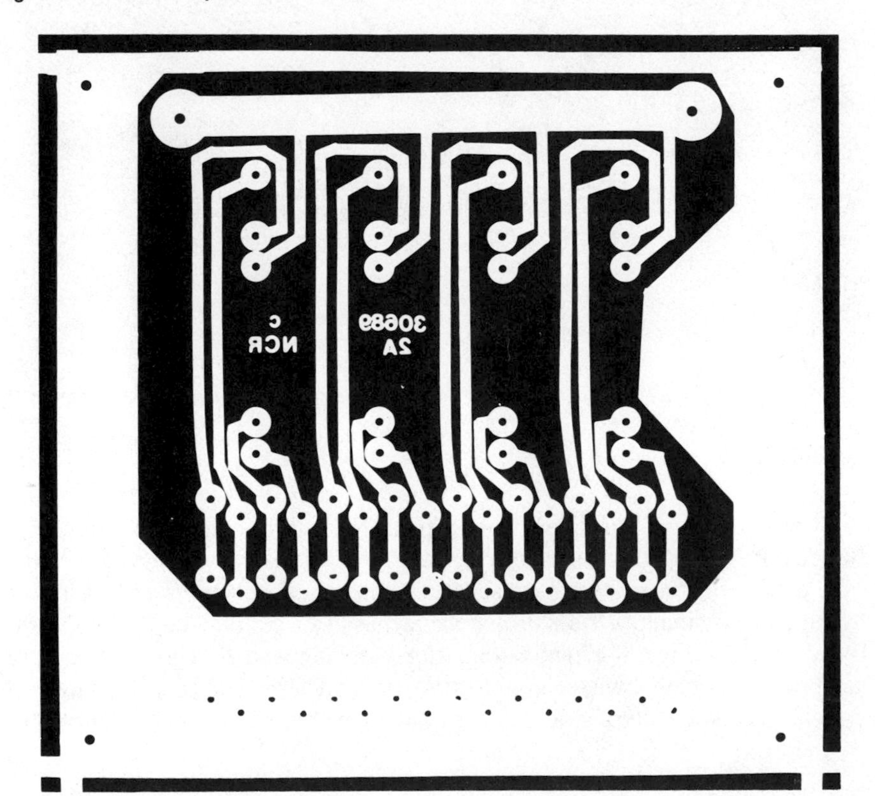

Fig. 6-48. PC board layout for display.

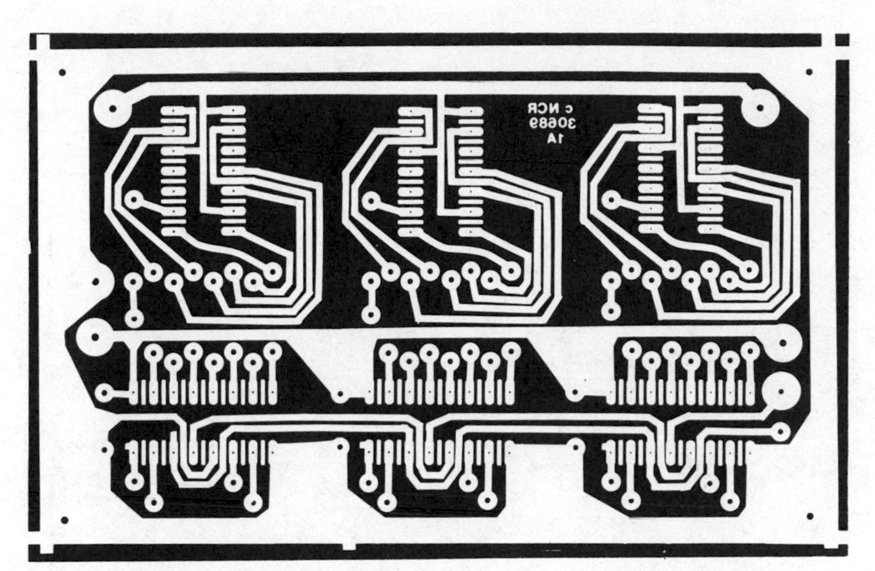

Fig. 6-49. Thumbwheel switch board.

resistors are still necessary and may be hardwired to the switches. See Fig. 6-45 for a schematic of the switchboard. Use ribbon cable for connections between PC boards for a neat job.

Assemble the PC boards by first inserting resistors and capacitors. Next, transistors. Do not install the opto isolator. Use DIP sockets for all IC devices if you can—it makes it a lot easier if you have either a defective device or accidentally blow a device during testing.

Check all wiring, then test for shorts and openings. Do not yet insert IC devices. Next, connect the primary of T1 to the ac line. Turn on S3 and check for +17 volts at the junction of D1 and D4. Check for +5V at all IC pins connected to +5V. With a scope, check for a 60 Hz square wave of 0 to +5V swing at the collector of Q4. Check for 0 to 12 volts at the emitter of Q1. Vary R26 throughout its range—it should vary the voltage from 0 to 12V at the emitter of Q1. Now remove power. Set the time switches to 999.

Insert all IC devices and reapply power. The LEDs should light (vary R26) and read 999, which was programmed by the switches. Try all three digits in all positions 0 through 9. The LEDs should indicate in each case the digit set on the switch. Next, place S4 in the open position and depress S1. The counter should slowly count down, once a second. Closing S1 should result in a ten times faster count. Verify with a watch or clock.

Connect a 100W lamp to J1, depress S1, the timer should start up and the lamp should light. At the count of zero the lamp should go out and the programmed time should reappear on the LEDs.

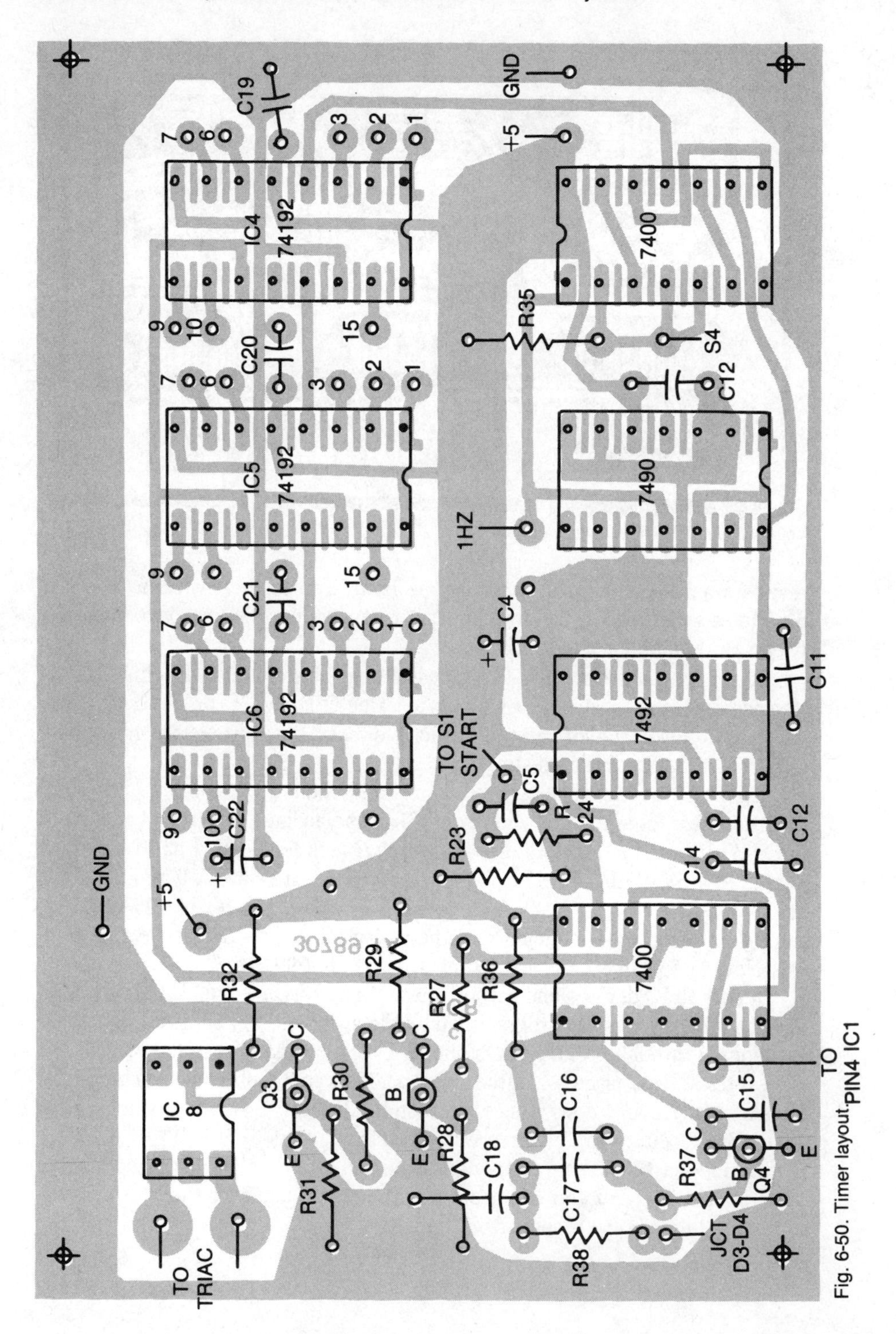

Fig. 6-50. Timer layout.

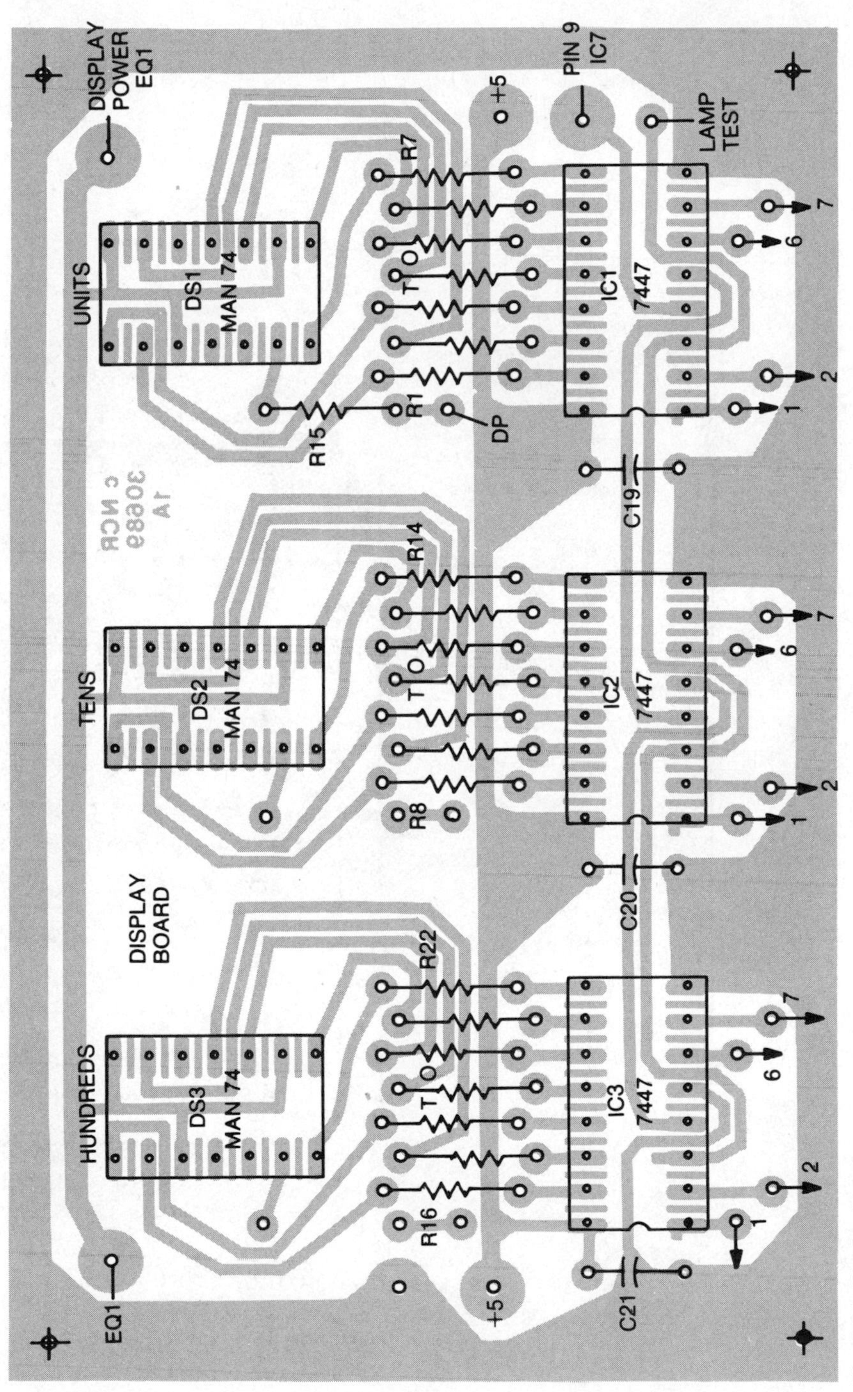

Fig. 6-51. Display layout.

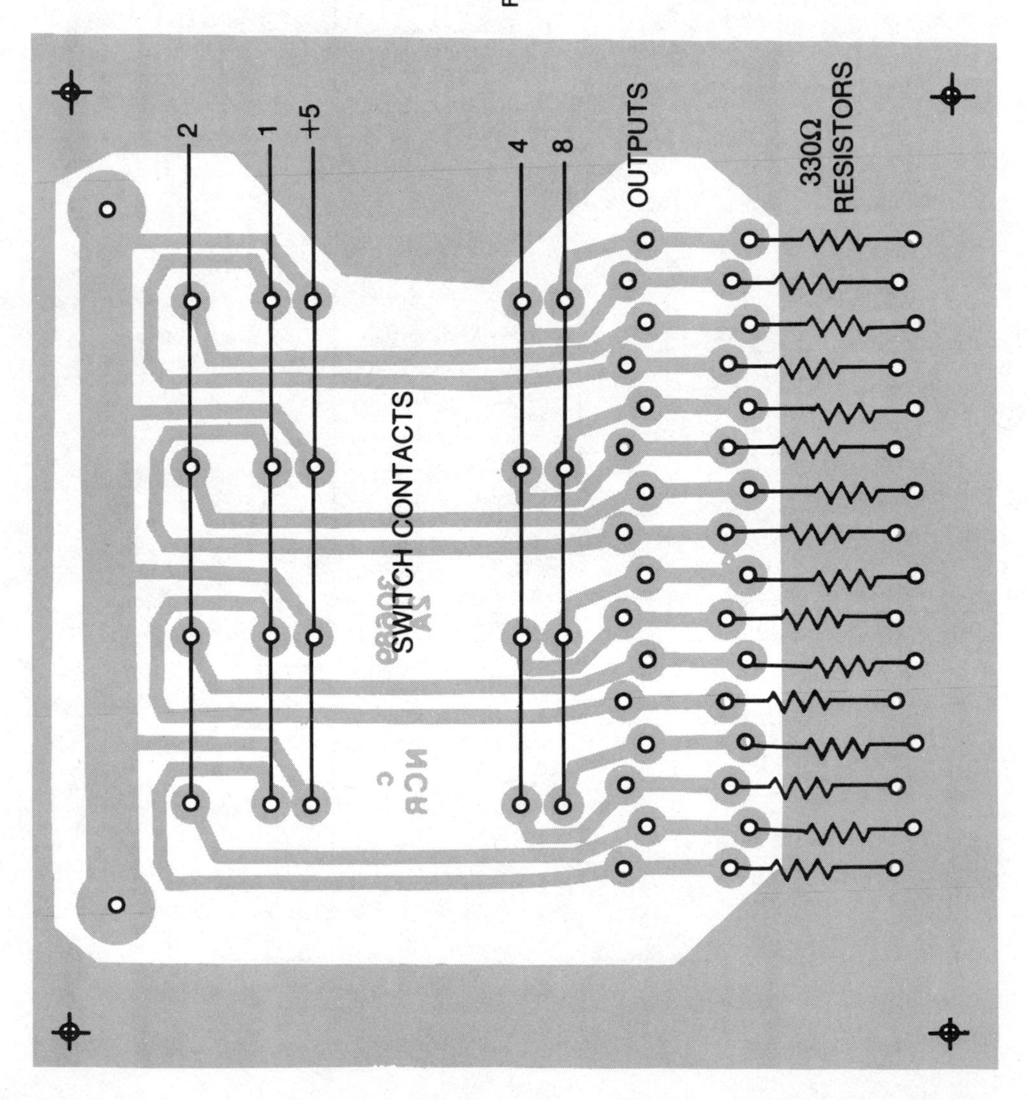

Fig. 6-52. Display layout.

RESISTORS 1/4W + 10%

R1 – R22	680 ohm
R23, R35, R36	2.2k
R24	2.2 Meg
R25	470 ohm
R26	1k ohm pot linear taper
R27, R28	4.7k ohm
R29, R39 – R50	330 ohm
R30, R31	1k ohm
R32	150 ohm
R33, R34	100 ohm
R37, R38	10k ohm

DISPLAYS

DS1, DS2, D3	Common anode 7 segment LED similar to MAN 74 equivalent

MISCELLANEOUS

S1	NO pushbutton
S2	SPDT 6A at 120Vac toggle
S3	SPST 6A at 120Vac toggle
S4	SPST 0.1A at 5Vdc toggle
T1	120V:12V at 1A or more transformer
J1	Ac outlet 120V 15A w/o round 1
F1	1A fuse assembly
S5, S6, S7	Thumbwheel switches BCD output

Also: Cabinet of choice
Ac line cord and plug
PC boards
Hardware as required

CAPACITORS

C1, C2, C3, C5, C6, C8, C11, C12, C13, C14, C15, C19, C20, C21	.01 disc
C4, C9	10μF 16V Elec.
C10	4700μF 16V Elec.
C7	1μF Elec.
C16	270 pF disc
C17, C18	.1 Mylar 50V

SEMICONDUCTORS

Q1	TIP41A
Q2, Q3, Q4	2N3904
D1, D2, D3, D4	1N4007
Q5	C106B

ICs

IC1, IC2, IC3	7447
IC4, IC5, IC6	74192 or 74LS192
IC7, IC11	7400
IC8	MDC3010 (GE)
IC9	7490
IC10	7492
IC12	LM7805

Fig. 6-53. Project 18 parts list.

This completes check out. If not already done, final assembly into whatever case has been used can now be done. A suggested package is shown in Fig. 6-46, but you can use your own ideas if preferred.

Appendix Component Identification Data

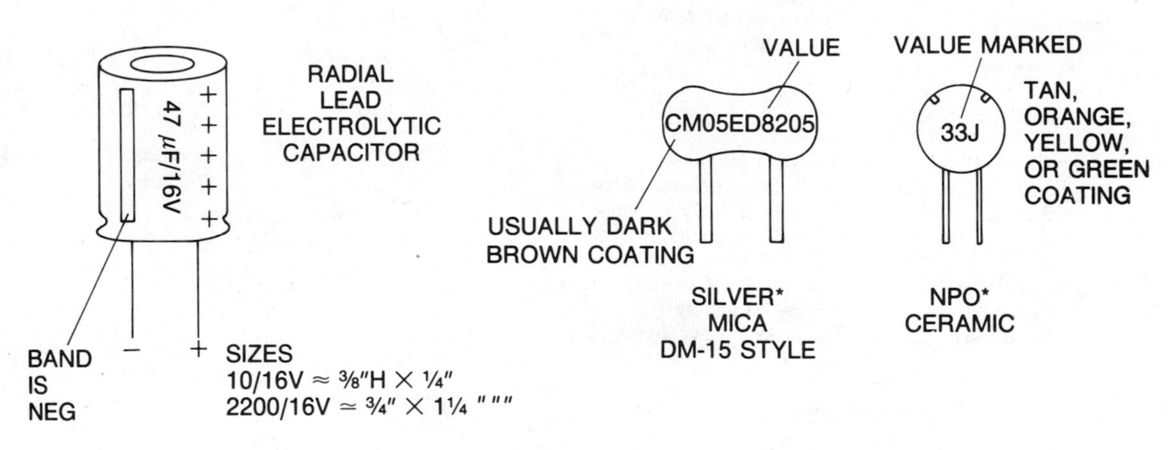

*NOTE—CAPACITOR VALUES SOMETIMES MARKED WITH FOLLOWING DESIGNATIONS

A) .01 disc—103Z, 103, .01, etc.
B) 470 disc—471
C) 330 pF MICA—CM05ED 331J
D) 330 pF NPO 331J
E) J = ±5% TOLERANCE
K = ±10% TOLERANCE
M = ±20% TOLERANCE
GMV = GUARANTEED MIN VALUE
(USUALLY NO LESS THAN. USED AS A
NONCRITICAL BYPASS, COUPLING, ETC.)

FOR EXAMPLE

471 MEANS 47 WITH 1 ZERO = 470 pF
331 MEANS 33 WITH 1 ZERO = 330 pF
102 MEANS 10 WITH TWO ZEROS = 1000 pF
820 MEANS 82 WITH NO ZEROS = 82 pF

F) NPO MEANS NEGATIVE-POSITIVE-ZERO (±30 PARTS PER MILLION PER °C) TEMPERATURE DRIFT—USUALLY VERY STABLE LOW VALUE (<1000 pF) DISC USED FOR TUNED CIRCUITS AND WHERE EXACT VALUE IS CRITICAL

VALUE DESIGNATION

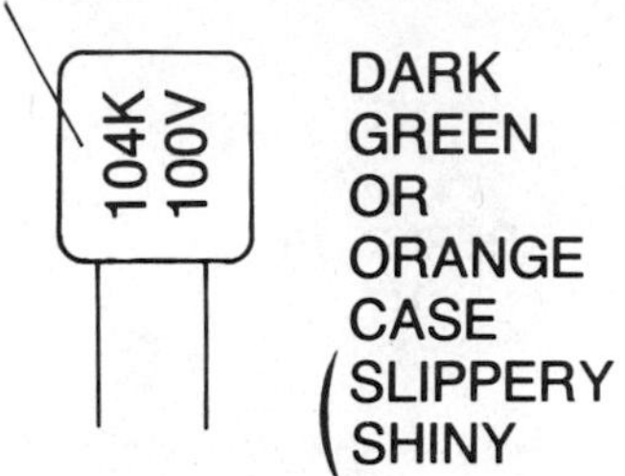

FILM CAPACITOR*
OR MYLAR CAPACITOR

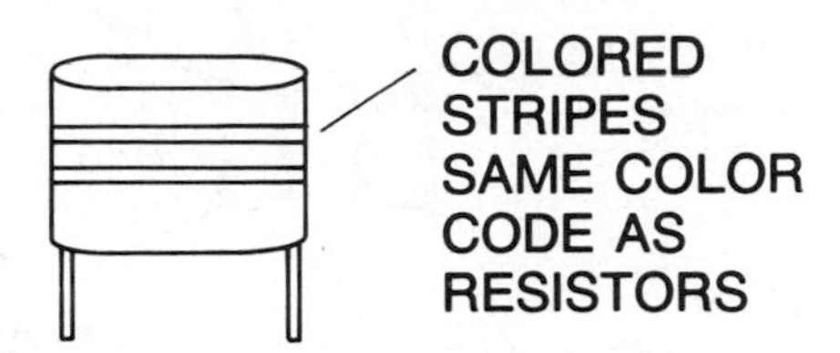

ALTERNATE
MYLAR CAPACITOR
("CANDY STRIPE" TYPE)

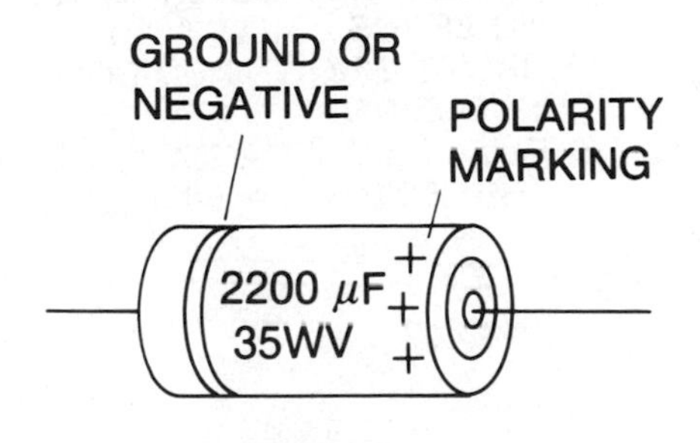

AXIAL LEAD
CAPACITOR (.001mF – .0022mF)

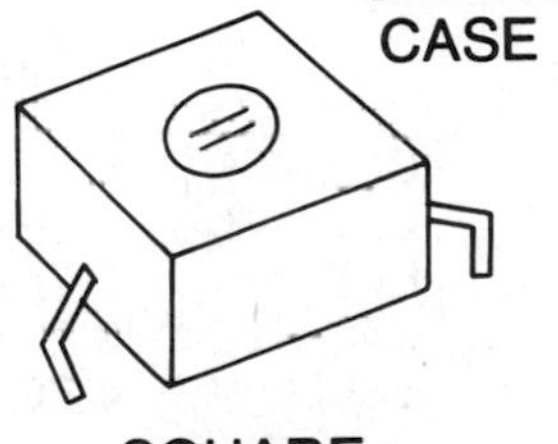

SQUARE
CERAMIC
TRIMMER
(7–50 pF)

*FILM OR MYLAR
VALUE DESIGNATIONS
SOMETIMES USED

102	1000 pF (.001)
222	2200 pF (.0022)
332	3300 pF (.0033)
472	4700 pF (.0047)
562	5600 pF (.0056)
103	.01 µF
223	.022 µF
473	.047 µF
104	.1 µF

LETTER
K = ±10% TOL
M = ±20% TOL

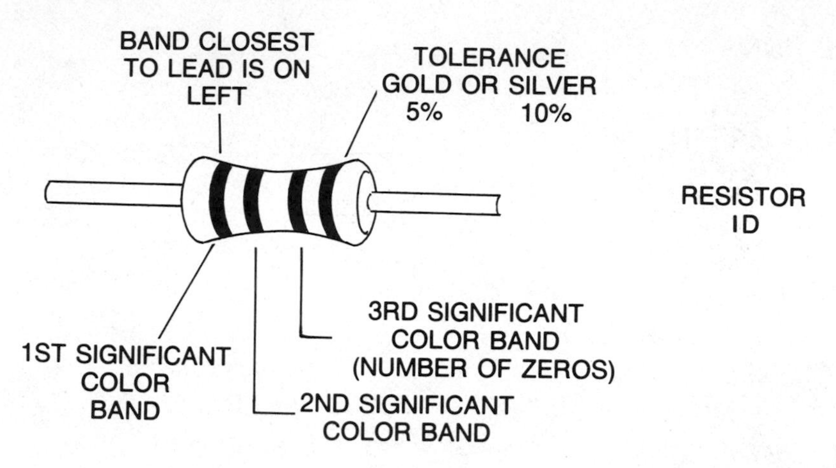

EXAMPLES

1) YELLOW-RED-BLACK-SILVER = 42 WITH NO ZEROS = 42Ω ± 10%
2) BROWN-BROWN-RED-GOLD = 11 WITH TWO ZEROS = 1100Ω ± 5%
3) BLUE-GRAY-ORANGE-GOLD = 68 WITH THREE ZEROS = 68000Ω ± 5% OR 68KΩ

DIGIT FIGURE	COLOR	
0	BLACK	(OR NO ZEROS)
1	BROWN	1 ZERO
2	RED	2 ZEROS
3	ORANGE	3 ZEROS
4	YELLOW	4 ZEROS
5	GREEN	5 ZEROS
6	BLUE	6 ZEROS
7	VIOLET	7 ZEROS
8	GRAY	8 ZEROS
9	WHITE	9 ZEROS

1/10 OR 1/8 WATT

1/4 WATT

APPROXIMATE SIZES

1KΩ = 1000Ω

1 MEG = 1,000,000Ω

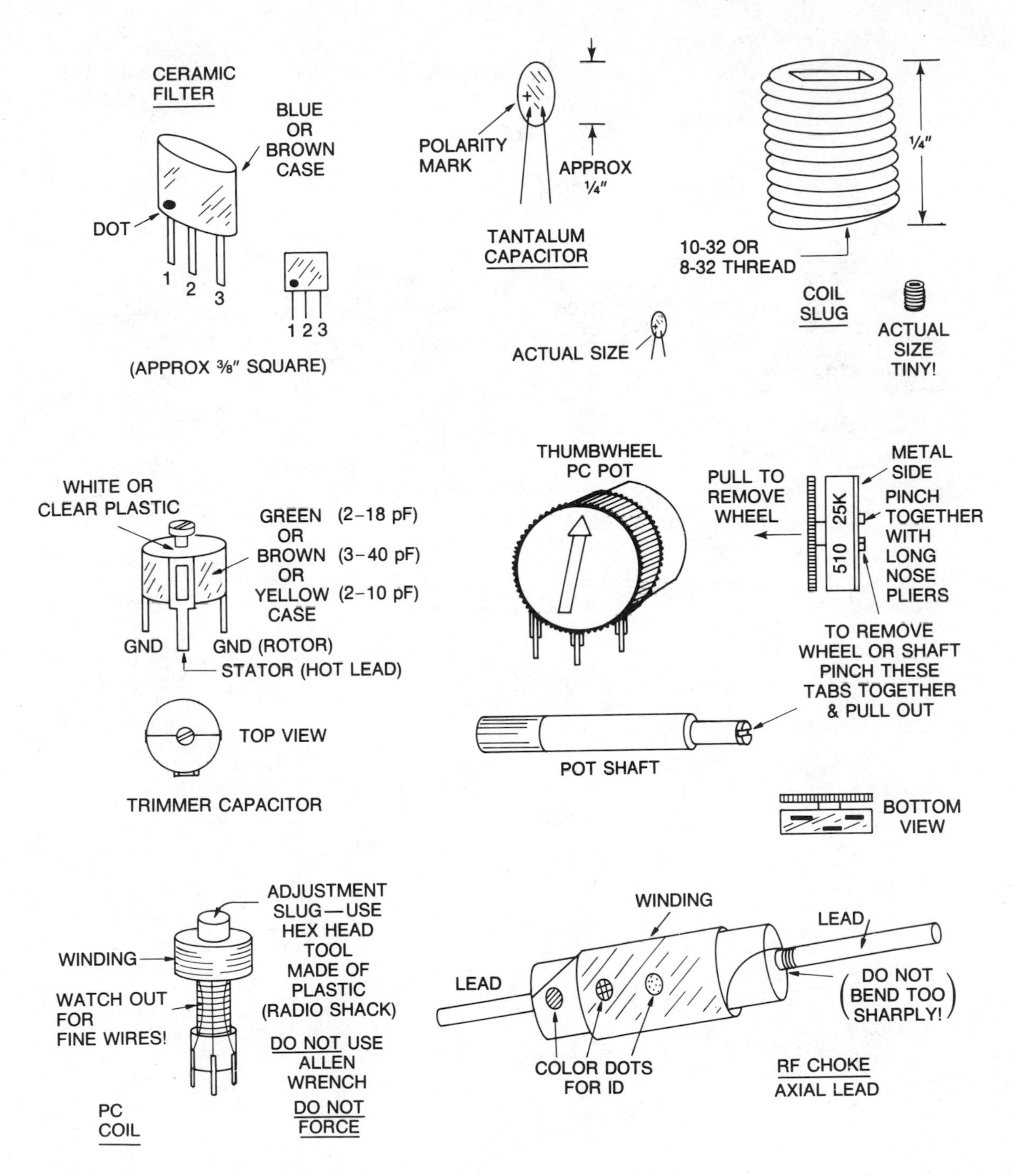
CERAMIC FILTER
BLUE OR BROWN CASE
DOT
1 2 3
1 2 3
(APPROX 3/8" SQUARE)
POLARITY MARK
APPROX 1/4"
TANTALUM CAPACITOR
ACTUAL SIZE
1/4"
10-32 OR 8-32 THREAD
COIL SLUG
ACTUAL SIZE TINY!
WHITE OR CLEAR PLASTIC
GREEN (2–18 pF) OR BROWN (3–40 pF) OR YELLOW (2–10 pF) CASE
GND
GND (ROTOR)
STATOR (HOT LEAD)
TOP VIEW
TRIMMER CAPACITOR
THUMBWHEEL PC POT
PULL TO REMOVE WHEEL
25K
510
METAL SIDE
PINCH TOGETHER WITH LONG NOSE PLIERS
TO REMOVE WHEEL OR SHAFT PINCH THESE TABS TOGETHER & PULL OUT
POT SHAFT
BOTTOM VIEW
ADJUSTMENT SLUG—USE HEX HEAD TOOL MADE OF PLASTIC (RADIO SHACK)
WINDING
WATCH OUT FOR FINE WIRES!
DO NOT USE ALLEN WRENCH
DO NOT FORCE
PC COIL
WINDING
LEAD
LEAD
DO NOT BEND TOO SHARPLY!
COLOR DOTS FOR ID
RF CHOKE AXIAL LEAD

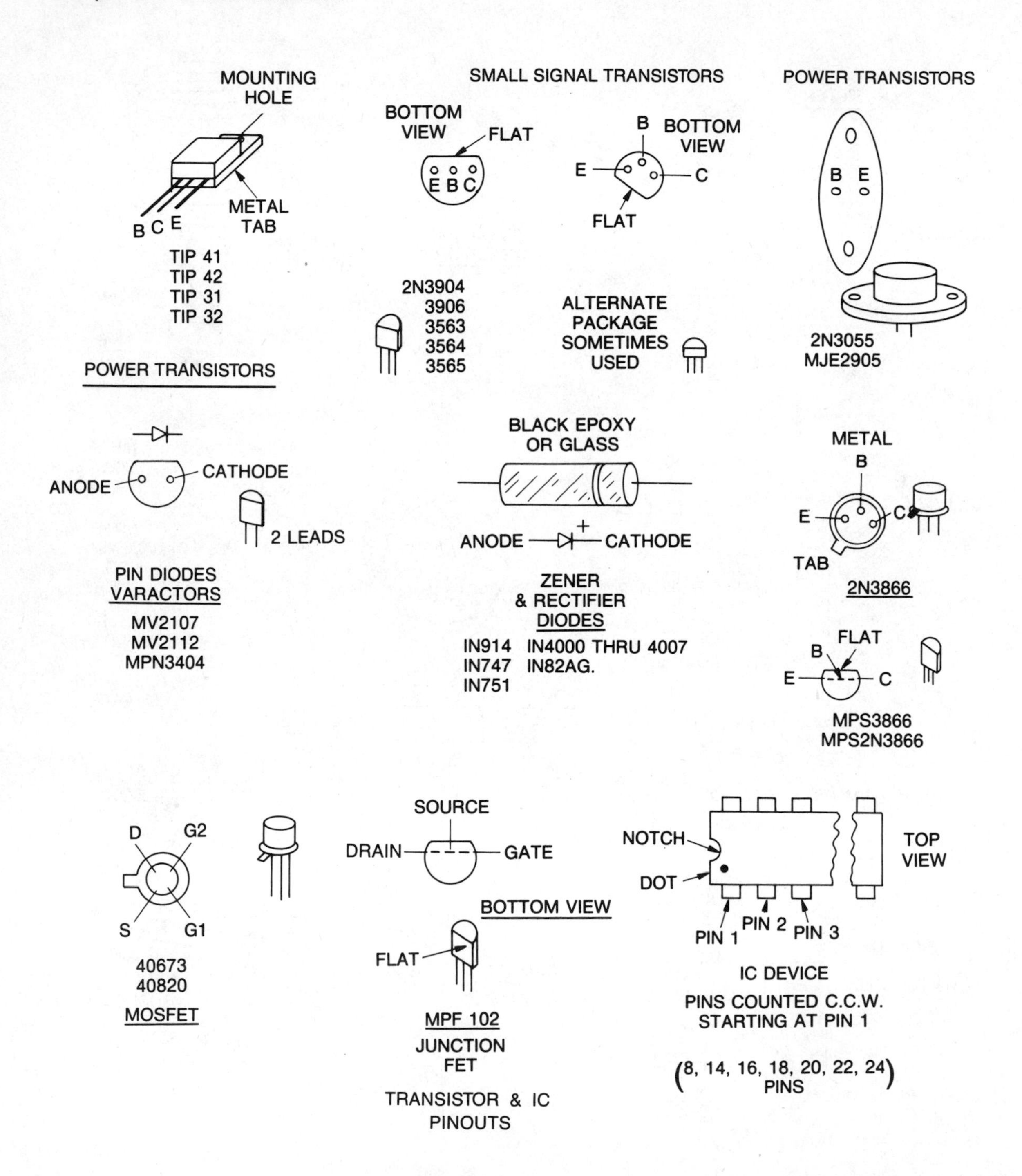

TRANSISTOR & IC PINOUTS

IDENTIFICATION OF RF CHOKES

CHOKE	COLOR DOTS—CODE	AVERAGE DC RESISTANCE—OHMS
1.8 μH	BROWN—GRAY—GOLD	0.8
5.6 μH	GREEN—BLUE—GOLD	4.5
18.0 μH	BROWN—GRAY—BLACK	2.0
47.0 μH	YELLOW—VIOLET—BLACK	6.5
68.0 μH	BLUE—GRAY—BLACK	7.9

CAUTION—THIS DATA IS FOR THE PURPOSE OF DISTINGUISHING BETWEEN RF CHOKES IN CASE OF DIFFICULTY IN READING THE COLOR CODES. A VARIATION OF UP TO ÷20% IN RESISTANCE READINGS MAY BE EXPECTED.

THIS DATA IS NOT MEANT TO BE A SOLE MEANS OF IDENTIFICATION, BUT AS SUPPLEMENTAL INFORMATION WHEN THE SUSPECTED INDUCTANCE IS KNOWN TO BE ONE OF SEVERAL VALUES. IT ONLY APPLIES TO INDUCTORS OF THE KIND SUPPLIED WITH OUR KITS.

Sources

All Electronics Corp.
P.O. Box 567
Van Nuys, CA 91408

Parts

B & H Photo
119 West 17th Street
New York, NY 10011

Photo Supplies,
Camera Equipment,
Darkroom Equipment

Bishop Graphics Inc.
5388 Sterling Ctr. Drive
West Lake Village, CA 91359
(818) 991-2600

Drafting Aids,
Precut Shapes

Chartpak
1 River Road
Leeds, MA 01053
(413) 584-5446

Drafting Aids,
Precut Shapes

Digi-Key Corporation
701 Brooks Avenue South
P.O. Box 677
Thief River Falls, MN 56701-0677

Parts

Freestyle Sales 5120 Sunset Blvd. Los Angeles, CA 90027 (800) 292-6137	*Photo Films, Developers, Sheet Film, Lith Film*
Jameco Electronics 1355 Shoreway Road Belmont, CA 94002	*Parts*
JDR Microdevices 110 Knowles Drive Los Gatos, CA 95030	
KEPRO Circuit Systems 630 Axminster Drive Fenton, MO 63026 (314) 343-1630	*PC Matls: Etching Fluid, Developer, PC Board Stock*
Mouser Electronics 11433 Woodside Avenue Santee, CA 92071	*Parts*
or	
Mouser Electronics 12 Emery Avenue Randolph, NJ 07869	*Parts*
North Country Radio P.O. Box 53E Wykagyl Station New Rochelle, NY 10804	*Kits*
Solid State Sales P.O. Box 74 Somerville, MA 02143	*Semiconductors*

Index

Index

R

Other Bestsellers of Related Interest

PRACTICAL ANTENNA HANDBOOK—by Joseph J. Carr

Build your own working antennas with the down-to-earth construction advice in this handbook. The emphasis is on practical! You'll get serviceable information on element length plus the theoretical material to go even further. Explore radio propagation, transmission lines, waveguides, microwave antennas, practical and shortwave antennas, and antennas for emergency communications. Use the projects as is or build your own specialized designs. Plus you get 22 computer programs written in BASIC for designing antennas. 416 pages, 351 illustrations. Book No. 3270, $21.95 paperback, $31.95 hardcover

THE BEGINNER'S HANDBOOK OF AMATEUR RADIO—2nd Edition—by Clay Laster, W5ZPV

Get all the information you need to pass the FCC licensing exam. Completely updated, the second edition covers newer FCC regulations including "Novice Enhancement" which gives new operators more privileges than ever before. You even get actual questions from the upgraded Novice Class License Exam and a sample test. You'll study Morse Code, ham radio operating procedures, how-to's for setting up your own radio station, and more. 432 pages, 305 illustrations. Book No. 2965, $16.95 paperback, $24.95 hardcover

GENERAL RADIOTELEPHONE OPERATOR'S LICENSE HANDBOOK—5th Ed.—by Harvey F. Swearer and Joseph J. Carr

290 pages, 144 illustrations. Book No. 3156, $14.95 paperback, $23.95 hardcover

GENERAL RADIOTELEPHONE OPERATOR'S LICENSE STUDY GUIDE—2nd. Ed.—by Thomas LeBlanc, NX7P

246 pages, 125 illustrations. Book No. 3118, $15.95 paperback, $23.95 hardcover

You'll pass the FCC's General Radiotelephone Operator's License test with flying colors when you prepare with this study team. Use the Handbook as a tutorial to gain the comprehensive knowledge you'll need to pass the test and start a profitable career. Brush up on electron theory, vector use, semiconductor theory and more. The Study Guide helps you prepare for the actual test with sample FCC practice exams and exam-taking tips. Plus it covers rules and regulations such as allowable frequency deviation, silent period, station logs, and more.

CET COMMUNICATIONS EXAM BOOK—by Ron Crow and Dick Glass

Increase your earnings. Become certified at the Senior or Journeyman level in the communications option offered by the ETAI/ISCET. Or just learn more about communications electronics. This comprehensive and practical text hows you how. It covers:

- FCC regulations
- transmitter operation
- troubleshooting communications systems
- cellular phones, modems, radar, satellite, and other communications techniques
- reducing radio interference
- satellite antennas

Plus it covers the full scope of the CET exam including practice tests and answers. 336 pages, 245 illustrations. Book No. 2910, $15.95 paperback, $24.95 hardcover

Other Bestsellers of Related Interest

TRANSMITTER HUNTING: Radio Direction Finding Simplified—by Joseph D. Moell (KODV) and Thomas N. Curlee (WB6UZZ)
"deserves a front row, center, seat on every radio experimenter's book shelf." **—Monitoring Times**

Join the increasing number of amateur radio operators who enjoy very weak or very strong signals, "sniff out the bunny," equip your vehicle, plan hunts, close in on a transmitter, build and use noise meters for additional RDF sensitivity, hunt, and more. Plus you et instructions for building your own doppler for less than $50, and two BASIC computer programs for triangulation. 336 pages, 252 illustrations. Book No. 2701, $18.95 paperback only

TROUBLESHOOTING AND REPAIRING VCR'S—by Gordon McComb

This bestseller enables you to repair all types of videocassette recorders including over 100 different brands of beta, VHS, and 8mm, as well as camcorders. You get solutions to all common and not-so-common ailments. . .troubleshooting flow-charts. . .scores of diagrams and instructive photographs. . .names and addresses of VCR manufacturers. .specs charts. . .and more. McComb, a certified Electronics Technician, owns of a successful VCR and electronics repair business. 336 pages, 354 illustrations. Book No. 2960, $17.95 paperback, $27.95 hardcover

TROUBLESHOOTING AND REPAIRING SOLID-STATE TVs—by Homer L. Davidson
". . .troubleshoot and repair almost any problem you come across in solid-state TVs, such as faulty remote control, high- or low-voltage power supply, defective horizontal-sweep circuits, brightness and picture tube problems, defective tuners, etc." **—Hands-On Electronics**

Through case studies, Davidson describes solid-state TV warning symptoms and the probable circuits and individual components that most often malfunction. 448 pages, 506 illustrations. Book No. 2707, $17.95 paperback, $26.95 hardcover

TROUBLESHOOTING AND REPAIRING COMPACT DISC PLAYERS—by Homer L. Davidson

Discover the latest technology used in all types of CD players including tabletop, portable, and automobile models. Step-by-step you'll examine each section, circuit, and component, and how they work together. Plus you get service literature and the actual schematics used in CD player production. You'll troubleshoot and repair servo control loops, remote control systems, optical lenses, laser assemblies, and crucial IC processors. 352 pages, 428 illustrations. Book No. 3107, $17.95 paperback, $26.95 hardcover

TROUBLESHOOTING AND REPAIRING AUDIO EQUIPMENT—by Homer L. Davidson

Learn the ABCs of audio component repair from renowned troubleshooter Homer Davidson. From telephone answering machines and compact disc players to turntables and car stereos, you'll learn which tools to use, how to conduct various tests, and how to locate defective parts. You get schematics, manufacturers' literature, and case histories for repairing name brand equipment. 336 pages, 354 illustrations. Book No. 2867, $17.95 paperback only

THE COMPLETE SHORTWAVE LISTENER'S HANDBOOK—3rd Edition—by Mark Bennett, Harry L. Helms, and David T. Hardy
". . .a comprehensive guide. . .explains everything you need to et started in (SWL) and other types of radio monitoring. . .in a way that is clear and easy to understand." **—Eugene R. Reich, V.D.A.**

Intercept conversations of Air Force One and pirate stations. This guide covers it all: receivers, antennas, frequencies, radio wave propagation, Q-codes, SWL terms, log books, selectivity, image rejection, station identification, DX clubs, and more. 304 pages, 96 illustrations. Book No. 2655, $16.95 paperback only